Surface Effects in
Magnetic Nanoparticles

Nanostructure Science and Technology

Series Editor: David J. Lockwood, FRSC
National Research Council of Canada
Ottawa, Ontario, Canada

Current volumes in this series:

Alternative Lithography: Unleashing the Potentials of Nanotechnology
Edited by Clivia M. Sotomayor Torres

Interfacial Nanochemistry: Molecular Science and Engineering at Liquid–Liquid Interfaces
Edited by Hiroshi Watarai, Norio Teramae, and Tsuguo Sawada

Nanoparticles: Building Blocks for Nanotechnology
Edited by Vincent Rotello

Nanostructured Catalysts
Edited by Susannah L. Scott, Cathleen M. Crudden, and Christopher W. Jones

Nanotechnology in Catalysis, Volumes 1 and 2
Edited by Bing Zhou, Sophie Hermans, and Gabor A. Somorjai

Polyoxometalate Chemistry for Nano-Composite Design
Edited by Toshihiro Yamase and Michael T. Pope

Self-Assembled Nanostructures
Jin Z. Zhang, Zhong-lin Wang, Jun Liu, Shaowei Chen, and Gang-yu Liu

Semiconductor Nanocrystals: From Basic Principles to Applications
Edited by Alexander L. Efros, David J. Lockwood, and Leonid Tsybeskov

Surface Effects in Magnetic Nanoparticles
Edited by Dino Fiorani

A Continuation Order Plan is available for this series. A continuation order will bring delivery of each new volume immediately upon publication. Volumes are billed only upon actual shipment. For further information please contact the publisher.

Surface Effects in Magnetic Nanoparticles

Edited by

Dino Fiorani

ISM-CNR
Area della Ricerca di Roma
Rome, Italy

 Springer

Library of Congress Cataloging-in-Publication Data

Fiorani, D.
  Surface effects in magnetic nanoparticles/Dino Fiorani.
      p.   cm. — (Nanoscience science and technology)
  Includes bibliographical references and index.
  ISBN 0-387-23279-6
      1. Nanotechnology.   2. Nanostructures.   Surfaces (Physics).   I. Title.   II. Series.

  T174.7.F56 2005
  620'.5—dc22

                                                                              2004060656

ISBN 0-387-23279-6

©2005 Springer Science+Business Media, Inc.

Printed in the United States of America.

9 8 7 6 5 4 3 2 1

springeronline.com

Preface

This volume is a selected collection of articles on different approaches to the investigation on surface effects on nanosized magnetic materials, with special emphasis to magnetic nanoparticles. The book is aimed to provide an overview of progress in the understanding of surface properties and surface driven effects in magnetic nanoparticles through recent results of different modelling, simulation and experimental investigations. Its intended audience is Ph.D. students and researchers in materials science.

Magnetic nanoparticles have been the subject of continuous and growing interest, from both fundamental and technological points of view, in the last 50 years, since the pionering work of Louis Néel. Nanoparticles are unique physical objects with remarkable magnetic properties which differ greatly from their parent massive materials. They are due to finite size effects of the magnetic core, related to the reduced number of spins cooperatively linked within the particle, and to surface and interface effects, related to the lack of coordination for the surface ions, inducing broken exchange bonds which can result in frustration and spin disorder. As the particle size decreases, the surface and interface effects are enhanced since the surface to volume ratio becomes larger. This determines completely different magnetic properties raising up basic and challenging questions for their theoretical description, being the "giant spin" model, which assumes that all spins participate in a single domain and that the magnetic energy is only determined by the particle volume and shape, no longer applicable. Surface and interface driven properties play a dominant role in all applications of magnetic nanoparticles (e.g. magnetorecording, magnetic sensors, biomedicine).

In this volume, theoretical and experimental investigations on ferro-, ferri- and antiferromagnetic particles are reported, providing a comprehensive description of different surface and interface effects. The general electronic structure of nanosized materials is reviewed. Surface and exchange anisotropy in nanoparticles and nanogranular materials are investigated by means of Monte Carlo simulations and a number of experimental techniques (e.g. magnetization measurements, resonance and Mössbauer spectroscopy). The magnetization reversal process is investigated in individual nanoparticles.

I would like to thank all the authors for their efforts which made it possible to provide this book to the scientific community.

Dino Fiorani
Rome, Italy

Contributors

X. Battle
Departament de Fisica
Fonanmental
Universitat de Barcelona
Diagonal 647, 08028 Barcelona
Spain

A. E. Berkowitz
Physics Department & Center for
Magnetic Recording Research
University of California, San
Diego, La Jolla, CA 92093
USA

E. De Biasi
Centro Atómico Bariloche
8400 S.C. de Bariloche, RN
Argentina

L. Del Bianco
National Institute for the Physics
of Matter (INFM) c/o
Department of Physics,
University of Bologna
40127 Bologna
Italy

D. Fiorani
Istituto di Struttura della Materia,
CNR, C.P. 10
00016 Monterotondo Stazione
(Roma)
Italy

A.J. Freeman
Department of Physics and
Astronomy
Northwestern University
Evanston, 60208-3112 IL
USA

D.A. Garanin
Institut für Physik, Johannes-
Gutenberg-Universität
D-55099 Mainz
Germany

A. Hernando
Instituto de Magnetismo
Aplicado UCM-Renfe,
P.O. Box 155
E-28230 Las Rozas, Madrid
Spain

O. Iglesias
Departament de Fisica
Fonanmental
Universitat de Barcelona
Diagonal 647, 08028 Barcelona
Spain

H. Kachkachi
Laboratoire de Magnetism et
d'Optique
Université de Versailles St.
Quentin
45 av. Des Etats-Unis
78035 Versailles
France

R. H. Kodama
Department of Physics,
University of Illinois at Chicago
Chicago, IL 60607
USA

A. Labarta
Departament de Fisica
Fonanmental
Universitat de Barcelona
Diagonal 647, 08028 Barcelona
Spain

K. Nakamura
Mie University
Mie 514-8507
Japan

R. Perzynski
Laboratoire des Milieux
Désordonnés et Hétérogènes
Université Pierre et Marie Curie
4 Place Jussieu,775252 Paris
Cedex 05
France

Yu. L. Raikher
Institute of Continuous Media
Mechanics
Ural Division of the Russian
Academy of Sciences
1 Korolyov St, 614013 Perm
Russia

C.A. Ramos
Centro Atómico Bariloche
8400 S.C. de Bariloche, RN
Argentina

H. Romero
Facultad de Ciencias,
Departamento de Física,
Universidad de Los Andes, 5101
Mérida
Venezuela

K.N. Trohidou
Institute of Materials Science
NCSR Demokritos
153 10, Aghia Paraskevi
Attiki, Athens
Greece

W. Wernsdorfer
Lab. L. Néel – CNRS
BP 166, 38042 Grenoble Cedex 9
France

R. Wu
Department of Physics and
Astronomy
University of California
Irvine, 92697 CA
USA

R.D. Zysler
Centro Atómico Bariloche
8400 S.C. de Bariloche, RN
Argentina

E.D. Zysler
Centro Atómico Bariloche,
8400 S.C. de Bariloche, RN
Argentina

Contents

                           in Ferrimagnetic Particles   105
                           A. Labarta, X. Battle and O. Iglesias

              4.1   Frustration in ferrimagnetic oxides   105
              4.2   Glassy behaviour in ferrimagnetic nanoparticles   107
              4.3   Monte Carlo simulations   121
              4.4   Open questions and perspectives   136

CHAPTER 5.   **Effect of Surface Anisotropy on the Magnetic Resonance**
                           Properties of Nanosize Ferroparticles   141
                           R. Perzynski and Yu.L. Raikher

              5.1   Introduction   141
              5.2   Spin perturbations in fine particles. Interplay of the exchange
                     and surface energies   147
              5.3   Spin-wave resonance in the presence of a uniaxial
                     surface anisotropy   153
              5.4   Experimental   165
              5.5   FMR in a spherical particle with the Aharoni surface
                     anisotropy   172
              5.6   FMR in a spherical particle with rotable exchange
                     Anisotropy   177
              5.7   Concluding remarks   182

CHAPTER 6.   **Surface-Driven Effects on The Magnetic Behaviour of**
                           Oxide Nanoparticles   189
                           R.H. Kodama and A.E. Berkowitz

              6.1   Introduction   189
              6.2   Atomic-scale magnetic modeling   192
              6.3   Ferrimagnetic nanoparticles   198
              6.4   Antiferromagnetic nanoparticles   206
              6.5   Remarks   213

CHAPTER 7.   **Exchange Coupling in Iron and Iron /Oxide**
                           Nanogranular Systems   217
                           L. Del Bianco, A. Hernando and D. Fiorani

              7.1   Introduction   217
              7.2   Nanocrystalline Fe   219
              7.3   Fe/Fe oxide nanogranular system   228
              7.4   Conclusions   236

Modern Electronic Structure Theory for Complex Properties of Magnetic Materials

A. J. Freeman

Department of Physics and Astronomy,

Northwestern University, Evanston, IL, USA 60208

Kohji Nakamura

Mie University, Mie 514-8507,Japan

Ruqian Wu

Department of Physics and Astronomy,

University of California, Irvine, CA, USA 92697-4575

1.   Introduction

As is clear from many chapters in this book and elsewhere, the field of magnetism is in an ongoing (one might say feverish) state of excitement, with discoveries important for both basic science and technological/device applications. From the theory end, it has been increasingly recognized that state-of-the-art *ab initio* electronic structure calculations based on the density-functional theory (DFT) have contributed to, and at times have led these developments. Indeed, they have achieved great success in the exciting field of thin film magnetism, in both explaining existing phenomena and, more importantly, in predicting the properties of new systems [1]. For example, the prediction of enhanced magnetic moments with lowered coordination number at clean metal surfaces and interfaces has stimulated both theoretical and experimental investigations for new magnetic systems and phenomena in

1

man-made transition metal thin films, which has accompanied the renaissance of magnetism and has led to a major impact on the magnetic recording industry [2].

In conjunction with remarkable progress in materials synthesis and characterization techniques, applications of first principles approaches have become more reliable and predictive when employed in the search and design of nanostructured and other magnetic materials. Unusual magnetic orderings are found in reduced dimensions and scale, as illustrated by several recent experiments and theoretical calculations. The magnetism in small entities, where a large fraction of atoms are exposed to surfaces and interfaces, introduces new physics of a fundamental nature. In addition, the new field of spintronics where both the spin and charge of the electron are manipulated and exploited is also developing rapidly. Thus, for example, it is of great interest nowadays to make direct electrical or optical switching of magnetization in ferromagnetic (FM) layers for further integration of logic, display and storage functions.

In this chapter, after a short introduction to density-functional theory and its most precise implementation via the full-potential linearized augmented plane wave (FLAPW) method, we present a brief discussion on the general aspects of the spin orbit coupling (SOC)-induced magnetic phenomena and present applications to the determination of magnetocrystalline anisotropy energies of various systems. This serves to illustrate the great progress made to date in treating theoretically/computationally the complex SOC-induced phenomena. In the second part of this chapter, we demonstrate how the remaining unsolved challenging issue — the role of noncollinear magnetism (NCM) that arises not only through the SOC, but also from the breaking of symmetry at surfaces and interfaces — is resolved using an extension of the FLAPW methodology.

2. Density Functional Theory and the FLAPW Method

2.1. DENSITY-FUNCTIONAL THEORY

Modern electronic structure theory employs the density-functional theory, introduced by Hohenberg and Kohn [3] and Kohn and Sham [4]. The underlying (Hohenberg-Kohn) theorem on which this theory rests is that the total energy, E, of a material system can be expressed as a functional of its electron density, ρ, namely, $E = E[\rho]$, that E is at its minimum for the ground-state density, and is stationary with respect to first-order variations in the density.

Typically, the Born-Oppenheimer approximation [5] is employed which assumes that the motion of the nuclei are negligible with respect to those of the electrons. This implies that the electronic structure is calculated for a given atomic geometry; the nuclei are then moved according to classical mechanics.

2.1.1. The Kohn-Sham Equations

To obtain the ground-state density, the variational principle is applied with respect to the one-particle wave functions:

$$[-\frac{\hbar^2}{2m}\nabla^2 + V_{eff}(\mathbf{r})]\psi_i(\mathbf{r}) = \varepsilon_i\psi_i(\mathbf{r}), \tag{1}$$

where

$$V_{eff}(\mathbf{r}) = V_C(\mathbf{r}) + \mu_{xc}[\rho(\mathbf{r})] \tag{2}$$

is the effective potential and ε_i the effective one-electron eigenvalues. Equations 1 are the "Kohn-Sham equations" and the solutions, $\psi_i(\mathbf{r})$, form an orthonormal set, i.e., $\int \psi_i^*(\mathbf{r})\psi_j(\mathbf{r})\,d\mathbf{r} = \delta_{ij}$. The Coulomb or electrostatic potential is given as:

$$V_C(\mathbf{r}) = -e^2 \sum_\alpha \frac{Z_\alpha}{|\mathbf{r} - \mathbf{R}_\alpha|} + e^2 \int \frac{\rho(\mathbf{r}')}{|\mathbf{r} - \mathbf{r}'|}d\mathbf{r}' \tag{3}$$

which can also be calculated using Poisson's equation,

$$\nabla^2 V_C(\mathbf{r}) = -4\pi e^2 q(\mathbf{r}), \tag{4}$$

where $q(\mathbf{r})$ represents the electronic charge distribution *and* the positive point charges at position $\mathbf{R}_\alpha$. The exchange-correlation potential is given by

$$\mu_{xc} = \partial E_{xc}[\rho]/\partial\rho. \tag{5}$$

Because the exchange-correlation potential (and energy) are not known, approximations have to be made.

2.1.2. Spin-polarized Density Functional Theory

The generalization of density-functional theory to spin-polarized systems has been made within the local spin density approximation (LSD) [6, 7]. The important quantity, in addition to the electron density $\rho(\mathbf{r})$, is the spin density $\sigma(\mathbf{r})$ which is the difference between the spin-up and spin-down configuration densities $\sigma(\mathbf{r}) = \rho_\uparrow(\mathbf{r}) - \rho_\downarrow(\mathbf{r})$; the total density is given by $\rho(\mathbf{r}) = \rho_\uparrow(\mathbf{r}) + \rho_\downarrow(\mathbf{r})$. Because the exchange-correlation potential for spin-up and spin-down electrons is in general different, the spin-polarized form of the Kohn-Sham equations are:

$$[(-\hbar^2/2m)\nabla^2 + V_{eff}^\sigma(\mathbf{r})]\psi_i^\sigma(\mathbf{r}) = \varepsilon_i^\sigma\psi_i^\sigma(\mathbf{r}), \quad \text{where } \sigma = \uparrow \text{ or } \downarrow, \tag{6}$$

and

$$V_{eff}^\sigma(\mathbf{r}) = V_C(\mathbf{r}) + \mu_{xc}^\sigma[\rho(\mathbf{r}), \sigma(\mathbf{r})]. \tag{7}$$

Thus there are two sets of single-particle wave functions, one for spin-up (or "majority") electrons and one for spin-down ("minority") electrons, each with corresponding one-electron eigenvalues.

2.1.3.   Exchange-correlation Functions

Local-density approximation (LDA)

A very successful and widely used approximation for the exchange-correlation energy is the local-density approximation (LDA). Here the exchange-correlation energy is assumed to depend only on the local electron density in each volume element dr:

$$E_{xc}[\rho] \approx \int \rho(\mathbf{r})\varepsilon_{xc}[\rho(\mathbf{r})]d\mathbf{r}. \qquad (8)$$

$\varepsilon_{xc}[\rho]$ is the exchange-correlation energy per electron of a *homogeneous* electron gas and is expressed as an analytic function of the electron density, as is the exchange-correlation potential, μ_{xc}. There are various forms of the LDA in the literature; we refer to those of Hedin-Lundqvist [8] and Wigner (non-spin-polarized forms), and von Barth-Hedin [6] (spin-polarized form) since they are the ones implemented in the present FLAPW program.

Generalized gradient approximation (GGA)

In recent years, the generalized gradient approximation (GGA) is considered as a possible improvement over the LDA since it has been found to generally improve the description of total energies, ionization energies, electron affinities of atoms, the atomization energies of molecules [9–11] and some solid state properties [12–15]. Adsorption energies of adparticles on surfaces are also reported to be improved [16, 17], as are reaction energies [18, 19]. Furthermore, the GGA leads to significantly better activation energy barriers for H_2 dissociation [20, 21]; also, the relative stability of structural phases appears to be better described for magnetic [22] and non-magnetic systems [23, 24].

There are a number of GGA approaches in the literature. The generic form of the GGA exchange-correlation energy may be written as:

$$E_{xc}^{GGA}[\rho] = \int \rho(\mathbf{r})\varepsilon_{xc}^{GGA}(\rho(\mathbf{r}), \nabla\rho(\mathbf{r}))\, d\mathbf{r} \qquad (9)$$

so that it depends locally on the electronic density $\rho(\mathbf{r})$ and its gradient.

The GGA developed by Perdew and Wang (PW) [9] is derived essentially from first principles, combining the gradient expansions of the exchange and correlation holes of a nonuniform electron gas with real-space truncations to enforce constraints imposed by properties of the physical exchange-correlation hole. The GGA developed by Becke and Perdew [25] on the other hand, relies on fitted parameters. In the present version of the FLAPW program, the GGA implemented is that proposed by Perdew, Burke and Ernzerhof (PBE) [26]. This functional is regarded to be conceptually more concise than the PW GGA but appears to yield very similar results [27].

2.2. THE FLAPW BASIS SET

In the FLAPW method, the real space in a bulk material is partitioned into spherical regions around atoms ("muffin tins" or "atomic spheres") and interstitial regions between the spheres [28]. In the sphere region, the basis functions are products of radial functions and spherical harmonics, and in the interstitial region plane waves are used. For treating a film

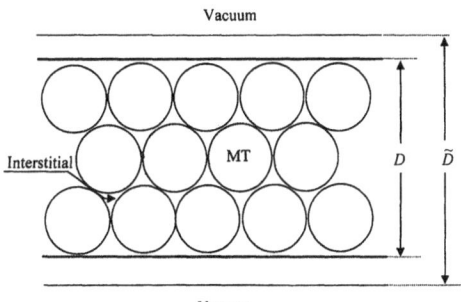

FIG. 1: Geometry for a film calculation showing division of space in a film geometry.

geometry, there are a number of atomic layers surrounded by vacuum; thus, in addition to sphere and interstitial regions, one defines a vacuum region, which starts at $\pm D/2$ and ends at $\pm \tilde{D}/2$ (see Fig. 1), where the wave functions are products of two-dimensional (2D) plane waves and z-dependent functions which are solutions of the one-dimensional Schrödinger equation of the (x, y)-averaged potential in the vacuum region.

Specifically, the FLAPW one-particle wave functions in the film geometry are:

$$\psi_i(\mathbf{r}, \mathbf{k}_\parallel) = \sum_j c_{ij}\phi(\mathbf{r}, \mathbf{K}_j); \quad \mathbf{K}_j = \mathbf{k}^\parallel + \mathbf{G}_j, \tag{10}$$

where $\mathbf{k}^\parallel$ is an arbitrary vector of the 2D Brillouin zone (BZ) and $\mathbf{G}_j$ is a three-dimensional (3D) reciprocal lattice vector (in the z-direction, an artificial periodicity between the boundaries at $\pm \tilde{D}/2$ is imposed). The basis functions are:

$$\phi(\mathbf{r}, \mathbf{K}_j) = \begin{cases} \Omega^{-1/2}e^{i\mathbf{K}_j\cdot\mathbf{r}} & \text{interstitial} \\ \sum_{lm}[A_{lm}^\alpha(\mathbf{K}_j)u_l(E_l^\alpha, r_\alpha) + B_{lm}^\alpha(\mathbf{K}_j)\dot{u}_l(E_l^\alpha, r_\alpha)]Y_{lm}(\hat{r}_\alpha) & \text{sphere} \\ \sum_q[A_q(\mathbf{K}_j)u_{kq}(E_\nu, z) + B_q(\mathbf{K}_j)\dot{u}_{kq}(E_\nu, z)]e^{i(\mathbf{k}^\parallel + \mathbf{K}_q^\parallel)\cdot\mathbf{r}} & \text{vacuum} \end{cases} \tag{11}$$

The Coulomb potential $V_c(\mathbf{r})$ is obtained by solving Poisson's equation in each of the three regions. At infinity the potential is gauged equal to zero in the case of the film geometry. The effective single-particle potential is constructed by adding the exchange-correlation potential, which is determined by the charge density in real space and its transformation into each spin representation. The core electrons are treated fully relativistically and are updated at each iteration using a scheme for *free* atoms (only the spherical part of the potential is used). The valence electrons are expanded in a variational basis set and are generally treated scalar-relativistically (i.e., neglecting the spin-orbit coupling),or when spin-orbit coupling is included, either perturbatively or fully self-consistently.

2.3. SPIN-ORBIT COUPLING

As is well known, SOC is the only term in the Hamiltonian that links the spatial and spin parts of the wave functions, and thus serves as the origin of many important phenomena such as magnetocrystalline anisotropy (MCA), magnetostriction, magneto-optical Kerr effects (MOKE), magnetic circular dichroism (MCD), and magnetic damping. In most magnetic systems, however, the SOC contributions are very weak compared to those from crystalline fields and exchange interactions and so need to be treated extremely carefully especially in magnetic transition metal systems. Significant progress has been achieved in the last decade to improve numerical stability and reliability in the determination of SOC-induced properties.

In a central potential $V(r)$, the SOC Hamiltonian, H_{SOC}, is given as

$$H_{SOC} = \xi\sigma \cdot L \ \text{ and } \ \xi = \frac{1}{4m^2c^2r}\frac{dV(r)}{dr}.$$

This format, however, has been utilized in almost all the cases since the major contribution is clearly from the near nuclear region where the dV/dr term is large and the spherical symmetry is retained. In addition, the value of ξ is 10-50 meV in 3d metals, and is much smaller than their crystalline fields. Therefore, H_{SOC} can be conveniently treated in a second variational approach, based on the ground state properties obtained from scalar-relativistic calculations.

In numerical calculations, one needs to construct the SOC matrix

$$\sigma \cdot L = \begin{pmatrix} \uparrow\uparrow & \uparrow\downarrow \\ \downarrow\uparrow & \downarrow\downarrow \end{pmatrix}$$

$$= \begin{pmatrix} \frac{A_+ + A_-}{2}\sin\theta + L_z\cos\theta & (A_-\cos^2\frac{\theta}{2} - A_+\sin^2\frac{\theta}{2} - L_z\sin\theta)e^{-i\phi} \\ (A_+\cos^2\frac{\theta}{2} - A_-\sin^2\frac{\theta}{2} - L_z\sin\theta)e^{i\phi} & -\frac{A_+ + A_-}{2}\sin\theta - L_z\cos\theta \end{pmatrix},$$

where θ and ϕ stand for the polar and azimuthal angles of the magnetic moment and $A_+ = e^{-i\phi}(L_x + iL_y)$ and $A_- = e^{i\phi}(L_x - iL_y)$. After re-diagonalizing the $N_e \times N_e$ eigenvalue problem,

$$\begin{pmatrix} \varepsilon(\uparrow) & 0 \\ 0 & \varepsilon(\downarrow) \end{pmatrix} + \xi\sigma \cdot L = \varepsilon' I,$$

the wave function, charge density, potential, energy band and total energy will be updated and become (θ, ϕ) dependent. Here $N_e = N_e \uparrow + N_e \downarrow$ is the number of states invoked in the variational procedure, while $\varepsilon(\uparrow)$ and $\varepsilon(\downarrow)$ are the eigen-energies in the majority and minority spin-channels, respectively. It has been shown that SOC alters the charge density, spin density and spin moment only negligibly in most 3d transition metal systems and thus it is not important to treat H_{SOC} self-consistently there.

3. Results and Discussion

3.1. MAGNETOCRYSTALLINE ANISOTROPY IN THIN FILMS

Magnetic anisotropy is one of the most important phenomena in low dimensional magnetism and has attracted extensive attention recently. It is important in various aspects of applications such as thermal stability, coercivity, and perpendicular magnetic recording, and is a challenge for fundamental understanding. There are three approaches currently to calculate the MCA energy (E_{MCA}), namely, force theorem, total energy and torque. All three have been shown to produce very close results in converged calculations for most 3d systems.

The force theorem assumes that the contribution from SOC-induced changes in the charge and spin density, is negligible in the total energy and thus E_{MCA} can be evaluated merely from the SOC-induced changes in the eigenvalues [29, 30],

$$E_{MCA} = E(\rightarrow) - E(\uparrow) = \sum_{occ'} \varepsilon_i(\rightarrow) - \sum_{occ''} \varepsilon_i(\uparrow) + O(\delta\rho^n), \qquad (12)$$

where ε_i stands for the band energy of the ith state and the arrows in the parentheses denote the magnetization directions. Strong numerical uncertainties have been encountered because the sets of occupied states, i.e., $\{occ'\}$ and $\{occ''\}$, were determined through the usual Fermi filling scheme which relies on the very limited information from the eigenvalues, ε_i[31]. Thus, one had to use a huge number of k-points ($>$10,000 in the two-dimensional Brillouin zone for thin films) to obtain reliable values of the E_{MCA} and thus only a few model systems (e.g., a free monolayer or simple alloys) could be treated. A simple solution for this problem is given in the state tracking (ST) approach in which the $\{occ'\}$ and $\{occ''\}$ states are determined according to their projections back to the occupied set of unperturbed states [32]. Since this procedure ensures a minimum change in the charge and spin densities, as required by the force theorem, and excludes possible randomness in the Brillouin zone (tracking at a given k-point), very stable MCA results were obtained with a relatively small number of k-points for magnetic thin films such as Fe, Co and Ni monolayers in both their free-standing case and on various substrates[33].

A torque (TQ) method can further depress the remaining uncertainties resulting from the SOC interaction between near-degenerate states around the Fermi level (so called surface

A.J. Freeman et al.

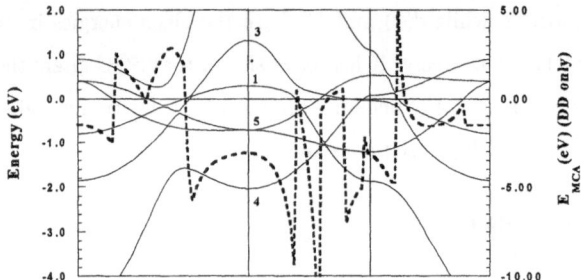

FIG. 2: The calculated band structure (thin lines) and MCA energy of a free standing Co monolayer (a=4.83 a.u.) along the high symmetric lines in the two-dimensional Brillouin zone.

pair coupling) [34]. We found for uniaxial systems that E_{MCA} can be directly evaluated through

$$E_{MCA} = \sum_{occ} < \psi'_i |dH/d\theta| \psi'_i > |_{\theta=45^\circ} = \sum_{occ} < \psi'_i |\partial H_{SOC}/\partial\theta| \psi'_i > |_{\theta=45^\circ}, \quad (13)$$

where ψ'_i is the ith perturbed wave function. The advantage of the torque method is obvious since E_{MCA} is expressed as the expectation value of the angular derivative of H_{SOC} and thus is much less sensitive to distortions of the Fermi surface — and only one Fermi surface needs to be determined.

To understand the behavior of MCA energies of magnetic films in different environments, it is instructive to analyze the electronic origin of the MCA energy for free-standing magnetic monolayers. The k-distributions of the MCA energy, $E_{MCA}(\vec{k})$, and the band structures of a Co monolayer, for example, are plotted in Fig. 2 along the high-symmetry directions in the 2D BZ. The obvious correlation between major E_{MCA} contributions and the locations of Fermi surfaces (where bands cross the Fermi level) indicates some simple physical insights for MCA, which once was thought to originate from very sophisticated interactions. Note that (i) only SOC interactions between pairs of states across E_F contribute to E_{MCA} and (ii) the pair with the same (different) magnetic quantum number(s), m, leads to a positive (negative) contribution to E_{MCA}.

From Fig. 2, the large negative MCA energy for the Co monolayer (-1.34 meV/atom) is mainly from the contributions around the $\overline{M}$ point. This is due to the SOC interaction between the occupied $d_{xz,yz}$ ($m = \pm 1$) states and the unoccupied d_{z^2} state ($m = 0$). By knowing detailed information about the origin of MCA for model systems, one can understand the MCA behavior of more complicated systems and furthermore tailor the MCA energy. For Co thin films, for example, one can enlarge the energy separation between the Co $d_{xz,yz}$ and d_{z^2} through the proximity effect of different substrates. For example, owing to say Co-Cu d-band hybridization, the $d_{xz,yz}$ states are split and diluted in a wide energy range for these systems. As a result, the negative contribution to E_{MCA} around $\overline{M}$ is drastically reduced. Further, the MCA energy for Cu/Co/Cu(001) becomes positive (0.54 meV/adatom, indicating a strong perpendicular anisotropy) [35], as observed experimentally [36, 37]. Many calculations have been done to study the effects of metal substrates or capping layers on the MCA of ultra-thin magnetic thin films. It is well established now that high quality first principles theory can yield quite satisfactory results for the uniaxial E_{MCA} of magnetic thin films [1] .

The E_{MCA} can also be tuned by exposing the films to different gaseous environments such as in H, O, or CO [38–40], in addition to capping the film with metallic overlayers. The presence of oxygen on Ni/Cu(001), for example, is found to reduce the surface E_{MCA} and hence to change the critical thickness of the spin reorientation transition in ultra-thin Ni films from 11 ML down to about 5 ML. Through first principles FLAPW calculations, we attribute these characteristics to modifications in the electronic structure of surface Ni atoms caused by the presence of oxygen and defects.

The spin magnetic moments of the surface Ni atom obtained through FLAPW calculations are 0.75-0.77 μ_B in clean and O-covered flat Ni/Cu surfaces. By contrast, the spin magnetic moment of the topmost Ni is drastically enhanced to 1.02 μ_B, along with a large orbital magnetic moment of 0.23 μ_B, if surface defects are introduced. Here, both O and Ni behave more in the vein of NiO except that the in-plane antiferromagnetic ordering was not allowed in the small unit cell. The induced spin magnetic moments of the O adatom are also large — 0.15μ_B on the flat and 0.26 μ_B on the defected Ni/Cu(001) surfaces, respectively.

We calculated the clean, Cu- and O-covered Ni$_n$ slabs (n= 5, 7, 9, 11 and 13 monolayers) for the determination of surface and bulk contributions to E_{MCA}. As plotted in Fig. 3, E_{MCA} displays a linear dependence on $1/d$ for films thicker than 9 layers. From the slopes, we found that the surface contributions are -140, -25 and 0.0 μ eV/atom for Ni, Cu/Ni and O/Ni surfaces, respectively. These results, together with the bulk contribution from the Ni atom in a tetragonally distorted lattice (-27 μeV/atom), agree very well with experimental data.

To elucidate the remarkable effect of O on the magnetic anisotropy energies, results of surface/interface density of states (DOS) are plotted in Fig. 4 for the Ni$_9$, Cu$_5$Ni$_9$, and O$_{0.5}$/Ni$_9$/O$_{0.5}$ slabs. The surface/interface effects are found to be limited mostly within the outmost two Ni layers. The DOS curves of the third Ni layer from the surface or interface (denoted as Ni(S-2)) are very close to each other in the three systems. The DOS of the

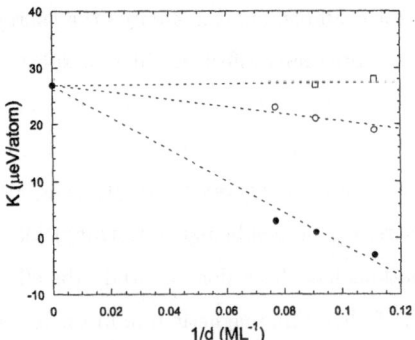

FIG. 3: The calculated magnetic anisotropy energies for clean Ni_n (filled circles) , Cu_5Ni_n (open circles) and $O/Ni_n/O$ (open squares) films plotted as function of the inversed Ni film thickness $1/d$.

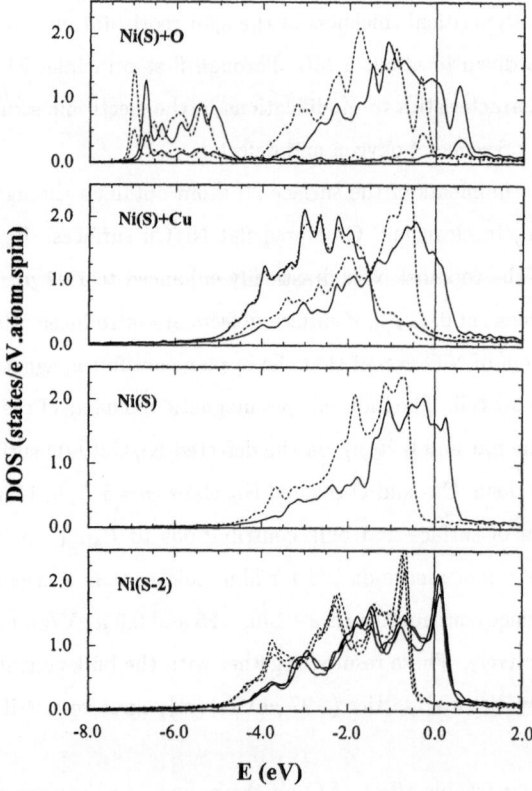

FIG. 4: The calculated density of states for Ni(S) (top three panels) and Ni(S-2) (the bottom panel) in Ni_9, Cu_5Ni_9 and $O/Ni_9/O$ films. The solid (dashed) lines are for the majority (minority) spin. Zero energy indicates the position of E_f. DOS curves of the interfacial O and Cu are also displayed in the top two panels, respectively.

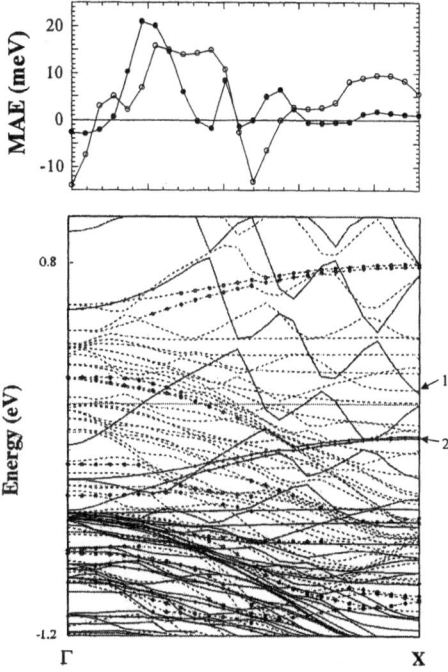

FIG. 5: The calculated bands (bottom panel) and distribution of MAE along the $\overline{\Gamma X}$ axis. In the bottom panel, the solid (dashed) lines are for the majority (minority) spin states, and the zero energy is the position of E_f. In the top panel, the line with filled circles is for the clean Ni_9 film. In the bottom panel, the bands with circles are those with more than 40% weight in the Ni(S) sphere.

surface Ni (denoted as Ni(S)) or even the subsurface Ni (not shown here), however, are significantly affected through hybridizations with Cu and O. In the top panel, one can find a pronounced antibonding O-Ni peak on top of the Ni d-band. This state, more clearly displayed in the bands of $O/Ni_9/O$ in Fig. 5(denoted as state "2"), is an O-induced surface state with the d_{xz} feature. The $\overline{\Gamma X}$ axis is chosen in Fig. 5 since analysis in the 2D BZ discloses that the effect of O on MAE of Ni films is mainly from the vicinity around the $\overline{X}$ point. This fact is also shown in the top panel of Fig. 5, where the MAE distribution of $O/Ni_9/O$ (the line with open circles) fluctuates around that of Ni_9 (the line with filled circles) but shows excessive positive contributions in the 1/3 of BZ near the $\overline{X}$ point. SO interaction between two states, namely, the empty minority spin state "1" (with mixed features of s and d_{z^2}) and the occupied majority spin state "2" (with the d_{xz} feature), is found to play a key role. Since they differ in magnetic quantum number by ± 1 and belong to different spin channels, the SO interaction between them leads to positive E_{MCA} [1]. This positive contribution is eventually overtaken by the others when the energy separation between "1" and "2" increases. Clearly, the O-induced surface state "2" is of a critical importance in the O-induced perpendicular magnetic anisotropy observed experimentally [41].

Finally, the presence of surface Ni vacancies was also found to promote positive MAE. Calculations for O on defected $Ni_4/Cu(001)$, for example, resulted in a MAE of +126 μeV per Ni atom.

3.2.   MAGNETO-CRYSTALLINE ANISOTROPY IN NANOWIRES

Magnetic nanowires fabricated on stepped surfaces [42] and in pinholes [43] have many significant applications such as spin filtering and high density magnetic recording. Monatomic chains of Co and Mn, the smallest possible nano-entities, were successfully fabricated through either self-assembled growth on Pt(997) [44] or tip manipulation in scanning tunneling microscope (STM) [45] experiments. On the ultra-small scale, magnetic properties display many peculiar behaviors that challenge theoretical explanations and predictions — especially for the size and shape dependence of the E_{MCA} that is crucial for the stability of magnetism in 1D systems.

Due to their intrinsic complexity and computational demands, very few first principles calculations for the determination of E_{MCA} in nanowires and nanoclusters have been presented so far. Tight-binding Hubbard model calculations revealed the complex behavior of E_{MCA} of unsupported small Fe clusters [46]. Lazarovits et al. calculated the magnetic properties of small Fe, Co, and Ni clusters on top of Ag(100) with the KKR Green's function method [47] and found that E_{MCA} for individual atoms strongly depends on their position [48].

In Table I, the calculated magnetic moments are given for three systems; free and supported (on both Cu(001) and Pt(001)) Co monatomic wires. The spin and orbital magnetic moments are 2.23 μ_B and 0.28 μ_B for a free-standing Co wire, respectively. Compared to the corresponding results for the free-standing Co monolayer with the same lattice constant (a=4.83 a.u.), the orbital magnetic moment of the 1D Co wire is more than doubled, whereas its spin moment is enhanced by only 10% [35]. Due to the proximity effect of Cu(001) the spin and orbital magnetic moments of Co are reduced roughly by 13 and 40%, respectively, compared to those in the free-standing case. However, on top of the Pt(001) substrate the reduction of the spin magnetic moment is less appreciable, whereas the decrease of the orbital magnetic moment is larger than in Co/Cu(001). This is rather surprising but not unreasonable since the large SOC of Pt and the strong Co-Pt hybridization affect $< L_z >$

TABLE I: Spin and orbital magnetic moments (in μ_B), and magnetic anisotropy energies (in meV/atom) for a free and supported monatomic Co wire

system	$< S_z >$	$< L_z >$	E_1	E_2	E_{shape}
Co	2.23	0.28	-1.29		
Co/Cu(001)	1.91	0.15	+0.35	-0.22	0.085
Co/Pt(001)	2.15	0.10	+1.02	-2.58	0.075
Mn/Cu(001)	3.73		+0.19	-0.29	
Mn/Pd(001)	4.05		-0.41	-0.39	
Mn/NiAl(110)	3.67		-0.25	-0.20	
Fe/NiAl(110)	2.82	0.09	-0.01	0.18	

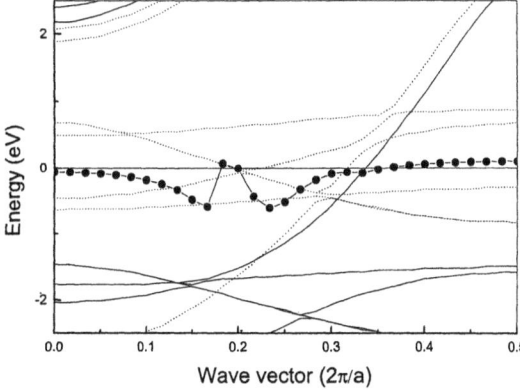

FIG. 6: Calculated band structure (thin solid lines) and E_{MCA} (solid circles, due to contributions of the minority spin states only) distribution along the one-dimensional Brillouin zone for a free Co wire

oppositely. The Co wire induces magnetic moments of 0.29 and 0.13 μ_B in the nearest and second nearest Pt surface atoms [49].

For nanowires, it is easy to show that the lowest non-vanishing energy terms can be expressed in the form

$$E = E_0 + \sin^2\theta[E_1 + E_2\sin^2\phi] \tag{14}$$

where θ is the polar angle of the magnetization away from the chain, while ϕ is the azimuthal angle in the plane perpendicular, measured from the surface normal direction for adsorption cases. For the free-standing cases, E_2 is zero.

The calculated magnetic anisotropy energies for free and supported Co monatomic wire are also listed in Table I. The negative E_1 of a free-standing Co monatomic wire means that the magnetic moment is aligned perpendicular to the wire axis. From the k-distribution of E_1 shown in Fig. 6, again, one can trace back an obvious correlation between E_{MCA} and the occupancies of a pair of states (d_{z^2} and d_{xz} if z-axis is set along the wire) in the minority spin channel (dashed lines). They are separated by the Fermi level and make large negative E_{MCA} in the first 60% of the one-dimensional Brillouin zone. When both of them are occupied around $k = 0.2\pi/a$, E_{MCA} change to the positive region.

For the Co wire on Cu(001), E_1 becomes positive with magnitudes of 0.35-0.39 meV/atom in both the (1×3) and (1×5) supercells. Clearly, E_2 appears to be more sensitive to the change of cell size [49]. Unfortunately, it is still a very hard task to explore the convergence with supercell size since the (1×5) cell is almost the maximum acceptable one for our current computing power. Nonetheless, the E_2 results from these calculations should still be relevant since the inter-wire separation for real applications is expected to be around 1 nm. From the results for E_1 and E_2, one can find that the easy axis in Co/Cu(001) turns along the wire. In Co/Pt(001), the magnitude of E_{MCA} is enhanced, primarily due to the strong SOC of the Pt atoms. The easy axis is perpendicular to the wire and is in the Pt(001) surface plane. No experimental data is available for direct comparison, at this time. The E_{MCA} of the Co wires in Co/Pt(997), with the easy axis also perpendicular to the chain, was measured to be -2.0 ± 0.2 meV/atom [44], a value that is close to our calculated $E_2 = -2.58$ meV/atom. This agreement, although encouraging, needs to be viewed cautiously since E_{MCA} is sensitive to subtle changes in the local environment.

If the size of the monatomic wire is further reduced to Co_2-Co_7 monatomic chains, $< L_z >$ and E_{MCA} are oscillatory while the spin magnetic moment is stable. For Co_2, Co_3, Co_4, Co_6 and Co_7, the nanochains have magnetization perpendicular to their axes, whereas Co_5 has parallel magnetization. As shown in Fig. 7, the total energy also displays a oscillatory tendency and indicates that Co_3, Co_4 and Co_6 are more stable compared to Co_2, Co_5 and Co_7. The energy corresponding to the two ends is 0.8 eV.

Note that neither the calculated E_{MCA} results for Co wires nor Co short chains are larger than those for Co thin films [1]. This indicates that one cannot achieve stable magnetic states in Co nanostructures at the ambient temperature by merely changing their size and shape. Instead, it was found to be possible to obtain large E_{MCA} by either placing Co on Pt surfaces [44, 50] or, furthermore, mixing with Pt to form binary chains [51].

3.3. HIGHHER-ORDER MAGNETO-CRYSTALLINE ANISOTROPY

The first principles determination of high-order E_{MCA} such as in cubic bulk magnetic Fe, Co and Ni metals or the in-plane anisotropy in a square lattice is still one of the challenging

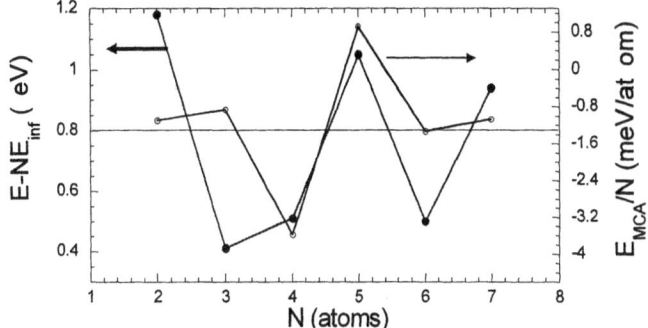

FIG. 7: The calculated total energy (red line) and anisotropy energy for the free standing Co_n chains with a fixed interatomic distance of 4.82 a.u. The horizontal line indicates the calculated E_{MCA} for an infinite Co monatomic wire with the same interatomic distance.

problems in condensed matter physics. The high-order E_{MCA} in these systems is extremely small – only about $1\mu eV/atom$. Quantitatively, such an energy difference scale is very close to or beyond the limit of precision of total energy calculations for most approaches. Our FLAPW calculations found that the in-plane coefficients [52] of E_{MCA} is extremely sensitive to the change in environment and to structural relaxation[35]. For cubic magnetic crystals, early calculations with the force theorem repeatedly gave the wrong sign for either Fe or Ni (or both). Using the LMTO-ASA approach, Daalderop et al. [53] obtained the right magnitude of E_{MCA}, but the wrong easy axis for hcp Co and fcc Ni. Guo et al. [54] and Strange et al. [55] also obtained the wrong easy axis or wrong magnitude for bulk Fe and Ni in their LMTO and KKR calculations. Based on the full potential LMTO method, Trygg et al. [56] treated the SOC Hamiltonian self-consistently but obtained (with $spd-$basis functions) almost the same results as Daalderop et al. The accuracy of total energy calculations was re-examined by Halilov et al [57], also using the LMTO-ASA method with combined corrections. They obtained the correct easy axis for all three metals through total energy calculations with a larger set of k-points. Beiden et al. [58] implemented a real-space locally self-consistent multiple scattering method. Again, they obtained the wrong easy axis for Ni and oscillatory results for hcp Co.

We extended the torque method [34] for the determination of E_{MCA} in cubic crystals, where the total energy can be well approximated in the form

$$E = E_o + K_1(\alpha_1^2\alpha_2^2 + \alpha_1^2\alpha_3^2 + \alpha_2^2\alpha_3^2) + K_2\alpha_1^2\alpha_2^2\alpha_3^2 \qquad (15)$$

Here α_1, α_2 and α_3 are directional cosines referred to the cube edges along the x, y and z

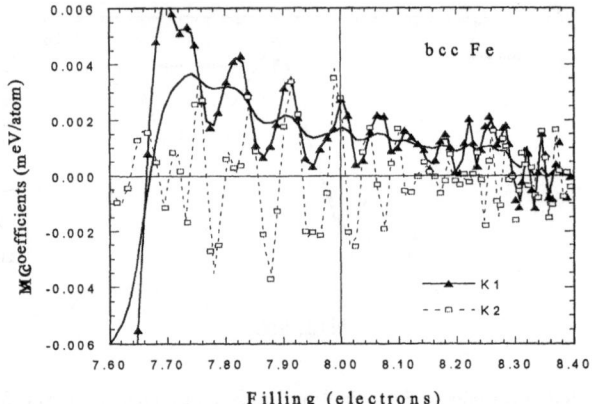

FIG. 8: MCA coefficients of the bulk bcc Fe with GGA. The bold solid lines are for k_1 (with either 0 or 15 meV Gaussian broadening) while the thin dashed line is for k_2.

axes. Clearly, the E_{MCA} can be evaluated from the K_1 and K_2 coefficients as

$$E_{111} - E_{001} = \frac{K_1}{3} + \frac{K_2}{27}; \qquad E_{110} - E_{001} = \frac{K_1}{4} \qquad (16)$$

As in Eq. 17 for thin films, the torque here, $T(\theta)$, is defined as the derivative of the total energy with respect to the polar angle away from the z-axis (denoted as θ below). To determine the values of K_1 and K_2, we focus on two special cases: (1) for $\alpha_2 = 0$ $(\alpha_1^2 = 1 - \alpha_3^2)$, we have

$$T_1(\theta) \equiv \frac{dE(\theta)}{d\theta}|_{\phi=0^\circ} = \frac{K_1}{2}\sin(4\theta), \qquad (17)$$

and (2) for $\alpha_1 = \alpha_2$ $(\alpha_1^2 = \alpha_2^2 = (1 - \alpha_3^2)/2)$

$$T_2(\theta) \equiv \frac{dE(\theta)}{d\theta}|_{\phi=45^\circ} = T_1(\theta) + \frac{\sin(2\theta)\sin^2(\theta)}{4}[2K_1 + K_2(3\cos^2\theta - 1)], \qquad (18)$$

where ϕ denotes the azimuzal angle in the xy plane. We have

$$T_1(\theta = 22.5^\circ) = K_1/2 \qquad\qquad T_2(\theta = 45^\circ) = (2K_1 + K_2/2)/8, \qquad (19)$$

and finally the MAE coefficients K_1 and K_2 can be evaluated very efficiently through

$$K_1 = 2T_1(\theta = 22.5^\circ) \qquad\qquad K_2 = 16T_2(\theta = 45^\circ) - 4K_1, \qquad (20)$$

The calculated K_1 and K_2 for bcc Fe with the FLAPW-GGA approach are given in Fig. 8. While K_1 is found to still be oscillatory but stable in sign with $70\times70\times70$ k-points in the full BZ for the cubic cell (with two atoms), K_2 changes its sign very rapidly. It appears that more k-points are needed to get a converged result for K_2. Fortunately, K_2 remains very small and does not affect the anisotropy energy too much (cf. the 1/27 scaling factor in Eq. 16 for $E_{111} - E_{001}$). As listed in Table II, the calculated E_{MCA} ($E_{111} - E_{001}$) for bcc bulk Fe is 0.9 (0.7) μeV/atom with GGA (LDA) formula (the (001) direction is the easy axis). The LDA value is very close to the results obtained by Daalderop et al. (0.5 μeV/atom, with the force theorem) [53], Trygg et al. (0.5 μeV/atom, with total energy) [56] and Beiden et al. (0.78 μeV/atom, with a real space approach). The discrepancy between theory (0.5-0.9 μeV/atom) and experiment (1.4 μeV/atom) appears not to be due to numerical problems, but to other physical reasons such as the neglected possible orbital polarization [56, 59].

For fcc Co and Ni, the E_{MCA} calculated results with the FLAPW approach are also very close to those obtained in previous density-functional calculations (cf. Table II). With $50 \times 50 \times 50$ k-points in the full BZ for the cubic cell (with four atoms), the correct easy axis is obtained for Co, but not for Ni. In addition, the theoretical E_{MCA} results are much smaller in magnitude than experiment. The failure of the density-functional description for E_{MCA} in bulk fcc Ni appears to be mainly due to $s - d$ charge transfer. Note that with a=6.66 a.u., the calculated spin magnetic moments of Ni are 0.62 and 0.67 μ_B with LDA and GGA, respectively. These values are markedly larger than the experimental result, 0.57 μ_B. Since the spin magnetic moment in Ni is almost equal to the number of holes in its minority spin d-band, this discrepancy indicates that the Ni-d band has about 0.05-0.10 fewer electron than what it should have. Such an error in band filling is enough to change the sign of E_{MCA}. Indeed, E_{MCA} ($E_{111} - E_{001}$) could change sign by moving only 0.03 electron from the Ni s-band to its d-band.

3.4.   MAGNETOSTRICTION

In general, the size of the magneto-elastic strain induced by rotation of the magnetization depends on the directions of the measured strain and of the spin moment with respect to the crystalline axes of the material. For a cubic material, the directional dependence of the fractional change in length can be expressed in terms of the direction cosines of the magnetization (α_i) and of the strain measurement direction (β_i) with respect to the crystalline axes [60]

$$\frac{\Delta l}{l_0} = \frac{3}{2}\lambda_{001}[\sum_{i=1}^{3} \alpha_i^2 \beta_i^2 - \frac{1}{3}] + 3\lambda_{111}\sum_{i \neq j}^{3} \alpha_i \alpha_j \beta_i \beta_j. \tag{21}$$

If the measurement is carried out along the (001) direction for example, $\beta_x = \beta_y = 0$ and $\beta_z = 1$, then Eq. 21 can be simplified as $\frac{\Delta l}{l_0} = \frac{3}{2}\lambda_{001}[\alpha_z^2 - \frac{1}{3}]$ or further, for systems with a

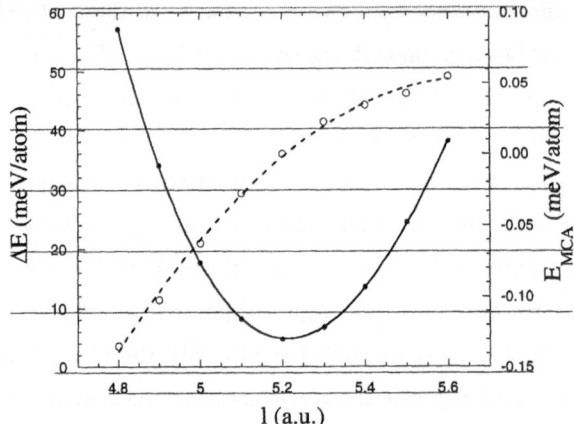

FIG. 9: The calculated total and MCA energies of the bulk bcc Fe (with LDA) with respect to the length of c-axis. The constant volume distortion mode was adopted.

single domain,

$$\lambda_{001} = \frac{2}{3} \frac{l_0(\alpha_z = 1) - l_0(\alpha_z = 0)}{l_0(\alpha_z = 1)}. \tag{22}$$

Clearly, λ_{001} represents the change in length along (001) when the magnetization turns from the x, y plane to the z direction.

The equilibrium length along the z direction, l_0, can be obtained by fitting the calculated total energy as a quadratic function of l

$$E(\alpha_z = 1) = al^2 + bl + c; \qquad E(\alpha_z = 0) = E(\alpha_z = 1) + E_{MCA}(l) \tag{23}$$

and so

$$\lambda_{001} = -2E'_{MCA}/3b. \tag{24}$$

Here $E'_{MCA} = dE_{MCA}/dl$, which is much smaller than the value of b. Note that b is always negative (since both a and l_0 are positive), and thus λ and E'_{MCA} have the same sign.

As an important benchmark test, the magnetostriction coefficients of cubic bulk magnetic transition metals are studied first. As shown in Fig. 9, the calculated MCA energy for bulk bcc Fe is a smooth monotonic function of the vertical strain. The slope of E_{MCA} and thus the magnetostrictive coefficient (λ_{001}) are found to be positive for Fe and Co, but negative for Ni[61]. This means that bulk Fe and Co (Ni) stretch (shrink) along the direction of magnetization, a conclusion that agrees well with experiment.

Quantitatively, the value of λ_{001} depends sensitively on the distortion mode (i.e., Poisson's ratio). As listed in Table II, the value of Poisson's ratio for Fe, Co and Ni optimized through total energy minimization is about 0.40, which is very close to that obtained using

TABLE II: Calculated equilibrium lattice constants a (in a.u.), Poisson's ratio (σ), spin and orbital magnetic moments (M_S and M_L, in μ_B), $E_{111} - E_{001}$ (E_{MCA}, determined with the experimental lattice constants, in μeV) and magnetostriction coefficients (in 10^{-6}) obtained with LDA and GGA corrections.

	a	σ	M_S	M_L	E_{MCA}	λ_{001}
bcc Fe						
LDA	5.20	−0.409	2.05	0.048	0.7	52
GGA	5.37	−0.486	2.17	0.045	0.9	29
EXP	5.41	−0.368	2.22	0.08	1.4	21
fcc Co						
LDA	6.48	−0.374	1.59	0.076	—	92
GGA	6.67	−0.396	1.66	0.073	−0.6	56
EXP	6.70	—	1.72	—	−1.8	79
fcc Ni						
LDA	6.46	−0.332	0.62	0.049	0.7	−63
GGA	6.64	−0.376	0.66	0.050	0.8	−56
EXP	6.66	−0.376	0.57	0.05	−2.7	−49

the measured elastic stiffness constants ($\sigma = -c_{12}/(c_{11} + c_{12})$) for bulk Fe and Ni (0.37-0.38). As a result, satisfactory quantitative agreement is achieved for λ_{001} between our (zero temperature) theory and experiment. The theoretical result can be further improved by using the GGA [26]. As seen in Table II, LDA leads to a 3% underestimation for the lattice constant and a more substantial difference for the spin magnetic moments at the equilibrium geometry. With GGA, most of the calculated values of the various magnetic properties are closer to experiment, especially for Fe in which the number of holes with majority spin is very sensitive to the change of environment. Note that Eq.24 can be extended for the determination of magnetostrictive coefficients along other orientations (e.g., (110) and (111)). Our recent calculations [62] found that the calculated λ_{111} for bulk Fe deviate significantly from the experimental data with either LDA or GGA approaches.

The inverse effect of magnetostriction is strain-induced uniaxial E_{MCA}, which plays a significant role in epitaxial magnetic films. Specially, due to the known deficiency of LDA, the interlayer distances are usually underestimated and thus it is vital to use GGA for the structural optimization. For example, with the in-plane lattice constant fixed (a=4.83 a.u.), the optimized Co-Cu interlayer distance in Co/Cu(001) and Cu/Co/Cu(001) is 3.44 a.u.

TABLE III: Calculated lattice constants a (in plane, in a.u.) and c (along z, in a.u.), magnetic moments (M, in μ_B), the Poisson ratio (σ), magneto-crystalline anisotropy energy (E_{MCA}, in μeV/cell) and magnetostriction coefficient (λ_{001}, in 10^{-6}). The corresponding experimental data are given in parentheses

	FeCo	FeCo$_3$	FeNi	FeNi$_3$	CoNi	CoNi$_3$
a	5.38 (5.39)	6.70	6.76 (6.76)	6.70 (6.71)	6.62 (6.67)	6.66 (6.65)
c	5.38 (5.39)	6.70	6.76 (6.76)	6.70 (6.71)	6.78 (6.67)	6.66 (6.65)
E_{MCA}	0	0	63	0	143	0
σ	−0.35	−0.36	−0.33	−0.35	−0.34	−0.36
λ_{001}	83(126)	-68	10(10-26)	27(13)	42 (40-100)	33

from GGA calculations [35], but it is only 3.08-3.11 a.u. if the LDA is adopted [63, 64]. The large difference in atomic arrangement, and as the gradient correction in exchange-correlation functionals, may strongly affect all the magnetic properties such as magnetic moments, magnetic ordering and E_{MCA} [65].

Now, Ni$_x$Fe$_{1-x}$ and Ni$_x$Co$_{1-x}$ magnetic alloys are widely used in magnetic recording technology and invar materials. The calculated magnetostrictive coefficients are listed in Table III. Reasonable agreement has been achieved for the alloys studied [66]. Our recent studies for FeGa alloys indicate that the observed giant magnetostriction is not necessarily due to the ground state. As shown in Fig. 10, while the B$_2$-like structure is unstable under tetragonal distortion and lies higher in energy that the L1$_2$ and DO$_3$ structures, it provides positive magnetostrictive coefficients (as observed in experiments) while the other two phases give negative magnetostriction. Owing to the reduced hybridization between Fe atoms, the magnetostrictive coefficient, λ_{001}, is enhanced by a factor of 10-20 compared to that for bulk bcc Fe.

Further, rare-earth intermetallic compounds have attracted great attention since the late 1960's due to their extraordinary magnetic properties, especially their large magnetostrictive coefficients (10^{-3}) at room temperature [60, 67, 68]. While it was believed that the localized rare-earth $4f$ states play a dominant role in magnetization and magneto-elastic coupling, recent experiments found that the effects of itinerant states can be equally important [60, 69]. Although a phenomenological approach was developed long ago to describe the dependence of single crystal magnetostriction on magnetization and measurement directions, the magnetostrictive coefficient for a given material, especially the contribution of itinerant electrons, has never been accurately calculated [60].

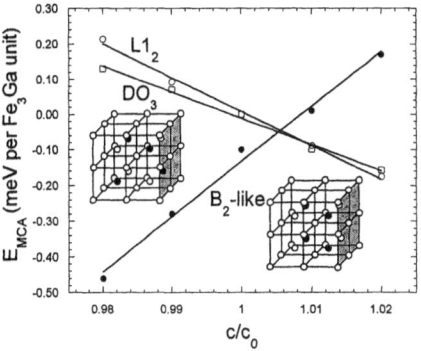

FIG. 10: Calculated strain-induced magneto-crystalline anisotropy energies of Fe_3Ga. The DO_3 and B_2-like structures are shown as insets; where the open circles represent Fe and the solid circle represent Ga.

TABLE IV: Calculated and measured (in parentheses) lattice constant, a, total magnetic moment per primitive unit cell, M, spin magnetic moment of itinerant electrons in rare-earth (M_{RE}) and transition metal M_{TM}, and magnetostrictive coefficient, λ_{001}, for different compounds.

	a (a.u.)	$M(\mu_B)$	$M_{RE}(\mu_B)$	$M_{TM}(\mu_B)$	$\lambda_{001}(10^{-6})$
GdCo$_2$	13.72(13.68)	4.99(4.9)	0.46	-1.24	-407(-1200)
NdCo$_2$	13.84(13.77)	5.28(3.8)	0.32	-1.14	-171
SmCo$_2$	13.74(13.71)	2.73(2.0)	0.52	-1.26	-290
ErCo$_2$	13.70(13.50)	7.06(7.0)	0.28	-1.10	-516(-1000)
GaFe$_2$	13.85(13.94)	3.85(2.8)	0.58	-1.96	44(39)

Very recently, we investigated the magnetostrictive properties of several rare-earth compounds in the C15 cubic Laves phase structure, a close-packed arrangement of spheres with two different sizes. As listed in Table IV, the calculated magnetostrictive coefficient agree with the experimental data available. Furthermore, the large values of λ_{001} indicate that the contribution from itinerant electrons is comparable to that from the f-shell. Using a rigid band picture, we found that the difference in the magnetostrictive behavior of GdFe$_2$ and GdCo$_2$ is mainly due to the change in band filling. A GdFe$_x$Co$_{2-x}$ compound (x=0.9-1.2) is predicted to be a strong magnetostrictive material with a positive sign for λ_{001} ($+450\times10^{-6}$)

4. Noncollinear Magnetism Phenomena at Surfaces and Interfaces

4.1. INTRODUCTION

The control and design of nano-scale structures and their magnetic properties are challenging issues in both basic and applied magnetism; in these, the complexity of the local magnetic structures such as surfaces, interfaces and domain walls plays a key role. To date, these artificial nanostructures – consisting, for example, of transition metal overlayers and superlattices – lead to perpendicular magnetocrystalline anisotropy (MCA)[70], exchange bias in ferromagnetic (FM) and antiferromagnetic (AFM) bilayers[71, 72], constricted domain walls as in a quantum spin interface[73, 74] and curling nano-scale dots as new candidates for nonvolatile memory applications[75].

As seen above, theoretical first-principles calculations for the prediction of these properties have also greatly developed. The significant progress obtained includes predictions of surface/interface magnetism such as enhanced magnetic moments and magnetic phenomena induced by the spin-orbit coupling (SOC)[1]. Indeed, very reliable results for the magnetocrystalline anisotropy (MCA) energy can now be obtained for most magnetic thin films (of order a few tenths of a meV/atom) and for magnetic cubic bulk materials (a few μeV/atom). However, one unresolved challenging issue remains to be resolved – the role of noncollinear magnetism (NCM) that arises not only through the SOC[76] but also from the breaking of symmetry at surfaces and interfaces[77, 78].

We have recently developed our highly precise bulk and film (single slab) full-potential linearized augmented plane wave (FLAPW) method[28] to treat NCM with no shape approximations for the magnetization[79]. To date, this new capability has proven highly successful in describing complex NCM phenomena in bulk such as: (i) canted ferromagnetism and its possible coexistence with superconductivity in $RuSr_2GdCu_2O_8$[80] and (ii) the stabilization of the 3D NCM structure of FeMn (a possibly useful candidate as an exchange bias material)[81].

In this section of the chapter, we focus on noncollinear magnetism phenomena induced at surfaces, interfaces, domain walls and in the magnetic vortex core of a quantum dot; we present some results obtained with our newly developed FLAPW method that includes intra-atomic NCM[79, 82, 83]. These results serve to illustrate the new level of performance of these modern computational simulations in making reliable predictions of complex materials/properties and to provide effective tools to help in the search for new magnetic materials with desirable properties for device applications.

4.2. IMPLEMENTATION OF FLAPW METHOD INCLUDING INTRA-ATOMIC NON-COLLINEAR MAGNETISM

In contrast with a collinear magnetic state that restricts magnetic moments to be parallel or antiparallel to a spin-quantization axis, a noncollinear magnetic state allows a full spatial variation of the spin directions. In this case, the electron density in density-functional theory (DFT) is treated with a 2×2 density matrix,[6, 84, 85]

$$\rho(\mathbf{r}) = \rho_0(\mathbf{r})\mathbf{I} + \mathbf{m}(\mathbf{r}) \cdot \boldsymbol{\sigma} \tag{25}$$

where I and $\boldsymbol{\sigma}$ are the unit matrix and Pauli spin matrix, respectively; $\rho_0(\mathbf{r})$ and $\mathbf{m}(\mathbf{r})$ are defined in a global coordinate reference and correspond to a charge density and a vector magnetization quantity. The off-diagonal elements in Eq. (25) give the magnetization perpendicular to the z axis. Following DFT, the variational procedure for the total energy functional with respect to $\rho(\mathbf{r})$ leads to the Kohn-Sham single-particle equations that determine a ground state,

$$\mathcal{H} = \{\mathcal{H}_{\mathrm{kin}} + V_0(\mathbf{r})\}\mathbf{I} + \mathbf{V}(\mathbf{r}) \cdot \boldsymbol{\sigma} \tag{26}$$

where $\mathcal{H}_{\mathrm{kin}}$ is a kinetic energy term for a noninteracting single particle and $V_0(\mathbf{r})$ and $\mathbf{V}(\mathbf{r})$ are nonmagnetic and magnetic parts of an effective potential (V_{eff}) given by external, Coulomb and exchange-correlation potentials. The potential is represented within the full-potential scheme prescribed by Weinert[86] and applied in the local spin density approximation (LSDA) using the Hedin-Lundqvist exchange-correlation[6].

In LSDA, where the exchange-correlation potential depends only on position, the effective potential may be evaluated by use of the spin $\frac{1}{2}$-rotation matrix $U(\theta(\mathbf{r}), \phi(\mathbf{r}))$[87], where $\theta(\mathbf{r})$ and $\phi(\mathbf{r})$ are the polar angles of the direction of magnetization $\mathbf{m}(\mathbf{r})$ at a point. The $U(\theta(\mathbf{r}), \phi(\mathbf{r}))$ transforms the global coordinate reference to a local one that parallels the z-axis magnetization direction. At a given $\mathbf{r}$, the $\rho(\mathbf{r})$ can be diagonalized as

$$\rho(\mathbf{r}) = U^\dagger(\theta(\mathbf{r}), \phi(\mathbf{r})) \begin{pmatrix} \rho_+(\mathbf{r}) & 0 \\ 0 & \rho_-(\mathbf{r}) \end{pmatrix} U(\theta(\mathbf{r}), \phi(\mathbf{r})), \tag{27}$$

and then the effective potential is given by

$$V_{\mathrm{eff}}(\mathbf{r}) = U^\dagger(\theta(\mathbf{r}), \phi(\mathbf{r})) \begin{pmatrix} v_{\mathrm{eff}}(\rho_+(\mathbf{r})) & 0 \\ 0 & v_{\mathrm{eff}}(\rho_-(\mathbf{r})) \end{pmatrix} U(\theta(\mathbf{r}), \phi(\mathbf{r})), \tag{28}$$

where $v_{\mathrm{eff}}(\rho_\pm(\mathbf{r}))$ is the effective potential exactly the same as that in collinear ferromagnetic electron gas theory[6]. The transformation is performed over all space, in contrast to the atomic sphere approximation[84, 85] where a single local spin direction with a spherically averaged magnetization is defined in each sphere.

The basis functions are specified with the LAPW basis[28]. In the noncollinear magnetic system, the plane wave is augmented with a spin-independent LAPW basis in order to avoid discontinuity in augmenting the basis functions at the MT and vacuum boundaries. Thus, the radial functions $u_\ell(\mathbf{r})$ and $\dot{u}_\ell(\mathbf{r})$ are derived from the radial Schrödinger and the energy derivative equations at energy parameters E_ℓ, using the nonmagnetic part of the spherical potential, $V_{0,\ell=0}(r)$. For the vacuum region, the basis function is augmented by solutions of the one-dimensional (out of surface-plane) Schrödinger and energy derivative equations at E_{vac}, using the nonmagnetic part of the potential in the vacuum region, $V_{\mathrm{vac},k_\parallel=0}(z)$. The coefficients $A_{\ell m}$ and $B_{\ell m}$ for the MT sphere and A_{vac} and B_{vac} for the vacuum are determined by requiring that the functions are continuous and differentiable at their boundary. The eigenstate at $\mathbf{k}$ in the first BZ is now expressed by

$$\Psi_{\mathbf{k}}(\mathbf{r}) = \sum_{\mathbf{K}=\mathbf{k}+\mathbf{G}}^{\mathbf{G}_{\max}} \begin{pmatrix} C_{\mathbf{K},\chi_+}\psi_{\mathbf{K}}(\mathbf{r}) \\ C_{\mathbf{K},\chi_-}\psi_{\mathbf{K}}(\mathbf{r}) \end{pmatrix}. \tag{29}$$

In the relativistic case, the corrections are first achieved in the scalar relativistic approximation[88–90] as done in the collinear calculations. This is performed by replacing the radial Schrödinger equation for obtaining the radial functions u_ℓ and $\dot{u}_\ell$ by the corresponding Dirac equation excluding the SOC term, using the non-magnetic part of the spherical potential $V_{0,\ell=0}(r)$. Thus, both the large and small components are used to evaluate the scalar relativistic Kohn-Sham Hamiltonian[88] ($\mathcal{H}_{scalar}$) matrix without the SOC term ($\mathcal{H}_{soc}$). The coefficients $A_{\ell m}$ and $B_{\ell m}$ in the LAPW basis are determined by enforcing the matching at the MT boundary using the large component and its derivative since the relativistic effects in the interstitial region are assumed negligible. This has an advantage for the implementation of our existing programmed code with minimal changes without great effort. The $\mathcal{H}_{soc}$ is then incorporated, which is approximated by using the non-magnetic part of the spherical potential, $V_{0,\ell=0}(r)$, in the MT sphere.

Having generated the full Hamiltonian matrix, $(\mathcal{H}_{scalar} + \mathcal{H}_{soc})_{\mathbf{K}',\mathbf{K}}$, and the overlap matrix, $S_{\mathbf{K}',\mathbf{K}}$, a diagonalization is carried out in the whole spin-space since spin-up and spin-down wavefunctions are no longer independent. Now the calculations require full self-consistency for the density matrix, i.e., the magnetization direction as well as the magnitude is determined self-consistently. The density matrix for the iterative procedure is generated by summing over the resultant eigenvectors of the occupied states as

$$\rho(\mathbf{r}) = \sum_i^{occ} \Psi_i^\dagger \Psi_i. \tag{30}$$

4.3. APPLICATIONS OF FLAPW METHOD TO SRRFACES, INTERFACES, DOMAIN WALLS AND VORTEX CORE

4.3.1. Enhancement of Magnetocrystalline Anisotropy in Ferromagnetic Fe Films by Intra-atomic Noncollinear Magnetism

As discussed above, at surfaces and interfaces, the SOC interaction is enhanced due to

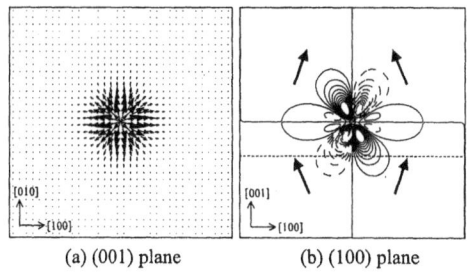

(a) (001) plane   (b) (100) plane

FIG. 11: In-plane components of the magnetization density, $m_{\parallel}(r)$, in a unit cell for a free-standing Fe(001) monolayer with an Ag (001) lattice constant of 5.45 a.u. when the average magnetic moment lies along the out-of-plane direction: (a) the $m_{\parallel}(r)$ on a (001) plane below $\frac{3}{10}a$ from the nucleus along [001], where the moment directions and magnitude are represented by arrows and their size, respectively; (b) a contour map of the $m_{\parallel}(r)$ on a (100) plane, where solid and broken lines indicate the moments canting in [100] and [$\bar{1}$00] directions from the z axis, respectively, as schematically illustrated by the arrows, and a horizontal broken line corresponds to the height of the (001) plane of (a).

a reduction of dimensionality, in which the MCA energy is as large as 10^{-4} eV. Recent first principles calculations[1] predict the MCA energy with the correct sign and value and reveal the origin of the MCA from detailed band structures. To determine the effects of intra-atomic NCM – also induced by the SOC – we applied our FLAPW method including intra-atomic NCM and SOC to the case of free-standing Fe monolayers with lattice constants matching fcc Ag(001) (a=5.45 a.u.) and Cu(001) substrates (a=4.83 a.u.).

Figure 11 shows the in-plane components of the spin magnetization density, $m_{\parallel}(r)$, on (100) and (001) planes for a=5.45 a.u. when the average moment parallels the out-of-plane direction. The intra-atomic NCM is clearly observed with a four fold symmetry around the z axis, and the components of the positive and negative parts cancel out when integrated over a unit cell. It is also found that the magnetization density loses mirror symmetry with respect to the layer-plane as a result of the SOC. The moments below the nucleus cant with respect to the in-plane direction on going away from the center of the atom (Fig. 11(a)), while those above the nucleus cant in the opposite direction, as schematically illustrated by arrows in Fig. 11 (b). The noncollinearity is emphasized near the nucleus where $m_{\parallel}(r)$ is close to 0.01 $\mu_B/(a.u.)^3$, due to the strong SOC effects arising from the large gradient of the potential.

Table V summarizes the calculated E_{MCA} and the integrated spin and orbital moments in the MT spheres when the average moment lies in the out-of-plane direction. For comparison, we present the E_{MCA} value determined self-consistently by the standard FLAPW method

TABLE V: E_{MCA} (in meV/atom) and integrated spin and orbital moments (m_{spin} and m_{orbit} in μ_B) in MT spheres for free-standing Fe(001) monolayers with lattice constants matching fcc Ag(001) and Cu(001) substrates.

	a=5.45 a.u. (Ag)			a=4.83 a.u. (Cu)		
	E_{MCA}	m_{spin}	m_{orbit}	E_{MCA}	m_{spin}	m_{orbit}
NCM	0.28	3.16	0.16	0.36	3.01	0.10
CM	0.24	3.16	0.16	0.30	3.01	0.10

for collinear magnetism (CM) combined with the second variational technique for the SOC. Both results show the spin and orbital moments to energetically favor pointing in the out-of-plane direction for both a=5.45 a.u. and 4.83 a.u. With decreasing lattice constant, the moments decrease due to greater hybridization while the E_{MCA} increases. The values of the spin and orbital moments in the NCM state are almost identical to those in the CM state. However, a small enhancement in E_{MCA} is observed (by 0.04 meV for a=5.45 a.u. and 0.06 meV for a=4.83 a.u.) which corresponds to an increase of 17-20% of the values in the CM states, caused by the presence of the intra-atomic NCM.

The calculated band structure along high symmetry directions in the 2D BZ for a=5.45 a.u., when the average magnetic moment lies along the out-of-plane direction, is shown in Fig. 12. The bands crossing E_F arise mainly from the minority-spin states. The majority-spin bands are almost fully occupied, and are located from -1 to -4 eV below E_F. Bands 3 ($d_{x^2-y^2}$) and 4 (d_{xy}) located above and below E_F, whose orbitals lie in the layer-plane, show a large dispersion due to the strong bonding between them: band 1, whose $d_{3r^2-z^2}$ orbital lies along the plane normal and shows a weak dispersion, is located above E_F except near $\bar{\Gamma}$; bands 5 and 5* (d_{xz} and d_{yz}) localize in crossing E_F, and are degenerate at $\bar{\Gamma}$ and $\bar{M}$ if the SOC is neglected.

In this band structure, the MCA originates mainly from the SOC between the minority-spin bands crossing E_F. According to perturbation theory[33], the SOC between occupied and unoccupied states with the same (different) m quantum number through the L_z (L_x and L_y) operator gives a positive (negative) contribution to the E_{MCA}. As seen in Fig. 12, in most of the BZ except near $\bar{M}$ and $\bar{\Gamma}$, the contribution is mostly positive from the SOC between the occupied band 5 and the unoccupied band 5*. However, near $\bar{M}$, although the SOC between bands 5 and 5* makes no contribution to the E_{MCA} since both bands are occupied, they are coupled to the unoccupied band 1, which gives a negative contribution so the moments prefer to cant relative to the in-plane direction. Near $\bar{\Gamma}$, the unoccupied

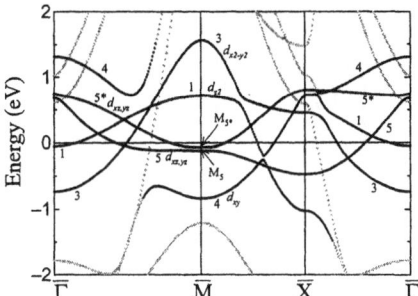

FIG. 12: Calculated band structures along high-symmetry directions for a free-standing Fe(001) monolayer with a lattice constant of 5.45 a.u. when the average magnetic moment lies along the out-of-plane direction. The bands 1, 3, 4 and 5 (5*) are composed of $d_{3z^2-r^2}$, $d_{x^2-y^2}$, d_{xy}, d_{xy} and $d_{xz(yz)}$ orbitals.

bands 5 and 5* couple to the occupied band 1, which also makes the moments cant relative to the in-plane direction. Through the SOC, these band effects lead to the intra-atomic noncollinear magnetism seen in Fig. 11.

The unperturbed degenerate states (i.e., at -0.1 eV below E_F at $\bar{M}$ and at 0.7 eV above E_F at $\bar{\Gamma}$) are split by the SOC, for which we found that an additional intra-atomic noncollinear magnetism is introduced. For example, we observed an energy gap of 54 meV for the noncollinear magnetic state and 53 meV for the collinear state, between the $\bar{M}_5$ and $\bar{M}_{5^*}$ states (Fig. 12). Using only the eigenvector for either the $\bar{M}_5$ state or the $\bar{M}_{5^*}$ state, we calculated the magnetization density. Figures 13 (a) and (b) show the in-plane components for the $\bar{M}_5$ and $\bar{M}_{5^*}$ states, respectively, on the (001) plane $\frac{3}{10}a$ below the nucleus along [001]. Interestingly, the moments from the $\bar{M}_5$ state cant going away from the center of the atom, while those from the $\bar{M}_{5^*}$ state cant in the opposite way, although both moments satisfy fourfold symmetry around the z axis.

4.3.2. Noncollinear Magnetic Structures at the Ferromagnetic NiFe and Antiferromagnetic NiMn Interface

Unidirectional magnetic anisotropy — the so-called exchange bias associated with anisotropy at an interface between ferromagnetic (FM) and antiferromagnetic (AFM) materials — holds promise as a useful phenomenon in controlling and designing magnetic applications such as storage and sensor devices. Theoretical investigations[91–96], mostly based on a model Hamiltonian or micromagnetic approach including effects of domain wall

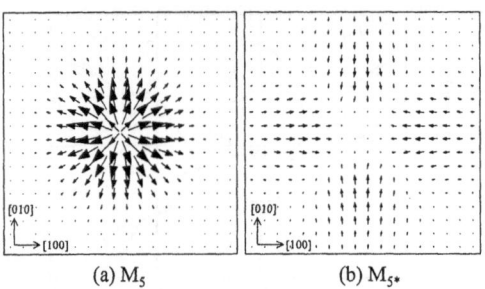

(a) M_5 (b) M_{5*}

FIG. 13: In-plane components of magnetic moments, calculated by using an eigenvector at either (a) the $\bar{M}_5$ state or (b) the $\bar{M}_{5*}$ state in Fig. 4, on a (001) plane below $\frac{3}{10}a$ from the nucleus along [001] for a free-standing Fe(001) monolayer with a lattice constant of 5.45 a.u. The moment directions are represented by arrows and the magnitude is in arbitrary units.

formation, roughness and a perpendicular (spin-flop) coupling at the FM/AFM interface, have been extensively performed to account for the exchange bias. Although these give intuitive pictures, little is known about the complexity of magnetic structures at FM/AFM interfaces quantitatively.

We applied the FLAPW method to determine magnetic structures including noncollinear magnetism at the compensated FM/AFM interface of NiFe/NiMn[97–101] of exchange bias films. Here, (001) and (111) interfaces are considered; the (111) interface was observed in experiments[99, 100]. As models, the $(NiFe)_2/(NiMn)_2$ superlattice structure with a (001) interface and the $(NiFe)_3/(NiMn)_3$ superlattice structure with a (111) interface were adopted, in which an $L1_0$ atomic ordering in the NiFe and NiMn layers was assumed. The lattice parameters of the NiMn layer were assumed to be those of the experiment. For the NiFe layer, the lattice parameters of the basal plane were assumed to match those of the NiMn layer, but with the c/a ratio chosen to preserve the experimental atomic volume of the $Ni_{0.5}Fe_{0.5}$ alloy. We confirmed that this structural approximation does not significantly alter the magnetic properties of NiFe found previously[102]. Note that ideally the (001) and (111) interfaces have no net magnetic moments on the AFM NiMn interface since the Mn moments align antiparallel to each other, i.e. compensated FM/AFM interfaces. The LSDA self-consistent calculations were carried out without SOC.

Figures 14 and 15 show the calculated magnetization density on the Fe, Ni and Mn layers for the (001) interface, and the NiFe and NiMn layers for the (111) interface, respectively. Here the moments lie parallel to their interface. In both cases, the average FM Fe moments

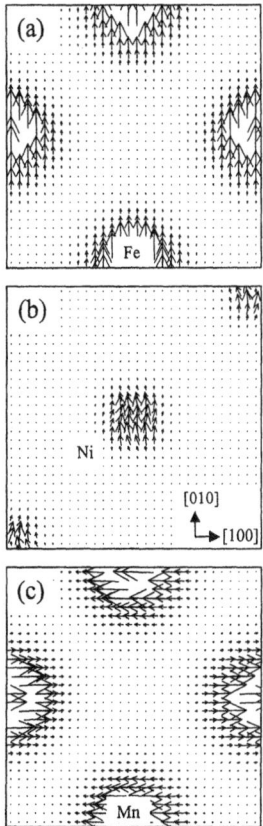

FIG. 14: Magnetization density on (a) FM Fe, (b) Ni and (c) AFM Mn layers at the (001) interface for the $(NiFe)_2/(NiMn)_2$ superlattice structure. The moment direction and the magnitude are represented by arrows and their size.

in Fig. 14 (a) and Fig. 15 (a) align perpendicular to the average AFM Mn moments in Fig. 14 (c) and Fig. 15 (b), respectively. The integrated Mn and Fe moments in the muffin-tin spheres at the (001) interface are 3.1 and 2.5 μ_B, and for the (111) interface, 3.0 and 2.5 μ_B, respectively. Those values are slightly smaller than those in their bulk[102]. The Fe and Mn moments cant slightly away from their FM or AFM axes. The canting of the Mn moments in the AFM interface for the (001) and (111) cases gives rise to a weak ferromagnetism with net moments of 0.07 and 0.20 μ_B, respectively.

In addition, the magnetization density at the Ni in Fig. 14 (b) and Fig. 15 (b) has a rather complicated pattern, with non-vanishing moments pointing perpendicular to the AFM Mn moments. Interestingly, the magnetization is also found to vary continuously on a

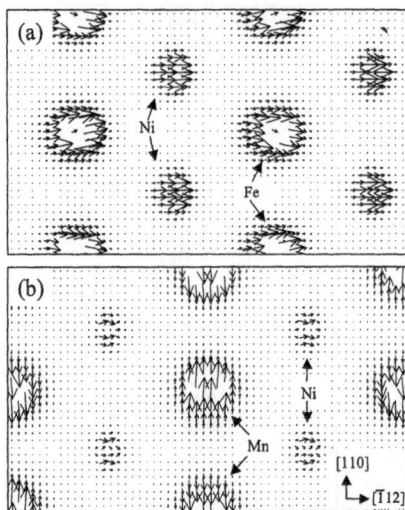

FIG. 15: Magnetization density on (a) FM NiFe, (b) AFM NiMn layers at the (111) interface for the $(NiFe)_3/(NiMn)_3$ superlattice structure. The moment direction and the magnitude are represented by arrows and their size.

smaller length scale inside the atom, i.e. *intra-atomic noncollinear magnetism exists.* As an example, for the (001) case, the moments on the right and left side portions of the Ni site in Fig. 14 (b) cant slightly toward the [100] direction away from the average Ni moments, and so correlate with the neighboring Mn moments [cf. Fig. 14 (c)] that point along the [100] direction; the moments on the top and bottom of the Ni site in Fig. 14 (b) cant toward the [$\bar{1}$00] direction, and correlate with the neighboring Mn moments [cf., Fig. 14 (c)] pointing along the [$\bar{1}$00] direction. Thus, the intra-atomic noncollinear magnetism is strongly affected by the coordination surrounding the atom; it is induced in the d_{xz} and d_{yz} orbitals, which are directly bonded to the neighboring FM Fe and AFM Mn moments and lead to a perpendicular coupling between them.

The total energies for both the (001) and (111) interfaces are found to be lower than those in the collinear magnetic states by 13 and 24 meV/(Mn-atom area), which correspond to 18 and 33 mJ/m^2 of biquadratic exchange energy (BEE). In the (111) case, the noncollinear magnetic structure is expected to be stable above room temperature because of the large BEE, leading to a high blocking temperature as observed in experiments[98, 101].

Since roughness at the interface such as steps, islands or point defects plays a key role in the exchange bias[94, 96, 103], we performed FLAPW calculations for the FM NiFe/AFM NiMn (001) interface including a line step defect along [010]. We employed a 12-layer superlattice structure assuming the bulk lattice parameters of AFM NiMn, as shown in Fig.16. We found that a collinear antiferromagnetic coupling at the FM/AFM interface is highly favored over a ferromagnetic one with a 54 meV/(Mn-atom area) difference in their total energies. Thus, the step defect produces an uncompensated FM/AFM interface across the step, in which the Fe moments align antiparallel to the Mn moments. Further, since the energy difference is larger than the BEE, the introduction of step defects in the flat interface (where the FM Fe moments align perpendicular to the AFM Mn moment) induces a torque that causes the Fe moments to align antiparallel to the Mn moments, leading to a unidirectional magnetic anisotropy of the exchange bias.

4.3.3. Noncollinear Magnetic Structures of Domain Walls in Ferromagnetic Fe and Antiferromagnetic NiMn

The domain wall structure, namely the change in magnetization orientation from one easy axis to another, is determined by a competition between the exchange energy and the anisotropy energy[104, 105]. The exchange energy tends to produce a slower variation of the magnetization whereas the anisotropy energy favors a rapid change from one easy axis to another, which leads to a stable domain wall thickness. Despite intensive research on domain walls based on phenomenological calculations made so far, little is known about it on an atomic scale.

As a first step to investigate the electronic and magnetic properties of a domain wall from first principles, we performed FLAPW calculations for 5, 9 and 17 layer-slabs of Fe(001) and NiMn(001) without constraints except that the magnetic orientations at both ends of the slab are fixed at 180° with respect to each other. The experimental values of the lattice parameters of bulk Fe and NiMn and the atomic $L1_0$ arrangement for NiMn were assumed. Since the MCA energy arising from the SOC is very small compared to the exchange energy for the wall thickness range of interest here, the SOC would not significantly modify the results obtained and so were not included.

The calculated magnetization densities for Fe(001) on each (001) layer plane from the end to the center of a 17 layer-slab are shown in Fig. 17, where the moments lie parallel to their planes. The moments at each atomic site almost orient to its average moment direction, although the small intra-atomic noncollinear magnetism is observed where the moment direction slightly cants away from the average direction. The integrated moment in the MT sphere on the center layer is 2.19 μ_B, which is close to the bulk value. Our self-consistent results clearly demonstrate that the average moments change from one orientation ([100]) to another ([$\bar{1}$00]) as expected in the 180° Bloch wall. It is also found that the

A.J. Freeman et al.

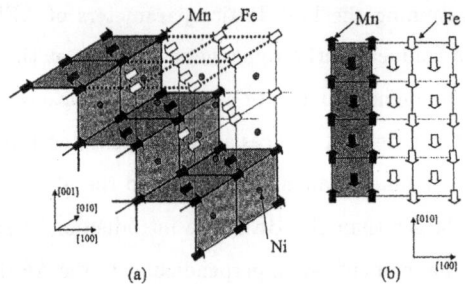

FIG. 16: (a) A model for the FM NiFe/AFM NiMn (001) interface including a line step defect along [010] and (b) the top view of the interface. Dark regions indicate the FM/AFM interface. Only Mn and Fe moments in (a) are given by arrows.

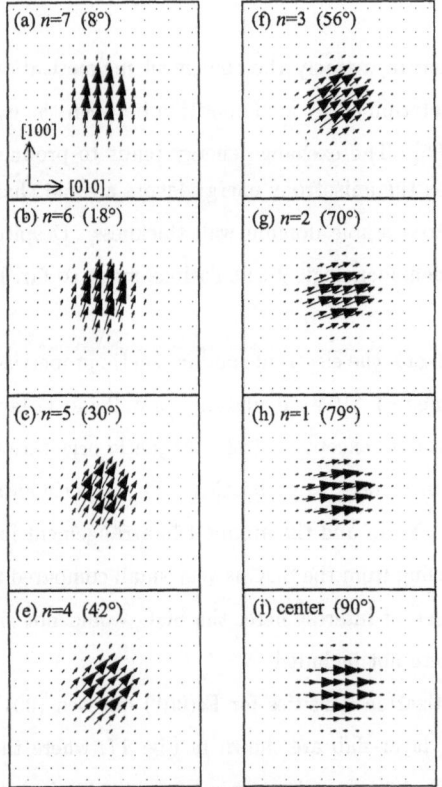

FIG. 17: Magnetization density on each (001) layer plane in the domain wall of the 17 layer Fe(001) slab. The magnetic moment directions and the magnitude are represented by arrows and their size, respectively, and (i) denotes the center layer in the domain wall, n indicates the layer index from the center layer and the angle of the average moment direction from [100] is in degrees. The Fe positions in (b), (e), (g) and (i) are shifted along $[\frac{1}{2}\frac{1}{2}0]$.

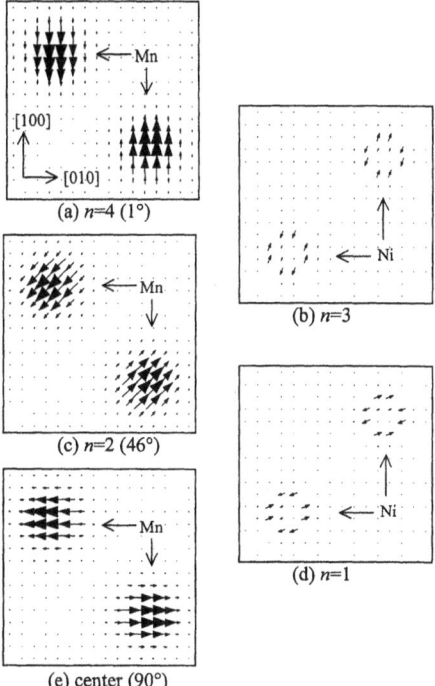

FIG. 18: The magnetization density on each (001) layer plane in the domain wall of the $L1_0$ 9 layer NiMn(001) slab; the size of the arrows for the Ni moments is enlarged five times, n indicates the layer index from the center layer and the angle gives the average moment direction from [100] in degrees.

orientation changes almost linearly, which implies that the exchange energy favors the slower variation of the magnetization.

Figure 18 shows the magnetization density on the (001) planes at the Mn and Ni sites of the 9 layer NiMn(001) slab. The Mn moments at each site almost orient to its average direction. The integrated Mn moment is 3.09 μ_B at the center layer, which is close to the calculated bulk value[102]. As seen in the Fe case, the average Mn moments change from one orientation to another while keeping the antiparallel alignment of the neighboring Mn moments on each layer. At the Ni site, although the integrated moment is zero, small moments are induced, which correlate with the neighboring Mn moments, as seen in Fig. 18 (b) and (d).

The formation energy of the domain wall (ΔE_{DW}) was calculated as the total energy difference between the systems with and without the domain wall, which is shown in Fig. 19 as a function of the inverse of the wall thickness. The ΔE_{DW} decreases significantly when

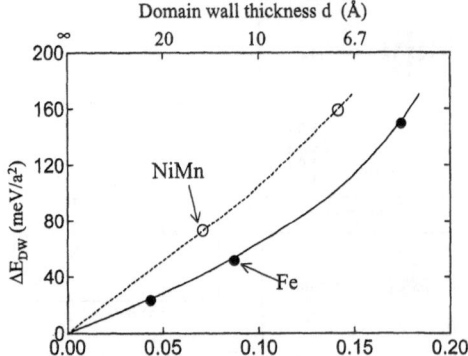

FIG. 19: Formation energy of the domain wall (ΔE_{DW} in meV/a^2), defined as the total energy difference between the systems with and without the domain wall, as a function of the inverse of wall thickness for FM Fe(001) and AFM NiMn(001), where a is the lattice constant.

the wall thickness increases: the ΔE_{DW} for both Fe and NiMn is 60-100 meV/a^2 at a wall thickness of about 10 Å and is reduced to less than 20-40 meV/a^2 when the wall thickness is more than 20 Å. The ΔE_{DW} is found to be approximately proportional to the inverse of the wall thickness, as expected from phenomenological micromagnetic calculations[104, 105]. Note that the agreement is better as the wall thickness is increased, as seen in the Fe case.

We estimated the exchange stiffness parameters, A, with the phenomenological relation $\Delta E_{ex} = \int A\theta'^2 dz$, assuming the angle θ between the magnetization and the x axis to follow a linear magnetization profile. This results in 1.13×10^{-11} J/m for Fe when the thickness is more than 10 Å, which is smaller than the prediction from phenomenological calculations, 1.49×10^{-11} J/m[104]. We also obtained A=1.43×10^{-11} J/m for NiMn, which is larger than that of Fe.

We also performed FLAPW calculations including the SOC with a perturbation treatment. For the wall thickness range of interest here, they confirmed that the anisotropy energy arising from the SOC is very low compared to the exchange energy, and so the SOC effects do not significantly modify the results obtained. However, we find that by introducing the SOC the Bloch wall structure is stabilized compared to the Néel one, since the domain walls breaks the spatial translation symmetry in crystalline solid. The energy difference between both structures was estimated to be about 0.1 meV/a^2 in case of the 17 layer slab (wall thickness is about 23 Å). Thus, the SOC effects give rise to not only a magnetic anisotropy but also to a breaking of the degeneracy of the Bloch and Néel type structures.

4.3.4.   Curling Spin and Orbital Structures in the Vortex Core of an Fe Magnetic Quantum Dot

In ferromagnetic dot nanostructures[105, 106], curling magnetic structures which determine their properties are known to form in order to reduce the demagnetization energy of the
dot volume, and the magnetization close to the center of the dot (so-called Bloch point) will
turn up along the perpendicular orientation to the curling plane. Experimentally, the vortex
core was inferred by magnetic force microscopy (MFM)[75], in which the magnetization near
the center assumes a perpendicular orientation, and the core structures were observed in a
4 or 5 nm radius by spin-polarized scanning tunneling microscopy (SP-STM)[107]. Understanding the nature of the Bloch point from a theoretical point of view, which is thought to
specially link classical and quantum magnetism[106], represents a severe challenge since so
little is known about it.

We have performed FLAPW calculations for the Fe dot, modeled as a repeated disk (rod)
structure with 29 Fe atoms in a unit cell, with the rods separated by vacuum regions; the
experimental lattice parameters of bulk bcc Fe were assumed. The self-consistent calculations were performed without any constraints except that the spin magnetic orientations in
the outer atoms of the disk are fixed along the tangential directions. As a reference, we
performed the calculations for a system without the vortex [i.e., a collinear ferromagnetic
(FM) dot] with the same lattice and computational parameters.

The spatial distribution of the calculated spin magnetization, $\mathbf{m}(\mathbf{r})$, and the integrated
out-of-plane components in the MT spheres, $M_\perp = \int_{\mathrm{MT}} \mathbf{m}_\perp(\mathbf{r}) d\mathbf{r}$, are shown in Fig. 20 and
21 (with solid circles), respectively. The spin moments aligning almost collinearly at each
atom are localized near the nucleus, and clearly orient in the tangential directions around the
center of the dot. Upon moving to the center, the in-plane components gradually decrease
while the out-of-plane components increase, so that the spin directions continuously turn
up along the perpendicular orientation to the curling plane, as predicted from classical
micromagnetic calculations [75, 105].

By forming the vortex core, the magnitude of the spin moments is reduced, however, from
that in the FM dot as a result of changes in the electronic structure: Fig. 21 (open circles)
shows the difference in magnitude of the total moments in the MT spheres between systems
with and without the vortex core, ΔM. The reduction of the moments in the center region
reaches the sizeable value of 0.14 μ_B.

Furthermore, a rather complicated pattern of the spin magnetization density is found near
the center in which the moment directions vary continuously on a smaller length scale inside
the atoms. Figure 22 shows the in-plane components, $\mathbf{m}_\|(\mathbf{r})$, of the spin magnetization

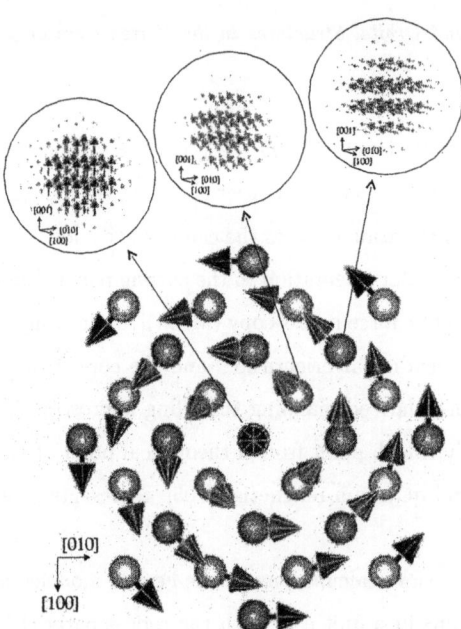

FIG. 20: Spatial distribution of the calculated spin magnetization density, $\mathbf{m(r)}$, in a magnetic vortex Fe dot. The large arrows on the atom sites show the average direction in the MT spheres, which turns up along the out-of-plane orientation from the in-plane orientation,on moving to the center of the dot. The blow-ups show details of the intra-atomic spin density distributions on selected sites.

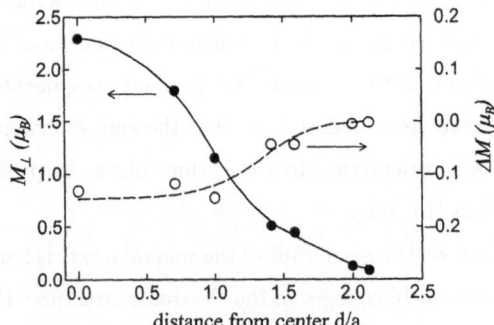

FIG. 21: (A) Out-of-plane components of spin moments inside the muffin-tin spheres, $M_\perp$, and (B) the difference in the magnitude of the total moments in the MT spheres between systems with and without the vortex core, ΔM, as a function of the distance d/a from the center of the dot, where a is the bcc lattice constant of bulk Fe.

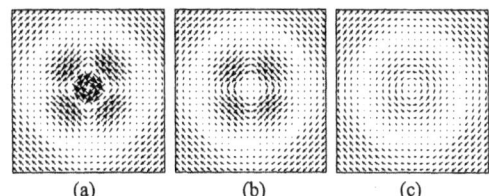

FIG. 22: In-plane components of magnetization densities, $\mathbf{m}_{\|}(\mathbf{r})$, on (A) the (001) plane at the center atom, and on parallel planes (B) $\frac{1}{20}a$ (a is the bcc lattice constant) and (C) $\frac{1}{10}a$ above the nucleus, where the magnitude of the moments in the interstitial region are enlarged three times that in the atom sites. The average moment orients in an out-of-plane direction. The area shown is $\frac{a}{2} \times \frac{a}{2}$.

density on (a) the (001) plane through the nucleus of the center atom and on parallel planes (b) $\frac{1}{20}a$ (a = bcc lattice constant) and (c) $\frac{1}{10}a$ above the nucleus, Surprisingly, the spin moments in the outer portions directed toward the nearest neighbor Fe atoms, which directly hybridize with $d_{xz,yz}$ orbitals, are predicted to curl in a counterclockwise direction while those near the nucleus cant in the opposite (clockwise) direction [cf. Fig. 22(a)]. An oppositely directed curling spin density is also found throughout the outer interstitial region due to the negative polarization of the delocalized s and p electrons[108], but with a very small moment.

The formation energy of the vortex core, ΔE_{vc}, calculated as the total energy difference between the systems with and without the vortex core, is found to be 88 meV/cell (3 meV/atom) – which roughly corresponds to the demagnetization energy of a 4 nm radius FM dot found by phenomenological calculations[105]. This may indicate that the vortex core will be stabilized when the radius becomes of this magnitude – which agrees with the SP-STM observations[107]. In determining more quantitatively these structures, of course, further investigations will be necessary.

The curling of the spin magnetization produces rotation structures in the effective LSDA potential, which lifts the orbital degeneracy and leads to a non-zero orbital angular momentum. Indeed, we observed orbital magnetic moments oriented perpendicular to the curling plane in which no in-plane components were observed. The out-of-plane components of the orbital moments, $M_{\perp}^{\mathrm{orbit}}$, calculated within the scalar relativistic approximation without SOC, are shown in Fig. 23 as solid circles. Orbital moments near the center, 0.03~0.06 μ_B, are clearly induced in the perpendicular orientation to the curling plane. Thus, the rotation properties in the spin density couple to the orbital motion, and may lead to a uniaxial

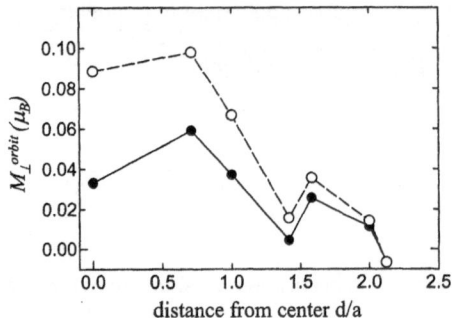

FIG. 23: Out-of-plane components of orbital moments, $M_\perp^{\mathrm{orbit}}$, as a function of the distance d/a from the center of the dot. Solid and open circles represent results without and with spin-orbit coupling.

magnetic anisotropy in the perpendicular direction.

Now, when the SOC is introduced, the vortex core structure is stabilized by 1 meV/cell in the ΔE_{vc}; the spin density structures are not significantly changed. Since the SOC induces the atomic orbital moments along the spin moment orientations, the orbital moments also curl around the center of the dot, in which the moments turn up along the perpendicular orientation at the center. However, due to the interaction with the curling spin density which induces the orbital moments along the perpendicular direction, the orbital moments do not coincide with the spin directions. The out-of-plane components are enhanced by about a factor of two from the states without SOC, as seen in Fig. 23. The magnitude at the center atom, 0.089 μ_{B}, is larger than that in the FM dot, 0.055 μ_{B}.

----

[1] R.Q. Wu and A.J. Freeman, J. Magn. Magn. Mater. **200** (1999) 498; **100** (1991) 497.

[2] See review articles in volume 200 of J. Magn. Magn. Mater.

[3] P. Hohenberg and W. Kohn, Phys. Rev. **136** (1964) B864.

[4] W. Kohn and L. J. Sham, Phys. Rev. **140** (1965) A1133.

[5] M. Born and R. Oppenheimer, Ann. Phys. Lpz. **84** (1927) 124.

[6] J. von Barth and L. Hedin, J. Phys. C **5**, 1629 (1972).

[7] O. Gunnarsson, B. I. Lundqvist, and S. Lundqvist, Solid State Commun. **11**, 149 (1972).

[8] L. Hedin and B. I. Lundqvist, J. Phys. C **4**, 2064 (1971).

[9] J. P. Perdew, J. A. Chevay, S. H. Vosko, K. A. Jackson, M. R. Pederson, D. J. Singh, and C. Fiolhais, Phys. Rev. B **46**, 6671 (1992).

[10] B. G. Johnson, P. M. W. Gill, and J. A. Pople, J. Chem. Phys. **98**, 5612 (1993).

[11] A. D. Becke, in *The Xhallenge of d and f Electrons*, ACS Symposium Series No. 394, edited by D. R. Salahub and M. C. Zerne (American Chemical Society, Washington, DC, 1989), p. 165.

[12] A. Garcia, C. Elsaesser, J. Zhu, S. Louie, and M. L. Cohen, Phys. Rev. B **46**, 9829 (1992).

[13] Y.-M. Juan, E. Kaxiras, and R. G. Gordon, Phys. Rev. B **51**, 9521 (1995).

[14] A. Dal Corso, A. Pasquello, A. Baldereshi, and R. Car, Phys. Rev. B **53**, 1180 (1996).

[15] M. Körling and J. Häglund, Phys. Rev. B **45**, 13293 (1992).

[16] P. H. T. Philipsen, G. te Velde, and E. J. Baerends, Chem. Phys. Lett. **226**, 583 (1994).

[17] P. Hu, D. A. King, S. Crampin, M.-H. Lee, and M. C. Payne, Chem. Phys. Lett. **230**, 501 (1994).

[18] D. Porezag and M. R. Pederson, J. Chem. Phys. **102**, 9345 (1995).

[19] J. Baker, M. Muir, and J. Andzelm, J. Chem. Phys. **102**, 2063 (1995).

[20] B. Hammer, K. W. Jacobsen and J. K. Norskov, Phys. Rev. Lett. **70**, 3971 (1993).

[21] B. Hammer, M. Scheffler, K. W. Jarcobsen, and J. K. Norskov, Phys. Rev. Lett. **73**, 1400 (1994).

[22] T. C. Leung, C. T. Chan, and B. N. Harmon, Phys. Rev. B **44**, 2923 (1991).

[23] N. Moll, M. Bockstedte, M. Fuchs, E. Pehlke, and M. Scheffler, Phys. Rev. B **52**, 2550 (1995).

[24] D. R. Hamann, Phys. Rev. Lett. **76**, 660 (1996).

[25] A. D. Becke, Phys. Rev. A **38**, 3098 (1988); J. P. Perdew, Phys. Rev. B **33**, 8822 (1986); Phys. Rev. B **34**, 7406 (1986).

[26] J. P. Perdew, K. Burke, and M. Ernzerhof, Phys. Rev. Lett. **77**, 3865 (1996).

[27] M. Fuchs, M. Bockstedte, E. Pehlke, and M. Scheffler, Phys. Rev. B 57, 2134-2145 (1998).

[28] E. Wimmer, H. Krakauer, M. Weinert, and A. J. Freeman, Phys. Rev. B **24**(1981) 864; M. Weinert, E. Wimmer, and A. J. Freeman, Phys. Rev. B **26** (1982) 4571, and references therein.

[29] G.H.O. Daalderop, P.J. Kelly, and M.F.H. Schuurmans, Phys. Rev. B **42** (1990) 1533; A.R. Mackintosh and O.K. Andersen, in *Electrons at the Fermi Surface*, edited by M. Springford (Cambridge University Press, Cambridge, 1980).

[30] X.D. Wang, D.S. Wang, R.Q. Wu, and A.J. Freeman, J. Magn. Magn. Mater. **159** (1996) 337.

[31] J.G. Gay and R. Richter, Phys. Rev. Lett. **56** (1986) 2728; C. Li, A.J. Freeman, H.J.F. Jansen, and C.L. Fu, Phys. Rev. B **42** (1990) 5433; G.Y. Guo, W.M. Temmerman, and H. Ebert, J. Phys. C **3** (1991) 8205; J. Magn. Magn. Mater. **104-107** (1992) 1772.

[32] D.S. Wang, R.Q. Wu, and A.J. Freeman, Phys. Rev. Lett. **70** (1993) 869; **71** (1993) 2166.

[33] D.S. Wang, R.Q. Wu, and A.J. Freeman, Phys. Rev. B **47** (1993) 14932; J. Magn. Magn. Mater. **129** (1994) 327; Phys. Rev. B **48** (1993) 15883.

[34] X.D. Wang, R.Q. Wu, D.S. Wang, and A.J. Freeman, Phys. Rev. B **54** (1996) 61.

[35] V.G. Gavrilenko and R.Q. Wu, Phys. Rev. B **60** (1999) 9539.

[36] P. Krams, F.Lauks, R.L.Stamps, B.Hillebrands, and G.Güntherodt, Phys. Rev. Lett. **69** (1992) 3674.

[37] B. N. Engel, M. H. Wiedmann, and C. M. Falco, J. Appl. Phys. **75** (1994) 6401.

[38] S. Hope, E. Gu, B. Choi, and J.A.C. Bland, Phys. Rev. Lett. **80** 1750 (1998); M. Tselepi, P.J. Bode, Y.B. Xu, G. Wastlbauer, S. Hope, and J.A.C. Bland, J. Appl. Phys. **89**, 6683 (2001).

[39] R. Vollmer, Th. Gutjahr-Loser, J. Kirschner, S. van Dijken, and B. Poelsema, Phys. Rev. B **60**, 6277 (1999).

[40] S. van Dijken, R. Vollmer, B. Poelsema and J. Kirschner, J. Magn. Magn. Mater. **210** 315 (2000).

[41] J.S. Hong, R.Q. Wu, J. Lindner, E. Kosubek, and K. Baberschke, Phys. Rev. Lett., submitted.

[42] Dongqi Li, B. Roldan Cuenya, J. Pearson, S.D. Bader, and W. Keune, Phys. Rev. B **64**, 144410 (2001).

[43] N. García, M. Muñoz, and Y.-W. Zhao, Phys. Rev. Lett. **82**, 2923 (1999).

[44] P. Gambardella, A. Dallmeyer, K. Maiti, M.C. Malagoli, W. Eberhardt, K. Kern, and C. Carbone, Nature **416**, 301 (2002).

[45] N. Nilius, T.M. Wallis, and W. Ho, Science **297**, 1853 (2002); and private communications.

[46] G.M. Pastor, J. Dorantes-Dávila, S. Pick, and H. Dreyssé, Phys. Rev. Lett. **75**, 326 (1995).

[47] B. Lazarovits, L. Szunyogh, and P. Weinberger, Phys. Rev. B **65**, 104441 (2002).

[48] M. Weinert and A.J. Freeman, J. Magn. Magn. Mater. **38**, 23 (1983); M. Weinert, W. Yang and A.J. Freeman, ICM2003 Proc. accepted.

[49] Jisang Hong and R.Q. Wu, Phys. Rev. B **67**, 020406(R) (2003).

[50] P. Gambardella, S. Rusponi, M. Veronese, S. S. Dhesi, C. Grazioli, A. Dallmeyer, I. Cabria, R. Zeller, P. H. Dederichs, K. Kern, C. Carbone, and H. Brune, Science **300**, 1130 (2003).

[51] Jisang Hong and Ruqian Wu, to be published.

[52] For systems with a fourfold symmetry with respect to the surface normal, their magnetic anisotropy energies, E_{MCA}, can be expressed in the lowest non-vanishing order of the polar and azimuth angles (θ and ϕ) as $E_{MCA} = K_1 \sin^2\theta + K_2 \sin^2(2\phi)\sin^4\theta$, where K_1 and K_2 are coefficients of the leading uniaxial and in-plane contributions, respectively.

[53] G.H.O. Daalderop, P.J. Kelly and M.F.H. Schuurmans, Phys. Rev. B **41** (1990) 11919 .

[54] G.Y. Guo, W.M. Temmerman and H. Ebert, Physica B **172** (1991) 61.

[55] P. Strange, J.B. Sraunton, B.L. Gyorffy and H. Ebert, Physica B **172** (1991) 51.

[56] J. Trygg, B. Johnansson, O. Eriksson and J.M. Wills, Phys. Rev. Lett. **75** (1995) 2871.

[57] S.V. Halilov, A.Ya. Perlov, P.M. Oppeneer, A.N. Yaresko, and V.N. Antonov, Phys. Rev. B **57** (1998) 9557.

[58] S.V. Beiden, et al, Phys. Rev. B **57** (1998) 14247.

[59] O. Eriksson, et al, Phys. Rev. B **41** (1990) 7311; **42** (1990) 2707.

[60] J. R. Cullen, A.E. Clark and K.B. Hathaway, in *Materials Science and Technology*, edted by R.W. Cahn, P. Hasen and E.J. Kramer, Vol. IIIB (1994) 529.

[61] R.Q. Wu, L.J. Chen, A. Shick and A.J. Freeman, J. Magn. Magn. Mater. **177-181** (1998) 1216.

[62] M. Fähnle, M. Komelj, R.Q. Wu and G.Y. Guo, Phys. Rev. B **65**, 144436 (2002).

[63] R.Q. Wu and A. J. Freman, J. Appl. Phys. **79** (1996) 6500; R.Q. Wu, L.J. Chen, and A.J. Freeman, J. Magn. Magn. Mater. **170** (1997) 103.

[64] A.B. Shick, D. L. Novikov, and A. J. Freeman , Phys. Rev. B **56** (1997) R14259; J. Appl. Phys. **83** (1998) 7258.

[65] V.G. Gavrilenko and R.Q. Wu, J. Appl. Phys. **87** (2000) 6098.

[66] *Data in Science and Technology: Magnetic Properties of Metals*, edited by H.P.J. Wijn (Springer-verlag, Berlin, 1986).

[67] A.E. Clark, in *Ferromagnetic Materials*, Vol. 1, edited by E.P. Wohlfarth (Amsterdam, North-Holland), page 531.

[68] K.N.R. Taylor, Advance in Physics, **2**, 551 (1971).

[69] K. Hathaway and J. Cullen, J. Phys. Condens. Matter **3** (1991) 8911.

[70] *Ultrathin Magnetic Structures*, edited by J. A. C. Bland and B. Heinrich (Springer, Berlin, 1994), Vols. 1, 2, and references therein.

[71] J. Nogués and I. K. Schuller, J. Magn. Magn. Mater. **192** (1999) 203.

[72] A. E. Berkowitz and K. Takano, J. Magn. Magn. Mater. **200** (1999) 552.

[73] G. Tatara and H. Fukuyama, Phys. Rev. Lett. **78** (1997) 3773.

[74] J. Prieto, M. Blamire, and J. E. Evetts, Phys. Rev. Lett. **90** (2003) 27201.

[75] T. Shinjo, T. Okuno, R. Hassdorf, K. Shigeto, and T. Ono, Science **289** (2000) 930.

[76] L. Nordström and D. J. Singh, Phys. Rev. Lett. **76** (1996) 4420.

[77] D. Hobbs, G. Kresse, and J. Hafner, Phys. Rev. B **62** (2000) 11556.

[78] P. Kurz, G. Bihlmayer, and S. Blügel, J. Appl. Phys. **87** (2000) 6101.

[79] K. Nakamura, A. J. Freeman, D. S. Wang, L. Zhong, and J. Fernandez-de-Castro, Phys. Rev. B **65** (2002) 12402; Phys. Rev. B **67** (2003) 14420.

[80] K. Nakamura and A. J. Freeman, Phys. Rev. B **66** (2002) 140405.

[81] K. Nakamura, A. J. Freeman, L. Zhong, and J. Fernandez-de-Castro, Phys. Rev. B **67** (2003) 14405.

[82] K. Nakamura, A. J. Freeman, L. Zhong, and J. Fernandez-de-Castro, J. Appl. Phys. **93** (2003) 6879.

[83] K. Nakamura, T. Ito, and A. J. Freeman, Phys. Rev. B **68** (2003) 180404.

[84] J. Kübler, K. -H. Höck, J. Sticht, and A. R. Williams, J. Phys. F **18** (1988) 469.

[85] L. M. Sandratskii, Adv. Phys. **47** (1998) 91.

[86] M. Weinert, J. Math. Phys. **22** (1981) 2433.

[87] M. E. Rose, *Relativistic Electron Theory* (John Wiley and Sons, New York, 1961).

[88] C. Li, A. J. Freeman, H. J. F. Jansen, C. L. Fu, Phys. Rev. B **42** (1990) 5433.

[89] D. D. Koelling and B. N. Harmon, J. Phys. C **10** (1977) 3107.

[90] A. H. MacDonald, W. E. Pickett, and D. D. Koelling, J. Phys. C **13** (1980) 2675.

[91] W. P. Meiklejohn and C. P. Bean, Phys. Rev. **102**, 1413 (1956); Phys. Rev. **105** (1957) 904.

[92] R. L. Stamps, J. Phys. D **33** (2000) R247.

[93] D. Mauri, H. C. Siegmann, P. S. Bagus, and E. Kay, J. Appl. Phys. **62** (1987) 3047.

[94] A. P. Malozemoff, Phys. Rev. B **35** (1987) 3679.

[95] N. C. Koon, Phys. Rev. Lett. **78** (1997)4865.

[96] T. C. Schulthess and W. H. Butler, Phys. Rev. Lett. **81** (1998) 4516.

[97] T. Lin, D. Mauri, N.Staud, C. Hawng, J. K. Howard, and G. L. Gorman, Appl. Phys. Lett. **65** (1994) 1183.

[98] S. Mao, S. Gangopadhyay, N. Amin, and E. Murdock, Appl. Phys. Lett. **69** (1996) 3539.

[99] B. Y. Wong, C. Mitumata, S. Prakash, D. E. Langhlin, and T. Kobayashi, IEEE Trans. Magn. **32** (1996) 3425.

[100] X. Portier, A. K. Petford-Long, and T. C. Anthony, IEEE Trans. Magn. **33** (1997) 3679.

[101] Z. Qian, J. M. Sivertsen, J. H. Judy, B. A. Everitt, S. Mao, and E. S. Murdock, J. Appl. Phys. **85** (1999) 6106.

[102] K. Nakamura, M. Kim, A. J. Freeman, L. Zhong, and J. Fernandez-de-Castro, IEEE Trans. Magn. **36** (2000) 3269.

[103] K. Takano, R. H. Kodama, A. E. Berkowitz, W. Cao, and G. Thomas, Phys. Rev. Lett. **79** (1997) 1130.

[104] S. Chikazumi, *Physics of Ferromagnetism* (Wiley, New York, 1964).

[105] A. Hubert and R. Schäfer, *Magnetic Domains* (Springer-Verlag, Berlin, 1998).

[106] J. Miltat and A. Thiaville, Science **298** (2002) 555.

[107] A. Wachowiak, J. Wiebe, M. Bode, O. Pietzsch, M. Morgenstern, and R. Wiesendanger, Science **298** (2002) 577.

[108] S. Ohnishi, A. J. Freeman, and M. Weinert, Phys. Rev. B **28** (1983) 6741.

A. Hubert and R. Schäfer, Magnetic Domains: Springer-Verlag, Berlin, 1998.

MONTE CARLO STUDIES OF SURFACE AND INTERFACE EFFECTS IN MAGNETIC NANOPARTICLES

K.N. TROHIDOU
Institute of Materials Science, NCSR Demokritos
153 10, Aghia Paraskevi, Attiki, Athens, Greece

1. Introduction

The study of magnetism in nanoparticles saw considerable developments in the late forties and fifties largely through the interest in paleomagnetism. The basic theory is due to Néel [1]. The magnetization of a ferromagnetic particle is assumed to be the same as that of the bulk material. At a certain temperature, the blocking temperature T_B, a particle's thermal energy approaches its anisotropy energy. Above T_B the particles are treated as a paramagnetic gas. The phenomenon, known as superparamagnetism [2], is described by the Langevin theory.

A microscopic treatment of the magnetization of ferromagnetic nanoparticles was developed later by Binder and co-workers [3-5] using Monte Carlo techniques. An important demonstration of the work was the reduction of the magnetization near the surface of the particle. Clearly this was to be expected because a surface spin has a smaller number of neighbors than it would have in bulk and, hence, experiences a reduced mean field. For very small particles (less than say 5 nm) the proportion of surface spins is such that they will make a major contribution to the magnetization. As a result, the magnetization will decrease with temperature over a range where the bulk magnetization is roughly constant and deviations from Curie-law behavior in the susceptibility are to be expected. In the period following the Monte Carlo work cited above, interest has developed in finite-size scaling, and it is in this context rather than through superparamagnetism per se that subsequent advances [6] in the nanoparticle magnetism have occurred.

The early experimental work [7] displayed conflicting results with some workers reporting Curie behavior while others observed considerable deviations. Bean and Livingston [7] proposed an argument, which led to the conclusion that the reduction in particle magnetization due to the surface effect did not occur. The treatment was based on a Heisenberg model and considered the exact behavior of 1, 2, and 3 spin clusters. Unfortunately, extrapolating to larger clusters is only possible with an infinite range

45

interaction and so the inferences they made are not valid. A decade later Hahn [8] readdressed the issue, but although he put it on a more formal basis, the questions raised have still not, to our knowledge, been satisfactorily resolved. It must of course be emphasized that, for ferromagnets, problems arise only with the extremely small particles, for which consideration other than statistical physics may be important. For example, at some point, electronic calculations for the cluster as a whole will become necessary.

Recent, increasing interest in the magnetic properties of nanoparticles comes from their important technological applications. Many experiments have been concentrated on the effect of the surface oxidation on the magnetic properties of ferromagnetic particles [e.g. 9-11]. In Fe particles coated by an antiferromagnetic iron oxide with thickness 1-2 nm, the coercivity showed a large increase in samples with iron core diameter below 6 nm reaching a value of about 2.7 kOe in a particle with core diameter 2 nm. In samples with a larger core diameter the coercivity was almost constant, ~250 Oe [11].

Néel first showed [12] that antiferromagnetic particles can exhibit superparamagnetic behavior if there is an imbalance in the number of spins on the 'up' and 'down' sublattices (uncompensated spins). Experimental work [13,14] on α-Fe_2O_3 particles in rocks and on Cr_2O_3 particles followed Néel's analysis. Subsequent work in antiferromagnetic particles in biological systems has attracted attention [15]. More recently, experimental work has focused on iron oxides, which are used as high-density recording media [see for example 16-18]. The magnetic properties of antiferromagnetic nanoparticles have been investigated by Monte Carlo simulations [19-22]. These simulations showed that the magnetization of a particle arises from the uncompensated spins that can be attributed to the surface of the particle, and that the temperature variation of the magnetization and the coercivity depends on particle size and the number of uncompensated spins.

In what follows the role of the surface and interface anisotropy in the magnetic properties of nanoparticles is investigated by the Monte Carlo simulation technique. Isolated ferromagnetic, antiferromagnetic and nanoparticles with ferromagnetic core surrounded by an antiferromagnetic shell, which will be called oxidized, are considered. Interparticle interaction effects, which are present in dense samples, are not taken into account. In all cases the results reported here are compared with analytical studies and experimental findings.

In a Monte Carlo (MC) simulation, the microstructure and the temperature are explicitly included. This method is therefore appropriate to study the magnetic behavior of nanoparticles and to elaborate the surface and interface effects. The Monte Carlo results cannot be drawn by simple mean field arguments, although qualitative agreement with experiment can be obtained in both cases [23,24]. The simulations discussed in this review show that the exchange anisotropy [25], due to the interaction between the ferromagnetic core and the antiferromagnetic coating of the oxidized particles and between the antiferromagnetic core and the ferromagnetic surface of the antiferromagnetic particles, plays a predominant role in the magnetic behavior of these nanoparticles [22]. The exchange anisotropy is not assumed beforehand in the simulations, but results from the microstructure that is absent in mean-field models.

2. The Model

We consider spherical nanoparticles that consist of an assembly of N classical spins placed on the sites of a three dimensional simple cubic (sc) lattice, within a radius of R lattice spacings of the central site. Assuming Heisenberg interactions between the spins, the energy of a particle in a magnetic field H is

$$E = -\sum_i \sum_{j \neq i} J_{ij} \vec{S}_i \cdot \vec{S}_j - \vec{H} \cdot \sum_i \vec{S}_i - \sum_i K_i S_i^2 \cos^2 \theta_i \tag{1}$$

where S_i is a vector that represents the atomic spin at site i, with magnitude set equal to 1. J_{ij} in the first term of the above equation is the nearest neighbour exchange coupling constant, which is equal to +J (J>0) or -J for the ferromagnetic and antiferromagnetic cases, respectively. The second term is the Zeeman energy. It is generally assumed that isotropic exchange forces dominate the thermodynamics, while anisotropy energy, which defines the easy directions of magnetization, determines the dynamics of relaxation near the blocking temperature. This anisotropy is described by the third term in Eq.(1), where, θ_i is the angle between S_i and the direction of the anisotropy easy axis at sites i.

We consider core/shell morphology for all the cases of nanoparticles in agreement with the experimental investigations [see for example 26-29], so in the simulations the bulk and surface anisotropy are included separately. As surface sites we consider those within the two outer shells of the ferromagnetic and antiferromagnetic nanoparticles, so the surface atoms are those lying between (R-2) and R of the center of the particle.

For nanoparticles, there is some evidence that the easy axis is along one of the crystallographic directions, even though in cubic bulk materials the easy axis is not uniaxial but along the three cubic axes [30]. The core anisotropy is, therefore, considered uniaxial along the z-axis with anisotropy coupling constant K_c. Many studies have shown that, due to the reduced symmetry of the surface, the surface crystal anisotropy is stronger than the bulk [31]. Moreover, calculations have shown that the easy axis direction is either perpendicular to the free surface of magnetic materials [32], known as radial anisotropy, or parallel to it. In order to account for these distinct features of surface anisotropy, we discuss the case of radial surface anisotropy with coupling constant K_s. Results of simulations will be presented without and with a distinct surface anisotropy for comparison. In the former case the surface spins experience the same uniaxial anisotropy along the z-axis as the core spins, this is the uniform anisotropy case.

In the case of the oxidized nanoparticles, the shell is considered as a layer surrounding the core and being four lattice spacings thick. The interface between the core and the shell is defined by the spins in the outer layer of the core being one lattice spacing thick. The surface of the particle is defined again by the spins in the two outer layers of the particle, with two lattice spacings thickness. Since the bulk anisotropy of the oxides is lower than that of the corresponding ferromagnetic materials, for illustration, we consider that the anisotropy coupling constant of the antiferromagnetic shell is again uniaxial with the same easy axis as in the core and with a coupling constant, K_{SH}, of magnitude one quarter of K_C. The reduced symmetry argument for the surface anisotropy holds obviously for the interface. In the simulation, the spins in the interface have the same type of anisotropy as in the core with an anisotropy coupling constant, K_{IF}.

The magnetization and the response to an external field have been studied for a range of parameters. In order to determine the coercive field, initially the spins are in the +z direction. A magnetic field is applied in the -z direction. The coercive field is defined as the magnetic field that reverses the magnetization of the particle in the simulation time, so that the z-component of the magnetization vanishes.

The Monte Carlo method for a Heisenberg system has been described many times [33]. In the presented simulations 18×10^4 Monte Carlo steps per spin (MCSPS) were found to be sufficient to produce results as compared with results for a number of MCSPS an order of magnitude higher. The results produced after an equilibration from the fully aligned state using 300 MCSPS. For each temperature, the amount of the change in the direction of a selected spin is chosen such that on the average one half of the selected moves are successful. Results have been checked by calculating the magnetization and the coercivity for different sequences of random numbers (ten runs). The statistical error found was very small even at high temperatures. Including the corresponding error bars in our figures would not affect the information obtained from them so they are omitted. In what follows, the temperature is measured in units of J/k_B, the magnetic field in units of $J/g\mu_B$ and the anisotropy coupling constants in units of J.

3. Ferromagnetic Nanoparticles

There have been many advances in the fabrication and characterization of fine magnetic structures with diameters even smaller than 10 nm [34-36] since the early nineties. The experimental results on the magnetic behavior of small Fe and Co clusters have been explained in terms of a superparamagnetic model [37-40]. However, additional novel features have been seen such as high coercivity of very small particles [41-44] that is attributed to the single domain structure of these particles and, also, to possible enhanced anisotropy relative to that of bulk magnetic materials. Moreover surface effects such as surface anisotropy and surface exchange interaction could be important and should be taken into account in order to explain the experimental observations, especially in the very small particle sizes. The coercive behavior of single domain ferromagnetic particles has been described in terms of mean field theory [45]. A modified mean field model has been formulated to accommodate surface effects and it gave qualitative agreement with the experimentally observed behavior [23]. The effect of surface anisotropy on the magnetic properties of ferromagnetic nanoparticles is discussed in Refs. [46-50].

An experimental probe for the magnetization behaviour of magnetic nanoparticles is small angle neutron scattering (SANS) [51-59]. SANS studies revealed the role of the surface in small Fe particles of radius $R \sim 1.5$ nm [51-53]. In bigger particles behaviour similar to that of a particle with uniform anisotropy was observed [54].

We review here the use of the MC simulation technique to investigate effects that emerge from the microstructure of the ferromagnetic particles. We discuss the size and temperature dependence of the magnetization and coercivity. Additionally, results on MC simulations for the scattering function of the SANS for ferromagnetic nanoparticles are reported.

3.1 MAGNETIZATION

First the magnetization of ferromagnetic particles in the absence of any surface anisotropy is presented. Representative behaviour is shown in Fig. 1 (full lines) for two spherical nanoparticles with radii $R=8$ (circles) and $R=12$ (triangles) atomic lattice spacings on a simple cubic (sc) lattice. The magnetization per spin is displayed; hence the magnetization is normalized to unity at zero temperature. The temperature is in units of T/T_c where T_c, the bulk Curie temperature, is 2.9 for the sc lattice [60,61]. In this figure the temperature dependence of the magnetization is shown for particles with uniform uniaxial anisotropy K_c with the easy axis along the z-direction, $K=K_c=K_s=0.1$ (full lines). For these ferromagnetic particles, the decrease in the magnetization with decreasing R is well known and it is ascribed to the increasing role played by the surface as R becomes smaller [19,20,35,46,47].

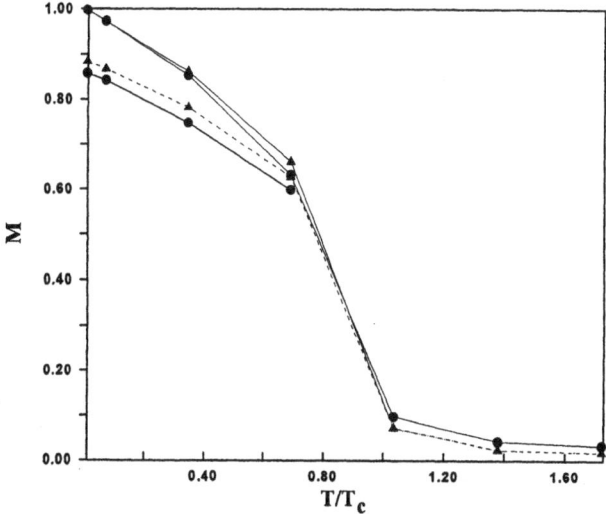

Figure 1. Magnetization versus temperature for ferromagnetic particles with sizes: $R=8$ (circles); $R=12.0$ (triangles) with uniform uniaxial anisotropy $K=0.1$ (full lines) and including radial surface anisotropy $K_s=10K_c$ (broken lines).

The effect of introducing radial surface anisotropy is next considered, for these two particles (broken lines). The surface anisotropy K_s is one order of magnitude higher than the core $K_c=0.1$, $K_s=10K_c$ with the easy axis normal to the surface at each site. We observe in this case a reduction of the magnetization due to the surface disorder introduced by the radial anisotropy and a more rapid fall of the magnetization with temperature. The radial direction of the easy axis orientation of the strong surface anisotropy when averaged over the whole surface of the spherical particle tends to eliminate the contribution to the magnetization from the surface layer. This behaviour is in agreement with experimental findings on $(Fe_{0.26}Ni_{0.74})_{50}B_{50}$ nanoparticles. [62], on PdFe particles [63], on metallic Fe nanoparticles [42] and Co nanoparticles [64].

3.2.   COERCIVITY

Next, the influence of the surface anisotropy on the coercive field is discussed. In figure 2 the coercive field versus temperature for two ferromagnetic particles of radius $R=6$ (circles) and $R=8$ (triangles) lattice spacings is displayed. Full symbols represent results for uniform z-axis anisotropy $K=K_c=K_s=0.1$ and open symbols for z-axis core anisotropy $K_c=0.1$ and radial surface anisotropy of size $K_s=10K_c$. The observed behaviour for the particles with uniform anisotropy is the predicted one from the phenomenological model of Kneller and Luborsky [45]. The bigger particle has the higher coercivity and this behaviour is valid for all temperatures.

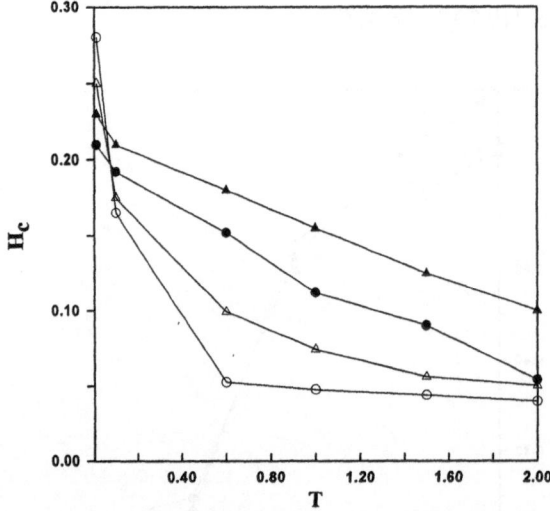

Figure 2. The coercive field versus temperature for particles of radii $R=6$ (circles) and $R=8$ (triangles), with uniform uniaxial anisotropy $K=0.1$ (full symbols) and including radial surface anisotropy $K_s=10K_c$ (open symbols).

The behaviour of the particles with the radial surface anisotropy is quite different. Increase of the low temperature coercivity for the particles with radius $R=6$ is observed and an inversion with the size. This behaviour is in agreement with experimental findings on Fe particles [42] and on Co particles [64] where an increase of the surface anisotropy with the decrease of the particle size was observed. Also the zero temperature analytical calculations of Refs. [46-48] are in agreement with the coercive behaviour presented by this model. As the temperature increases this behaviour is reversed, also we observe a steeper drop with temperature than in the uniform anisotropy case in agreement with Ref. [65]. This is due to the fact that the thermal fluctuations mask the surface contribution.

3.3   SMALL ANGLE NEUTRON SCATTERING

The structure factor observed with small angle scattering of neutrons (SANS) is

$$S(Q) = \sum \langle S_i S_j \rangle \exp iQ \cdot (r_i - r_j) \qquad (2)$$

Here Q is the scattering vector. In the MC simulations, ferromagnetic spherical nanoparticles with uniform anisotropy are considered. The quantity $S(Q)$ was evaluated for a particular configuration and then the configuration average was obtained [19]. This was done for Q along several different crystallographic directions and the results directionally averaged. The normalized scattering function is

$$I(Q) = \langle S(Q) \rangle_{av} / S(0) \qquad (3)$$

where the angular brackets denote the directional average. Equation (2) describes the Q dependence of magnetic neutron scattering. Nuclear scattering, of course, occurs as well, but the magnetic part can, in principle, be separated out by the use of polarized neutrons [53,54,59].

There are essentially two regimes of interest for ferromagnets. For temperatures sufficiently below T_c, fluctuations are negligible and $\langle S_i S_j \rangle \rightarrow \langle S \rangle^2$. The scattering function then becomes the standard expression for noninteracting spheres given by Rayleigh [66] as

$$I(Q) = \left[\frac{3[\sin(QR) - QR \cos(QR)]}{(QR)} \right]^2 \qquad (4)$$

The behavior in this regime is well demonstrated by the Monte Carlo simulation. The MC data for a particle of radius $R = 7.5$ and at two temperatures are plotted in Fig. 3.

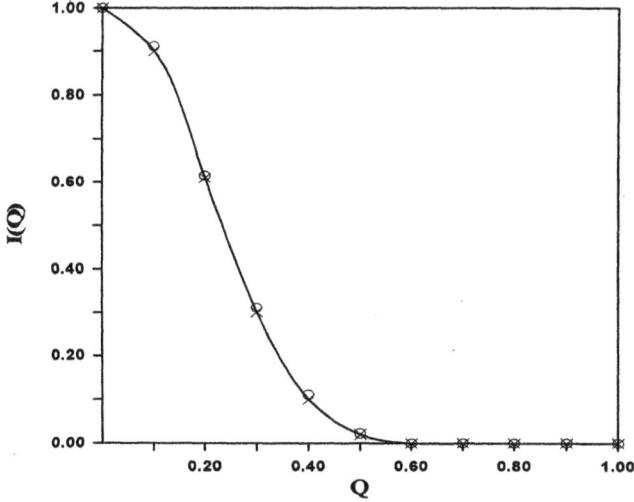

Figure 3. Scattering function versus Q, in units of lattice spacings, for a ferromagnetic particle with radius $R=7.5$ and temperatures $T=2.0$ (circles) and $T=3.0$ (crosses). The solid line gives the functional form Eq. (4).

The continuous curve represents Eq. (4) and the agreement between MC data and the equation is excellent. Notice that the actual radius of the particle is being used rather than an effective radius. Even though the particle contains less than 2000 atoms, the surface effects are rather small. This is consistent with Wildpaner's calculations on classical Heisenberg particles. He considered the effective radius rather than calculating the small angle scattering directly and found that it differed from the actual radius only when the particle was very small [5]. These results are also in agreement with the polarized small angle scattering of neutrons experiments (SANSPOL) in $Fe_{10}Al_{90}$ granular films [54,59].

Near T_c we are in the regime of critical scattering and as long as the correlation length is smaller than the size of the particle, the Ornstein-Zernike form [67] should apply:

$$I(Q) \sim (Q^2 \xi^2 + 1)^{-1} \qquad (5)$$

For the critical scattering, MC results are shown in Figs. 4(a) and 4(b) for particles of radii 7.5 and 12.0 respectively for temperature $T=4.5$ and 5.0. The bulk T_c is at 4.5126 [61]. $1/I(Q)$ is plotted against Q^2 and it can be seen that the MC data are well described by Eq. (5). From the slopes we can extract the correlation length ξ. For each particle we obtain $\xi \sim 1.5$ (lattice spacings) at $T=5.0$, and, for the larger particle, $\xi \sim 4.0$ at $T=4.5$. The straight line fit at $T=4.5$ is not as good for the smaller particle because ξ is comparable with the size of the particle in agreement with experimental finding of Ref. [53].

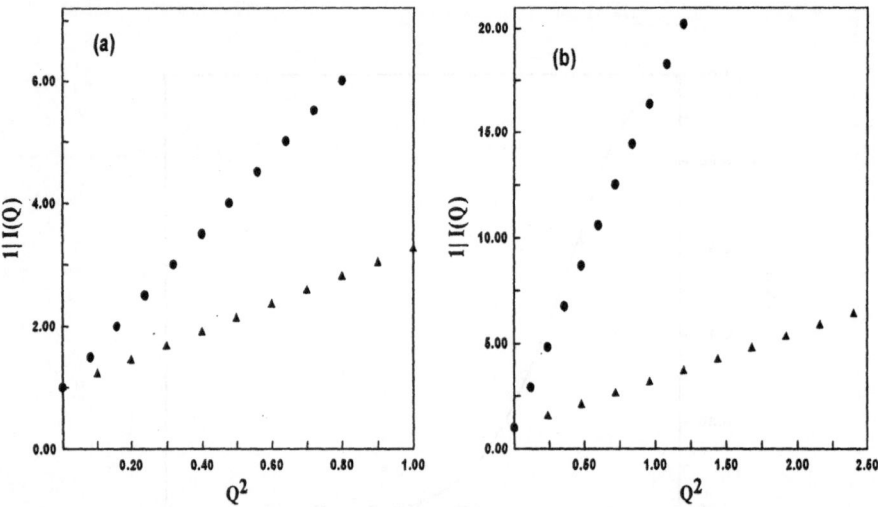

Figure 4. Inverse scattering function versus Q^2 for ferromagnetic particles with sizes (a) $R=7.5$ and (b) $R=12.0$, for temperature $T=4.5$ (circles) and $T=5$ (triangles).

4. Oxidized Nanoparticles

Recently, many experiments have been concentrated on the effect of oxidation on the magnetic properties of ferromagnetic particles [see 11,26,68-71]. The small particles are readily oxidized when they are exposed in the atmosphere. The complete oxidation can be avoided by placing the particles in controlled oxygen atmosphere. In this way, an oxide shell is developed around the ferromagnetic core. In samples where the ferromagnetic core diameters were smaller than 10 nm [72-74] enhanced coercivities have been measured after the oxidation, whereas in samples with core diameters of the order of 50 nm, the coercive fields measured before and after the oxidation were not very different [75].

In Fe particles coated by an antiferromagnetic iron oxide with thickness 1-2 nm, the coercivity showed a large increase in samples with iron core diameter below 6 nm reaching a value of about 2.7 kOe in a particle with core diameter 2 nm. In samples with a larger core diameter the coercivity was almost constant, ~250 Oe [11,73]. The surface and interface effects between the oxide shell and magnetically hard Fe core were considered responsible for the high coercivity values observed at low temperatures and its drastic temperature dependence. The highest coercivity obtained in Ref. [11] at room temperature was 1050 Oe for a particle with a 14 nm core diameter, and its value at 10 K was 1425 Oe, whereas in a sample with core diameter 2.5 nm the coercivity decreased from a value of 3400 Oe at 10 K to a negligible value at 150 K. Thus, in smaller particles the temperature dependence of the coercivity is much stronger than in bigger particles. In smaller particles the Fe core feels much more the effect of the Fe-oxide shell, due to higher Fe-oxide to Fe ratio. The strong decrease of coercivity with temperature can be explained by the superparamagnetic behavior of the Fe-oxide shell and its low blocking temperature. It is estimated that the Fe-oxide shell becomes superparamagnetic at T~10-50 K [72,73]. In all experiments on oxide-coated particles, shifted hysteresis loops have been obtained at low temperatures [10,71,74,76,77]. The loops become symmetric at temperatures above the blocking temperature of the oxide. The shifted hysteresis loop has been attributed to the exchange interaction between the ferromagnetic core and the oxide shell, which is unidirectional [25].

The coercivity of single domain ferromagnetic nanoparticles with uniform anisotropy is well described by the phenomenological theory of Kneller and Luborsky [45]. A modification of this theory was developed by Trohidou et al. [23] to incorporate the core-shell morphology and to explain the coercive properties of ultrafine iron particles coated by iron oxide. In the modified theory, the effect of a shell surrounding the core and of the interface and the surface, all with different anisotropy constants, is dealt with by substituting the magnetization and the anisotropy energy in the simple mean field theory by effective values that are determined by the structure of the particle. The model explained qualitatively the experimental data by considering one atomic layer thick interface with an effective anisotropy about ten times as large as that of the bulk Fe and a shell thickness of about 1.3 nm. In the mean field model the magnetic behaviour was attributed to the strong anisotropy at the interface between the core and the shell of the particle. A strong surface anisotropy is also considered as a candidate to explain the observed high coercivities. This model can only indicate the importance of the complex morphology of the particles without giving any insight in the physics originating from the ferromagnetic core-oxide shell morphology, which is thought to be responsible for the magnetic properties of the particles of interest. There are open questions about the effects due to the microstructure of the particle, the ferromagnetic interaction between the spins in the core, the antiferromagnetic interaction in the shell and particularly the exchange interaction at the interface of the particle, as well as about the effect of temperature.

We review in this chapter results on spherical magnetic nanoparticles with core/shell morphology using the MC simulation technique, where the microstructure and the temperature are explicitly included. The parameters used in the model are the particle size R expressed in lattice spacings, the exchange coupling J and the four anisotropy coupling constants K for different parts of the particle. Results are presented for a set of these parameters in order to discuss the physics emerging from some morphology characteristics of these systems. In order to distinguish the effects of interface and exchange anisotropy, the cases of a ferromagnetic shell and an antiferromagnetic shell are discussed separately.

In the experimental situation there are several cases where the ferromagnetic core is surrounded by a ferrimagnetic oxide [70,78]. In the present review only the case of an antiferromagnetic shell is considered. The effect of the exchange anisotropy is more pronounced in this case [25,79,80].

4.1   MAGNETIZATION AND COERCIVITY

First the case of spherical single domain particles composed of a ferromagnetic core and a ferromagnetic shell with anisotropy lower than that in the core, is considered. We will call them composite particles from now on. The parameters used in the simulations are $K_C=0.1$, $K_{IF}=K_C$ or $K_{IF}=10K_C$ (enhanced interface anisotropy), $K_{SH}=0.025$, $K_S=10K_{SH}$ or $K_S=10K_C$ in the case of simple ferromagnetic particles. In Figures 5(a) and 5(b) the coercivity as a function of temperature is shown for two composite particles with radii R = 8 lattice spacings (N=2109 spins) and 10 lattice spacings (N=4169 spins) (diamonds), respectively. The particles have a surface two lattice spacings thick and radial surface anisotropy.

In Figure 5, the coercive fields of two ferromagnetic particles (asterisks) of the same size with core anisotropy constant K_C =0.1 and radial surface anisotropy constant K_S =1.0 have also been plotted, for comparison. The coercive behavior of the simple ferromagnetic particles has been described in the previous section. Let us now turn to the composite particles. Initially an interface anisotropy equal to the core anisotropy (full symbols in Figure 5) was considered. In this case it can be seen in Figure 5 that the low anisotropy ferromagnetic shell results in a significant reduction of the coercivity for both particles compared to the simple particles. The coercive fields of the smaller composite particle are slightly lower than those of the bigger one. In both cases, the coercivity is weakly dependent upon the temperature. Next, an interface anisotropy one order of magnitude higher than that in the core (open symbols in the figures) is introduced in the model. As can be seen, in Fig. 5, the coercivity has increased in both nanoparticles, but more significantly in the bigger one. In the presence of an enhanced interface anisotropy, the coercive fields exhibit a considerable decrease with temperature. The coercivity of the simple ferromagnetic nanoparticles is slightly higher in the case of the enhanced interface anisotropy from that of the composite particles. If we could attribute an effective anisotropy to the composite nanoparticle, then obviously, this effective anisotropy would be much lower than that of the simple ferromagnetic nanoparticle. This fact can explain the lower coercive fields of the composite nanoparticle. Including the enhanced interface anisotropy we increase the effective anisotropy of the composite nanoparticle significantly, but it is still slightly lower than that of the simple one. The effect of the interface anisotropy is more pronounced in the bigger nanoparticle because of the larger number of sites on the interface.

   The case of the ferromagnetic nanoparticle coated by a ferromagnetic shell of lower anisotropy than the core has been examined, in order to distinguish the anisotropy effects from the exchange interaction effects. Now, we proceed with the experimentally interesting case of an antiferromagnetic shell. We will refer to these nanoparticles as oxidized nanoparticles. The difference between the composite and the oxidized nanoparticles is, therefore, the type of the exchange interaction in the interface and the shell of the nanoparticles. The results of the simulations for the coercivity of two oxidized nanoparticles with sizes R=8 and R=10 (circles) are also shown in Figures 5(a) and 5(b) respectively. We first considered the case of an antiferromagnetic shell with interface anisotropy equal to the core anisotropy (full circles in Figure 5). By comparing the coercive field values in this case with those of the simple nanoparticles, we see that there is a small difference in magnitude between them. The coercive fields of the oxidized nanoparticles are slightly higher and they exhibit a weaker temperature dependence. This effect has been produced by the presence of the antiferromagnetic shell itself, and it is not an effect of the reduced anisotropy of the shell, since it is not observed in the case of the ferromagnetic coating.

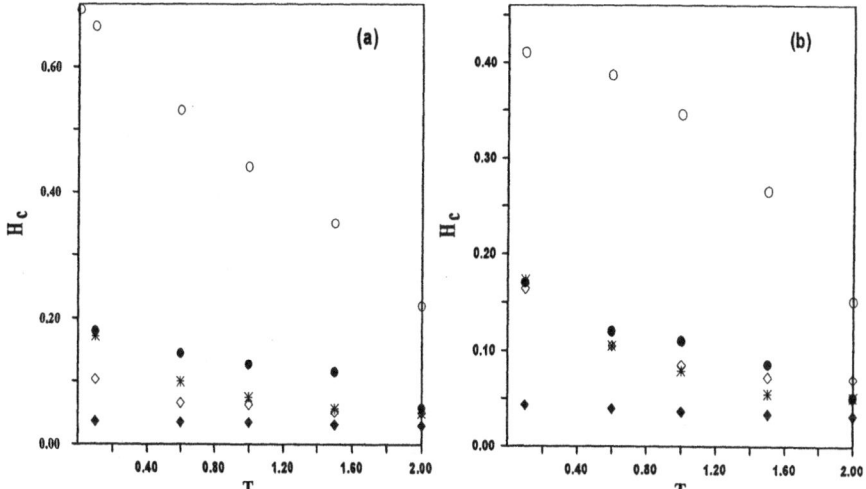

Figure 5. Coercive field versus temperature for composite spherical particles (diamonds) and for oxidized particles (circles) of radius R=8 (a) and R=10 (b), for $K_{IF}=K_C$ (full symbols) and $K_{IF}=10K_C$ (open symbols. The asterisks correspond to the simple ferromagnetic particles of the same size.

   From the simulation, it turns out that this effect should be attributed to the exchange interaction between the ferromagnetic core of the nanoparticle and the antiferromagnetic shell. This interaction has the same effect as an extra unidirectional interface anisotropy as has been described in several experimental works on Fe/FeO nanoparticles [10,11,68,78,81] on Ni/NiO [82,83] and Co/CoO [29,71, 74, 77,84], and is known as exchange anisotropy [25]. In Figures 5(a) and 5(b) we can also see the effect of the enhanced interface anisotropy (open circles). This anisotropy causes a considerable increase in the coercive fields of both nanoparticles. What is remarkable here is that there is a significant reversal in the size dependence of the coercivity. The smaller particle has higher coercive fields than the bigger. As can be seen in Figures 5(a) and 5(b), the effect on the smaller particle is more drastic, so that its coercivity is much higher than the coercivity of the big particle for all temperatures. The enhanced interface anisotropy causes an enhancement of the effect induced by the exchange anisotropy in the interface. They essentially result in a pinning of the magnetic moments in the interface. The interface anisotropy introduces a steep temperature dependence in agreement with the experimental findings for the coercivity of small Fe particles [10,11].

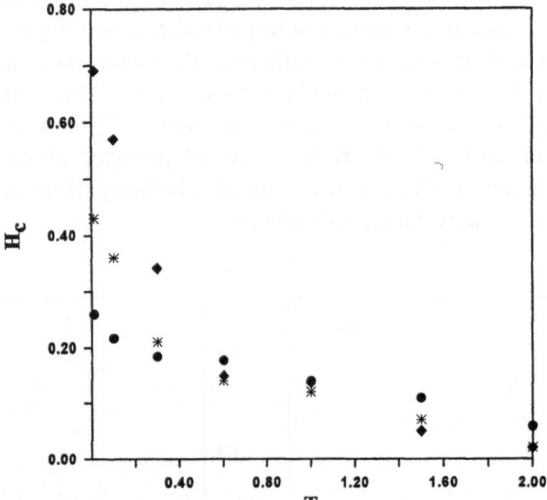

Figure 6. Coercive field versus temperature for oxidized particles of radii R=8 (diamonds),   R=10 (asterisks) and R=20 (circles). The anisotropy coupling constants of the core, interface, shell and surface are temperature dependent as described in the text.

Finally, the effect on the coercivity of the temperature dependence of the anisotropy coupling constant is discussed. The following expression for the temperature dependence of the anisotropy $K_l(T)=K_l(0)\,\hat{I}_{l+1/2}(\hat{I}^{-1}_{3/2}(m))$, where the $\hat{I}$'s are hyperbolic Bessel functions and m is the magnetization at temperature T, was proposed many years ago by Callen and Callen [85]. For the sc lattice, the values of the anisotropy follow this form for $l=4$. For this value a close to exponential fall-off of K_{SH} with T appears which becomes negligible for $T>1.0$. For K_C the temperature dependence is smoother [86]. The temperature dependence of the anisotropy and coercivity is in good agreement with the experimental findings [9,35,73,74]. The results are shown in Figure 6 for three nanoparticles with considerably different radii, R=8 (N=2109 spins), 10 (N=4169 spins) and 20 (N=33401 spins) (diamonds, asterisks and circles respectively), taking as zero temperature values for the anisotropy constants the ones described above (the enhanced K_{IF} case). The temperature dependence of the anisotropy results in a steeper decrease of the coercivity with temperature for the smaller particle. The coercive fields of the big nanoparticle vary less rapidly with temperature. The size dependence of the coercivity exhibits a low temperature cross-over, at $T\sim0.5$. Above this temperature, the coercive fields of the nanoparticle with R=8 are lower than those of the nanoparticles with R=10 and 20. Therefore the temperature dependence of the anisotropy, causes early disorder in the oxide shell, which gives the low cross-over temperature observed in experiments

[11,71]. Experimental investigations on Fe/FeO particles [27] have shown that the

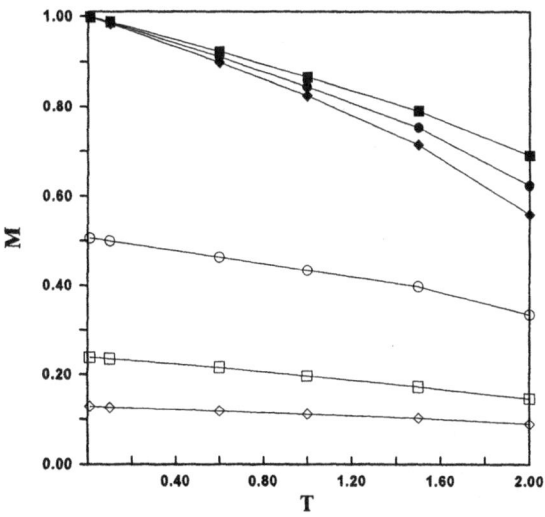

Figure 7. Magnetization per spin versus temperature for simple ferromagnetic particles (solid symbols)

increase of the shell thickness reduces the effect of the exchange anisotropy due to the increase of the surface disorder that masks it.

5. Antiferromagnetic Nanoparticles

Antiferromagnetic nanoparticles have finite magnetization due to the difference between the populations of the two sublattices of antiferromagnetism (uncompensated spins). Néel discussed how the superparamagnetic and antiferromagnetic susceptibilities are superimposed [12] and pointed out the predominant role of the surface in the magnetization and in the enhanced antiferromagnetic susceptibility ('superantiferromagnetism'). The experimental work was initially on α-Fe_2O_3 particles in rocks, on Cr_2O_3 particles [13,14] and later on ferritin [15,87-90]. Recently, much experimental activity has been devoted to the magnetic properties of metal oxide particles, which are used as high-density recording media [34,91-99]. From these works, it becomes apparent that the surface plays a predominant role in the observed superparamagnetic behavior and the low blocking temperatures of antiferromagnetic nanoparticles. High coercivity and shifted hysteresis loops are also reported in Refs. [98,100-103]. Deviations from a Langevin function fit to the superparamagnetic magnetization are attributed [87,99,100,104-106] to the surface moments. Surface spin disorder is assumed to explain the observed moments [91,102], the small magnetization and the high coercivities at low temperatures, found in several experimental investigations [76,97,107]. In Ref. [108] the magnetic properties of nanosized maghemite particles are attributed to the interparticle interactions that cause spin-glass-like ordering of the magnetic moments. It was also concluded [109] that small antiferromagnetic particles have small shape anisotropy compared with crystalline anisotropy. Quantum tunneling effects are more important in antiferromagnetic particles than in ferromagnetic particles [110,111] and can account for the magnetic behavior at temperatures as low as a few Kelvin [89,90,112,113]. Theoretical studies on very small Ising antiferromagnetic clusters (up to 30 spins in [114] and 561 spins in [115] have shown that the variation of magnetization with applied field and temperature depends on the geometry of the clusters. The variation of the magnetization of free and fixed-single-crystal antiferromagnets with the applied magnetic field is investigated in a macroscopic model in [20] and a method to determine the anisotropy strength is proposed. The magnetic properties of antiferromagnetic particles of sizes up to 10 nm have been studied by means of atomic scale modeling of the nanoparticles [102] and MC simulations [19-22]. In [19] Ising spins were considered for simplicity. The magnetization of a particle that arises from the uncompensated spins and, in particular, how the temperature variation of the magnetization depends on particle size and the number of uncompensated spins was investigated. In [20] a more realistic model with classical Heisenberg interactions between the spins was introduced, taking also into account the effect of the core and surface anisotropy.

Ferrimagnetic nanoparticles present common characteristics with the antferromagnetic ones. This is due to the antiferromagnetic ordering of the two sublattices. There is a lot of experimental work on ferrimagnetic nanoparticles and MC simulations [28, 116-121].

MC calculations on antiferromagnetic particles from size <300 spins up to several thousand spins are presented in this review. The magnetization for a range of parameters, the response to an external field and the magnetization reversal mechanism are examined. A two-sublattice model for antiferromagnetism is employed. N will denote the total number of spins in a particle and N_u will indicate the number of uncompensated spins (i.e. $N_u = N_1 - N_2$ where N_1 and N_2 are the number of sites on the two sublattices). In most of the simulations, the uniaxial core anisotropy along the z-axis is present; in these cases magnetization refers to the z component and the x and y components essentially average to zero. In cases where such anisotropy is absent we take the magnetization as the magnitude of the net particle magnetization vector and average it; without anisotropy arbitrary rotations of the magnetization vector can occur in the simulations.

5.1   MAGNETIZATION

5.1.1 Zero field magnetization

First the dependence of the magnetization of the antiferromagnetic particles on the particle size (radius R, total number of spins N) and on the number of uncompensated spins N_u, in the absence of any anisotropy is described. Representative behavior is shown in Figure 8 for three particles. The magnetization of identical ferromagnetic particles is also shown to provide a comparison.

For ferromagnetic nanoparticles, the total spin can undergo reversals on time scales of the order of the duration of the MC simulation. For this reason the magnetization, M, is generally defined as $< |S| >$ and this is the definition used in the current work. As a consistent definition for antiferromagnetic particles we have used $M = (< |S_1| > - < |S_2| >)/N_u$ where S_1 and S_2 are the sublattice magnetizations. The plots are normalized to 1.0 at $T=0$, so that it is the magnetization per spin that is displayed for the ferromagnets while, for the antiferromagnets, it is the magnetization per uncompensated spin that appears. The temperature is in units of T/T_N where T_N, the bulk Néel temperature, is 2.9 for the sc systems [61]. These values of T_N refer to bulk antiferromagnets without any anisotropy and they are used as reference temperatures for all figures.

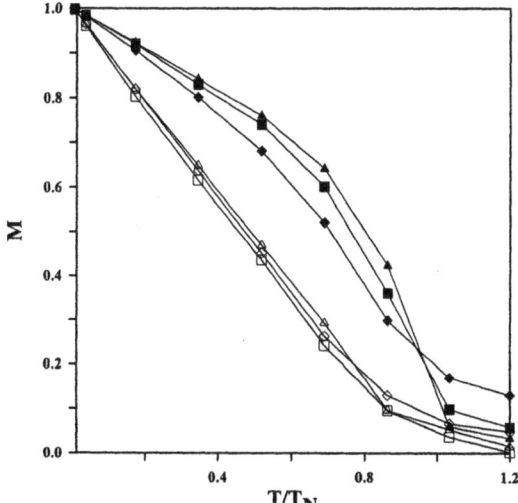

Figure 8. Magnetization per spin versus temperature for ferromagnetic particles (solid symbols) and antiferromagnetic particles (open symbols) for the following sizes $R=4.0$ ($N=251$, $N_u=19$) (diamonds), 7.75 ($N=1935$, $N_u=17$) (squares) and 11.58 ($N=6619$, $N_u=19$) (triangles) and for zero anisotropy.

The two most obvious features shown by the antiferromagnetic particles in contrast to the ferromagnetic case are the relative insensitivity of the magnetization to N, N_u and secondly, the linear temperature dependence of the magnetization. The size dependence in the ferromagnetic system, as is well known, arises from the increased importance of the low coordination number surface spins for smaller particles. In the antiferromagnetic case, we can regard the net superparamagnetic component due to the uncompensated spins as coming entirely from the surface irrespective of the size of the particle and thus the size dependence is much smaller. The temperature dependence of the magnetization can be deduced from spin wave theory. Again, noting that we are looking at a surface effect, consideration of surface spin waves would lead to a linear temperature dependence (as opposed to the 3/2 power law predicted for bulk ferromagnets). Spin wave predictions, of course, apply at low temperatures. It appears that good linear behavior actually operates over quite a wide temperature range. The contrast with Ising model behavior is familiar [19]; in that case there is a low temperature regime where deviations from full alignment require finite energy excitations and the variation with temperature is very small.

We now consider the effect of introducing anisotropy. Figure 9 shows the magnetization for a particle with radius $R=7.5$ lattice spacings Three anisotropy regimes are examined: (i) z-axis anisotropy ($K=K_c=K_s=0.5$) at all sites, (ii) z-axis core anisotropy ($K_c=0.5$) and small radial surface anisotropy ($K_s=1.0$), (iii) z-axis core anisotropy ($K_c=0.5$) and large radial surface anisotropy ($K_s=10.0$), this value was considered because early experiments indicated that, under certain conditions, the magnetization shows considerably less temperature dependence than could be attributed to the uncompensated spins. Pinning of moments at the surface is often invoked to explain this effect [17,122].

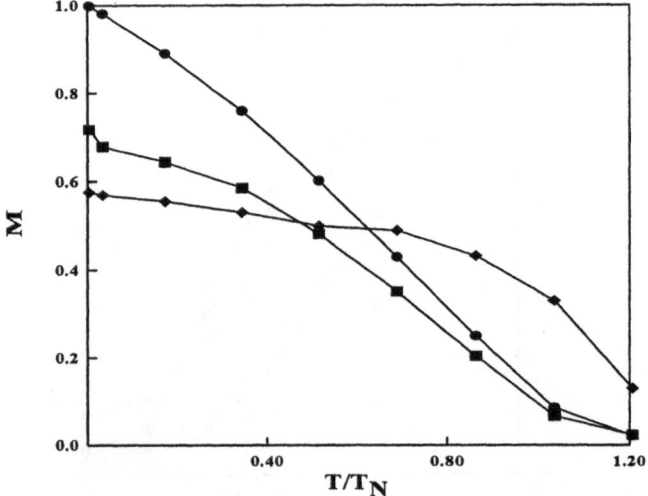

Figure 9. Magnetization versus temperature for a sc antiferromagnetic particle, R=7.5, N=1791, N_u=79. The anisotropies are K =0.5 (circles), K_c=0.5 and K_s= 1.0 (squares), 10.0 (diamonds).

In case (i) we notice the slower decrease of the magnetization with temperature than in Figure 8. The $K_c \rightarrow \infty$ limit is just the Ising model in which the variation with temperature is small for T/T_N less than about 0.3 [19]. The value of K_c (=0.5) used here is intermediate between the classical Heisenberg and the Ising limits. It is useful to consider the large K_s limit and regimes (ii) and (iii) are stages in the approach to that. The surface spins are pinned by K_s to lie parallel to a radius vector. We assume that the uncompensated spins are distributed reasonably uniformly over the surface of the particle, therefore, the magnetization (per uncompensated spin) is reduced to a factor that is the average of $|\cos\theta|$ over the surface, namely 1/2.

One can see from Figure 9 that the zero temperature magnetization is showing this trend with increasing K_s. Despite the fact that the magnetization is reduced the pinning also makes it more robust against the disordering effect of increased temperature. This is apparent in the high K_s plot, which lies above the others at higher temperatures. We have also considered the effect of randomness in the orientations of the surface anisotropy. The behavior is not significantly different from that just described, which is not too surprising. The average of $|\cos\theta|$ is still 1/2 and the pinning effects are similar but the orientation of the pinning is not related to the position on the surface.

A particular particle in Figure 9 is chosen, to display the effects of both kinds of anisotropy on the magnetization. The behavior is typical of all particles both sc and bcc as long as N_u is reasonably large. This condition is necessary because the generic behavior depends on the uncompensated spins being distributed reasonably uniformly over the surface of the particle. If N_u is small there is often sufficient uniformity in the distribution for the typical behavior represented in Figure 9 still to occur. However, the magnetization may show features that are not generic and the actual behavior has to be

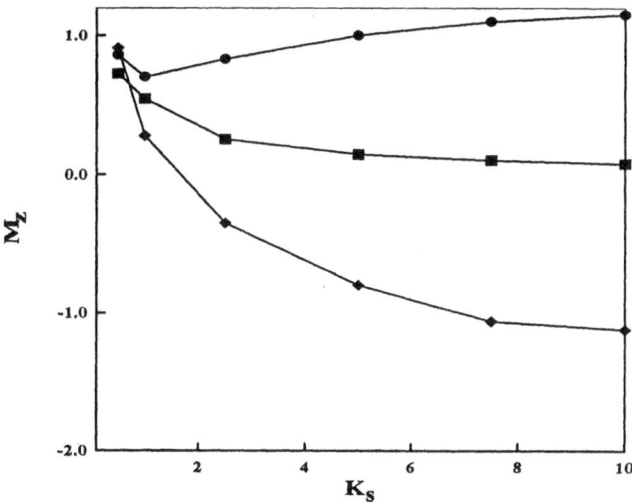

Figure 10. Zero temperature magnetization versus surface anisotropy K_s. The core anisotropy, K_c=0.5. Plots for three particles are shown: R=7.75, N=1935, N_u=17 (circles), R=10.75, N=5185, N_u=19 (squares); R=8.0, N=2103, N_u=7 (triangles).

determined by a study of the individual particle. Some examples are shown in Figure 10. The atypical features show up most strongly at low temperatures and so the plots display the magnetization as a function of K_s in the $T\rightarrow0$ limit. In one of the plots the magnetization falls below 1/2 for large K_s while, in another it rises above unity. The third plot show a case where the magnetization changes sign as K_s is increased. Figure 10 displays examples of behavior that can occur. These particular particles are not unique; they exemplify features that can occur and do occur in other particles as well. The origin of these features can be understood as follows. We have N_1 spins on the 'up' sublattice and N_2 spins on the 'down' one. Let us consider spins within planes perpendicular to the z-axis and in particular those planes that have z close to zero. These are the ones most strongly affected by the radial surface anisotropy, which tends to align the surface spins in these planes perpendicular to the easy z-axis of the particle. Now in the absence of distinct surface anisotropy N_u=N_1-N_2, and this determines the magnetization. The planes of spins close to z=0 may have an excess of 'up' or of 'down' sublattice sites. In the former case the introduction of radial surface anisotropy will tend to reduce the magnetization while in the latter case, the effect is to increase it. With a small N_u the reduction can be enough to produce a sign change whereas an increase will push it above unity. As temperature is increased the behavior is closer to standard behavior. For example, in cases which produce a sign change, the magnetization changes to the expected sign as the temperature is increased.

It is seen that, if no radial surface anisotropy is present, both large and small N_u particles behave in a very similar way. In the presence of the radial surface anisotropy, however, the small N_u particles are quite likely to display non-generic behavior that is specific to that particle. In experimental works on antiferromagnetic particles [88,102], the magnetization behavior is attributed to a random orientation of the surface spins.

Analogous non-generic behavior for small N_u is possible in the case of random anisotropy. In that case it would cause the deviations from the average in statistics on small samples (of uncompensated spins), whereas in our model it comes from particular geometric configurations.

5.1.2 Behavior in an applied magnetic field

A sc particle with size $R=7.5$ lattice spacings ($N=1791$ and $N_u=79$) is used to display typical magnetic behavior as a function of the applied magnetic field for a range of temperatures. Magnetization is plotted in Figure 11 against H. There is no distinct surface anisotropy, but a uniform anisotropy along the z-axis ($K=K_c=K_s=0.5$) is included. Over the range of fields relevant to the present discussion the field dependence is linear.

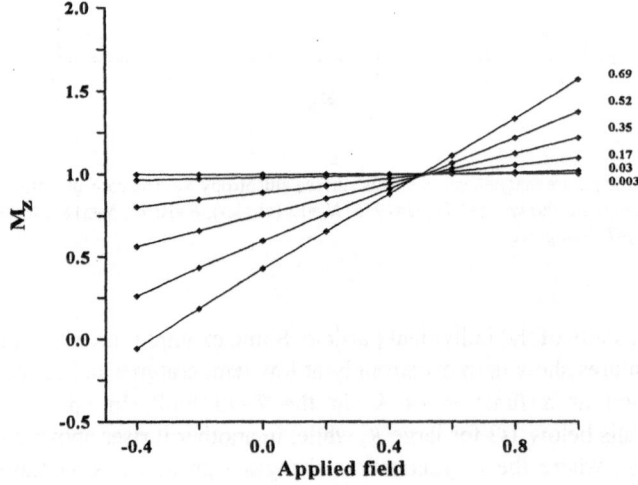

Figure 11. Magnetization versus applied field for sc particle with $R=7.5$, $N=1791$, $N_u=79$ and uniform anisotropy ($K=K_c=K_s=0.5$), for six temperatures. The figures on the right of the curves are the corresponding T/T_N.

Deviations from linearity become significant for $H<-0.5$. This behaviour is in agreement with experimental findings on ferritin [87,106,123] and α-Fe2O3 nanoparticles [99,105]. It is interesting to note that the different temperature plots cross at a very well-defined field which we will denote by H^*. The implications of this well-defined field H^* can be developed as follows.

The magnetization of a particle arises from the uncompensated spins and therefore changes due to increased temperature should scale with N_u. The effect of the field, on the other hand, will be felt by all the spins in the system and the susceptibility is expected to scale with N. We can write the magnetization per uncompensated spin as

$$M(T,H) = M(T,H=0) + \frac{N}{N_u}\chi(T)H \qquad (6)$$

where $\chi(T)$ is the susceptibility (per spin). Because of the well-defined H^*, one infers that the change in the zero field magnetization must also be proportional to $\chi(T)$, i.e. $M(T,H=0)-M(T=0, H=0)=F(T)$ with $F(T)=h\chi(T)$, where h is a constant of proportionality. This leads to the relation

$$M(T,H) = M(T=0, H=0) + \chi(T)\left[\frac{N}{N_u} H - h\right] \qquad (7)$$

the crossing point is then given by $H^* = h(N_u/N)$ so H^* should scale as the ratio N_u/N. We have plotted H^* against this ratio for particles with a range of values of N, N_u and both sc and bcc structures in Figure 12. For a rough argument it is seen that the scaling is

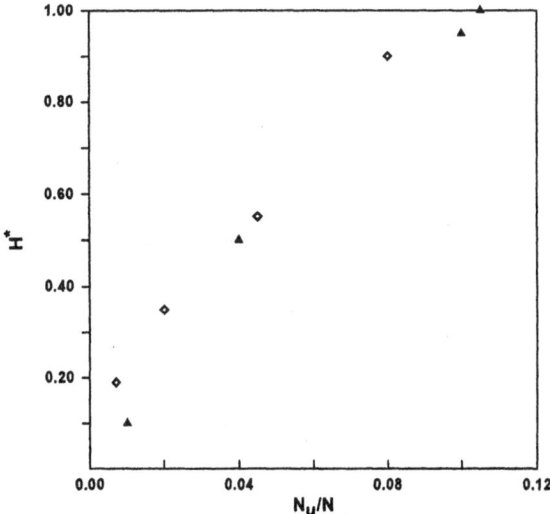

Figure 12. Crossing field H^* versus N_u/N for four sc (triangles) and four bcc (diamonds) particles. The points from left to right correspond to the particles: sc: $R=7.75$ ($N=1935$, $N_u=17$), $R=7.5$ ($N=1791$, $N_u=79$), $R=4.1$ ($N=257$, $N_u=25$), $R=3.1$ ($N=123$, $N_u=13$). bcc: $R=7.9$ ($N=4111$, $N_u=95$), $R=7.2$ ($N=3119$, $N_u=17$), $R=3.1$ ($N=259$, $N_u=13$), $R=4.1$ ($N=561$, $N_u=47$).

obeyed rather well. In the total absence of anisotropy the above argument can be justified with an $F(T)$ that is linear in T. This can be seen in Figure 8 where the magnetization decreases linearly with temperature to a good approximation. As far as the susceptibility, $\chi(T)$, is concerned we should use the parallel susceptibility for an antiferromagnet. This also varies close to linearly with temperature. Thus it is reasonable to take $F(T)=h\chi(T)$. When uniaxial anisotropy is introduced there are deviations from a linear behavior. The magnetization now falls less rapidly (Figure 9, circles). However, the anisotropy will affect the susceptibility in a similar way and so, at least phenomenologically, Eq.(6) will still represent the behavior to a fair approximation. Certainly the results of the simulation in Figure 12 and the well-defined value of H^* support this description.

Figures 13(a) and 13(b) show how the behavior displayed in Figure 11 is modified when the surface anisotropy is included ($K_s = 1.0$ and 10.0 respectively). There are two features that contrast with Figure 11. First, the near zero temperature plot is no longer field independent and, second, in Figure 13(b), the crossing point has become considerably blurred. The surface anisotropy tends to align the surface spins in a radial direction and, because of the exchange interactions, the spins in the vicinity of the

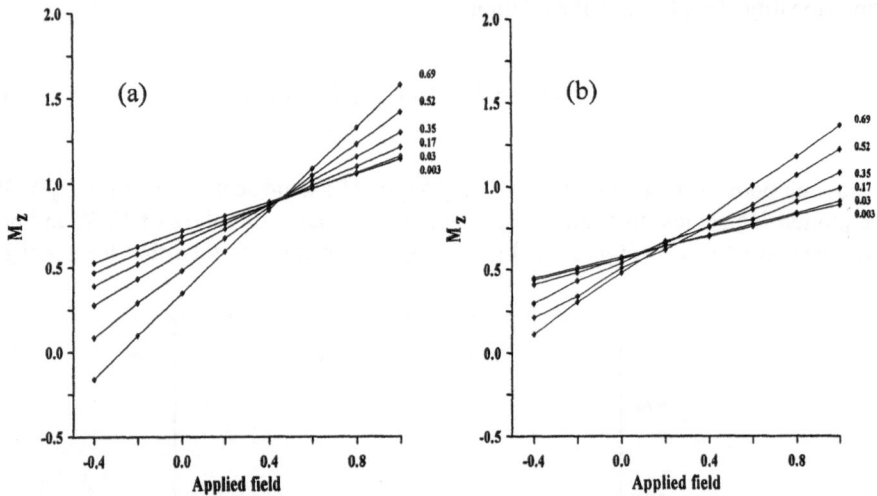

Figure 13. Magnetization versus applied field for particle with $R=7.5$, $N=1791$, $N_u=79$, sc, for six temperatures. The figures on the right of the curves are the corresponding T/T_N. Anisotropies are: $K_c=0.5$. (a)$K_s=1.0$ and (b)$Ks=10$

surface will try to follow in their alignment. The magnetic field will now have the additional effect of pulling these spins toward the z-direction. The result of this is a non-zero susceptibility even in the $T \rightarrow 0$ limit, thus accounting for the first of the new features. However, for the smaller value of the surface anisotropy (Figure 13(a)), a well-defined crossing point and value of H^* still occurs. As the surface anisotropy increases (Figure 13(b)), the crossing point becomes blurred. The clear division between surface and core spins does not apply and the simple argument that led us to a similar temperature dependence for $F(T)$ and $\chi(T)$ when interpreting the behavior in Figure 11 can no longer be used.

It is worth emphasizing that the arguments leading to Eq. (7) and a well-defined H^* depend only on the assumption that similar mechanisms operate in determining both the field and temperature dependent behavior. A particular particle has been used throughout to display the behavior, but the plots can be regarded as typical for all systems studied. We have mentioned caveats in predicting generic behavior for the magnetization for small N_u particles. These concerned the large K_s regime however. A well defined H^* occurs for small K_s and so the description given for this situation applies irrespective of N_u (both small and large N_u particles are included in Figure 12).

5.2   COERCIVITY

The response of the spins in spherical antiferromagnetic nanoparticles to a magnetic field applied in a direction opposite to the net magnetization is reported here.. The magnetization reversal is examined for surface anisotropy with easy axis: (a) along a fixed direction which is the same as in the core of the particle and (b) radial at each surface site. There are two magnetization reversal mechanisms for both types of anisotropy. In the first mechanism, the switching field reverses all spins in the particle and the magnetization relaxes to the opposite remanence after removing the applied field.

In the second mechanism, the particle magnetization vanishes by a switching field that distorts the surface spin alignment causing a decrease in the ferromagnetic component of the magnetization which arises from the uncompensated spins.

The two switching fields have distinct characters. The larger field causes a permanent reversal of the particle magnetization, i.e. when the magnetic field is switched off the remanent magnetization is opposite in sign to its original value. The smaller field causes zero magnetization in each individual particle but the magnetization of the particle returns to its initial remanence after switching off the field. To understand the physical origin of these two fields it is useful to consider the magnetization of the antiferromagnetic particles in the presence of a magnetic field as made up of two components, the ferromagnetic and the antiferromagnetic one. The ferromagnetic component is due to the uncompensated spins which rotate in order to align with the applied field in the $-z$ direction. The uncompensated spins rotate and due to the exchange interaction, their neighbors follow them. The z-component of the particle magnetization is therefore reduced. The antiferromagnetic particle acquires a net moment in the $-z$ direction in the presence of a field. This is strongly temperature dependent following the behavior of the parallel susceptibility (described in the previous section) and becomes more important as temperature increases. The reduction of the magnetization starts from the surface of the particle because the uncompensated spins belong to the surface. Moreover, the surface antiferromagnetic susceptibility is higher than the bulk one [12]. The applied field causes a distortion of the spin alignment at the surface of the particle. This distortion is extended deeper in the core of the particle for stronger applied fields. As the applied field increases, the magnetization of the particle decreases and eventually it vanishes. Since N_u is small compared to the total number of spins in the particle, it is possible to get zero magnetization solely due to the surface spin response to the switching field. In this case, after switching off the applied field the exchange interaction with the core spins forces the surface spins back to their original orientation and restores the initial remanence. To reverse the magnetization of the particle the applied field has to be stronger in order to affect the core spins.

In Figure 14 the coercive field as a function of temperature has been plotted for an

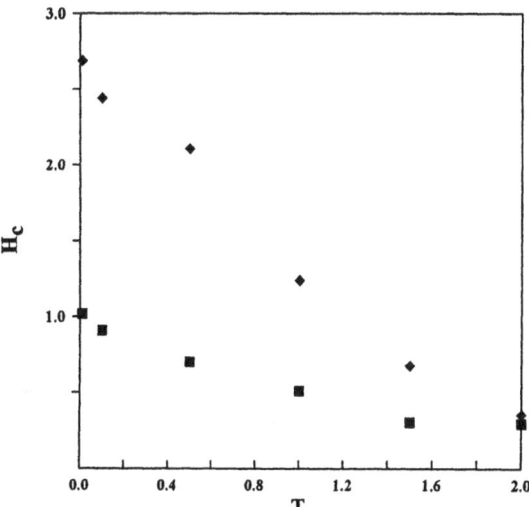

Figure 14. Coercive field versus temperature for a ferromagnetic particle of radius $R=7.5$ (squares) and an antiferromagnetic particle (diamonds) of the same size with z-easy axis uniform anisotropy $K=0.5$.

antiferromagnetic particle of radius $R=7.5$, together with that for a ferromagnetic particle of the same size, in both cases we consider uniform z-axis anisotropy. The coercive behavior of the ferromagnetic particle with uniform anisotropy is known. The coercive field of the antiferromagnetic particle is considerably higher. This is expected since in the antiferromagnetic particle we have contribution to the coercivity from both the antiferromagnetic core and the uncompensated spins on the surface.

In Figures 15, 16(a) and 16(b) the coercive field for the particle with radii $R=7.5$, 7.75 and 13 respectively have been plotted as a function of temperature for uniform anisotropy along the z-axis $K=0.5$(circles) and with $K_c=0.5$ along the z-axis including radial surface anisotropy with $K_S=1.0$ (diamonds) and radial surface anisotropy with $K_S=10$ (open diamonds). In order to demonstrate that the effect of random anisotropy is similar to that of radial surface anisotropy, in Figure 15 we have also plotted the coercivity as function of temperature for random axis surface anisotropy $K_S=10$ (asterisks). We observe that the random anisotropy results in smaller reduction of the coercivity than the radial surface anisotropy. It gives a higher coercivity at low temperatures and shows stronger temperature dependence but for a wide range of temperature is very similar to the coercivity for radial surface anisotropy of the same size. For a direct comparison of course we have to consider a specific system.

If we compare the coercivities of the particles in the case of a uniform anisotropy $K=0.5$ (circles in Figures 15, 16(a) and 16(b)) we can see that in the zero temperature limit they have the same value ~2.8 in our units. This value can be understood by examining the reversal mechanism of the spins in the zero temperature limit. If the anisotropy is small, the magnetization is inversed by coherent rotation of the spins and the coercivity is equal to $2K$. For larger values of the anisotropy constant, it might be energetically favorable that the surface spins respond first to the opposite magnetic field, because the exchange field is lower on the surface. In this case, the spins have to overcome the anisotropy energy barrier that gives a contribution $2K$ to the coercive field,

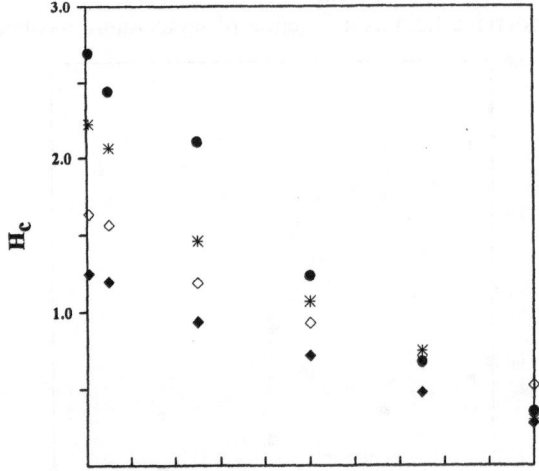

Figure 15. Coercive field versus temperature for an antiferromagnetic particle of radius R=7.5. The circles correspond to the particle with a z-easy axis anisotropy K=0.5, and the diamonds to the particle with z-easy axis core anisotropy $K_C=0.5$ and radial surface anisotropy $K_S=1.0$ (solid diamonds) and $K_S=20K_C$ (open diamonds). The asterisks correspond to random easy-axis surface anisotropy $K_S=10.0$.

and the surface exchange field H_{exc} that is opposite to the external field, in order to reverse the magnetization of the particle. Along these lines, we write the coercivity as the sum of the two contributions $H_c = 2K + aH_{exc}$, where the parameter a depends on the strength of the anisotropy. For a weak anisotropy, a is equal to 0, reflecting the inversion of the magnetization by coherent rotation of the spins in the particle. In the limit of a very strong anisotropy, each surface spin responds independently to the applied field and has to overcome the whole surface exchange barrier in which case a is equal to 1. For intermediate values of the anisotropy constant, a is smaller than 1, because more surface spins respond simultaneously to the applied field and the local exchange field is smaller. The average surface exchange field in the limit of a very strong anisotropy, is equal to $0.5 < z_s > S^2$ (in units of 2J), where $< z_s >$ is the average surface coordination number and is equal to ~4 for the sc particles. In Fig. 17 the parameter a is plotted as a function of the anisotropy and we see that in our case which lies in the high anisotropy regime a is equal to 0.9.

These arguments cannot be applied at higher temperatures. From Figures 16 (a) and (b), we see that for the particles with radii $R=7.75$ and $R=13$ in the uniform $K=0.5$ case, the coercivity over a wide range of temperature (1.0-2.0 in our units) is almost the same. These two particles have different sizes (1935 and 9092 spins respectively) and very different number of uncompensated spins (17 and 77 respectively), but they have almost the same ratio N_u/N (0.0087 and 0.0085 respectively). This ratio for the particle with radius $R=7.5$ in Figure 15 is five times bigger ($N_u / N = 0.044$) and the coercive field for the same range of temperature as previously is five times higher for the $K=0.5$ case. So the coercivity scales with the ratio N_u/N. At these temperatures, the coercivity is explained in [19,20] in terms of the susceptibility that it is found to scale with N_u/N. In the Heisenberg model, the scaling behavior holds for the component of the susceptibility

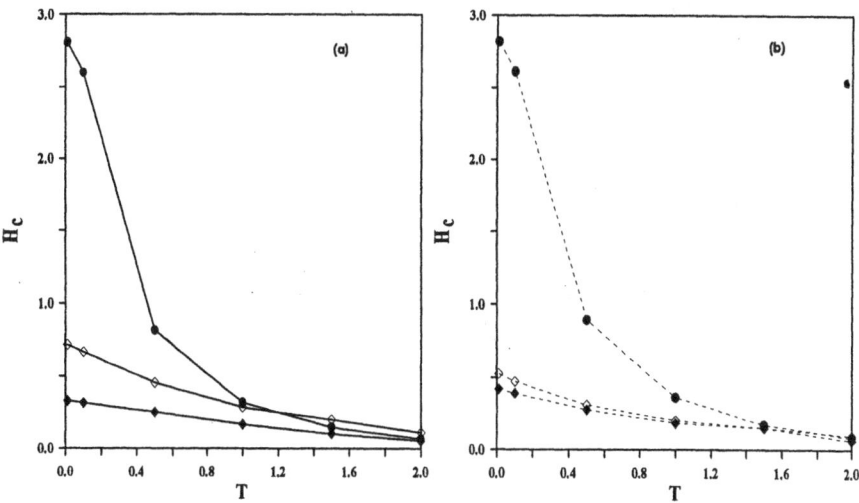

Figure 16. Coercive field versus temperature for an antiferromagnetic particles of radius *(a)* $R=7.75$ and (b) $R=13$. The circles correspond to the simple antiferromagnetic particle with a z-easy axis anisotropy $K=0.5$, and the diamonds to the particle with z-easy axis core anisotropy $K_C=0.5$, and radial surface anisotropy $K_S=1.0$(solid diamonds) and $K_S=10$ (open diamonds).

parallel to the magnetic field [20], which is what was simulated in the Ising system [19].

The radial surface anisotropy favors the radial orientation of the surface spins and lowers the barriers that they have to overcome in order to reverse the magnetization of the particle. At low temperatures, the coercivity is significantly smaller in the presence of the radial surface anisotropy, as can be seen in Figures 15, 16(a) and 16(b). In this case, we cannot apply the simple relation for the two contributions to the coercivity at zero temperature, because there is a distribution of energy barriers depending on the positions of the spins on the surface rather than a single barrier for the anisotropy and the exchange. The coercivity comes out from the averaging on all the surface contributions and it is therefore expected to be dependent on the number and the distribution of the uncompensated spins.

At higher temperatures, in the particles with the small N_u/N (Figures 16(a) and 16(b)), the thermal fluctuations screen the effect of the radial surface anisotropy and the values of the coercivity become comparable in all cases of anisotropy. In the particle with the bigger N_u/N (Figure 15), the anisotropy dominates over the thermal fluctuations up to the highest temperatures considered. For $K_s=10$, the coercivity has values higher than at the smaller K_s case. The increase is more significant in the particle with radius $R=7.5$ and big N_u/N. In the other two particles with radii $R=7.75$ and 13, which have the same N_u/N, the increase is bigger in the first particle which, as we discussed in the previous section, exhibits a significant increase of the magnetization for $K_s=10$ (Fig. 10) and it is smaller in the second particle which exhibits a decrease of the magnetization; arguments about the distribution of the uncompensated spins were used to explain this behavior. In the temperature range (1-2), the coercivity scales with N_u/N for the three particles in the case of $K_s=1.0$, but for $K_s=10$ the scaling does not hold for the particle with radius $R=7.75$. As we discussed, in the uniform $K=0.5$ case the scaling behavior of the coercivity holds for the range of temperatures that the zero field magnetization is independent of N_u and decreases linearly with temperature. Such a behavior can be seen for all cases except for the case of the particle with $R=7.75$ and $K_s=10$, as we discussed in Figure 10 and it was demonstrated subsequently in [87].

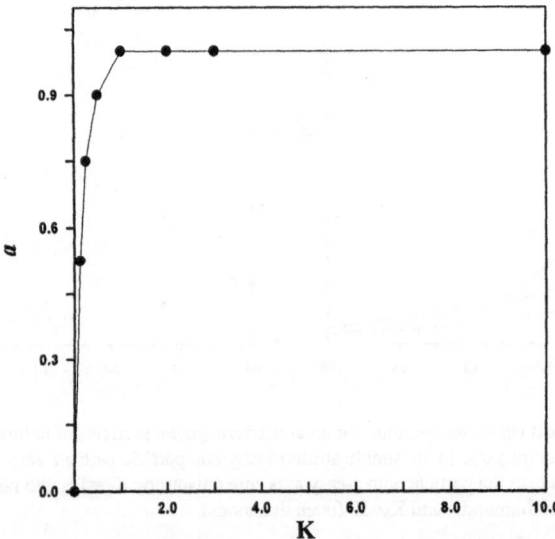

Figure 17. Parameter *a* versus anisotropy coupling constant K for spherical antiferromagnetic particles.

Our results are in agreement with experimental findings of the coercivity versus temperature on ferritin [87, 106,123] on Cr oxide particles [104] and on Co_3O_4 nanoparticles [101].

5.3.   SMALL ANGLE NEUTRON SCATTERING

It has been suggested for a long time that, in the case of antiferromagnetic particles, uncompensated spins reside at the surface [12,102,124,125]. To verify this expectation, the scattering function of SANS is obtained by MC simulations. This function will be compared with an analytical formula derived under assumption that all uncompensated spins are located at the particle's surface.

For antiferromagnetic particles SANS has the nice feature that the superparamagnetic component is distinct from the antiferromagnetic one, the former being determined by the static structure factor in the region around $Q = 0$. The dominant scattering, of course, is near the antiferromagnetic Bragg peaks rather than around $Q = 0$. There will, however, be a tail to the Bragg peaks which will be strongly temperature dependent, but weakly dependent on Q (over the Q range of interest near $Q = 0$). The scattering amplitude will scale with the number N of spins in the particle.

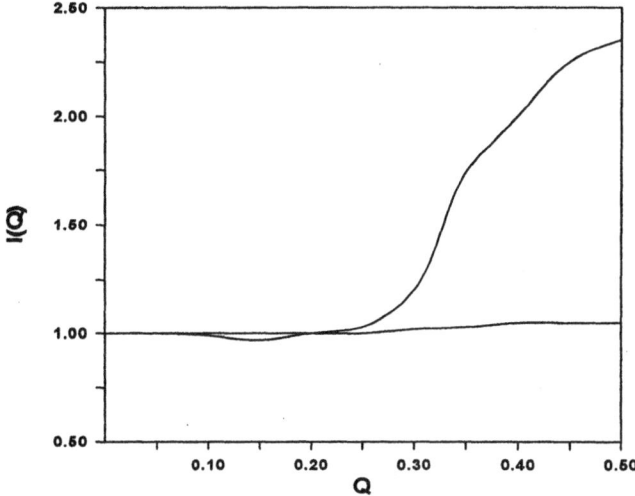

Figure. 18. Intensity versus Q for antiferromagnetic particles of radius $R = 12$ for temperatures $T=2.0$ {solid line) and $T=3.0$ (dashed line).

The scattering amplitude arising from the uncompensated spins will depend on $N_u<S>$, where $<S>$ is the mean magnetization (per uncompensated spin). The uncompensated spins can be viewed as effectively residing in the surface of the particle. Developing this picture, the scattering function is regarded as due, not to a sphere, but to a shell of radius R [19]. This argument leads to the following scattering function for antiferromagnetic particles at temperatures sufficiently below T_N:

$$\langle S(Q)\rangle_{av} = N^2 A(Q) + N_u^2 \langle S\rangle^2 \left[\frac{\sin(QR)}{QR}\right]^2 \qquad (8)$$

where $A(Q)$ represents the weakly Q-dependent scattering from the antiferromagnetic
Bragg tail. As stated earlier, at temperatures above about 70% of T_N the magnetization
of the particle becomes comparable with antiferromagnetic fluctuations. Following from
this, any critical scattering involving the uncompensated spins is completely masked by
the antiferromagnetic component and can be neglected.

It is proposed that Eq. (8) represents the only scattering for antiferromagnetic
particles around $Q = 0$. The MC procedure was used to test its validity. A particle of R
$=11.5$, $N = 6403$, and $N_u = 101$ was studied. The only unknown in Eq. (8) is $A(Q)$,
because $<S>$ can be obtained from the Monte Carlo simulation. $A(Q)$ was obtained from
calculations on a particle with $R = 12$ and $N = 7123$. N_u is 5 for this particle so the
second term in Eq: (8) is almost negligible.

The $I(Q)$ for the $R = 12.0$ particle is shown for two temperatures in Fig. 18. A very
small amount of scattering from the five uncompensated spins is apparent near $Q = 0$.
From Fig. 18, Eq. (8) was used to obtain $A(Q)$. Equation (8) is now an expression with
no free parameters with which one can predict the scattering from the $R = 11.5$ particle
which has a significant number of uncompensated spins. The results for $T = 2.0$ are
shown in Fig. 19(a). It can be seen that the MC data make a really excellent fit to Eq.
(8). A similar plot for $T=3.0$ is displayed in Fig. 19(b). Although not as precise as in Fig.
19(a), the agreement is still remarkably good. In [126] inelastic neutron scattering
experiments on AF particles also gave a ferromagnetic component from the surface.

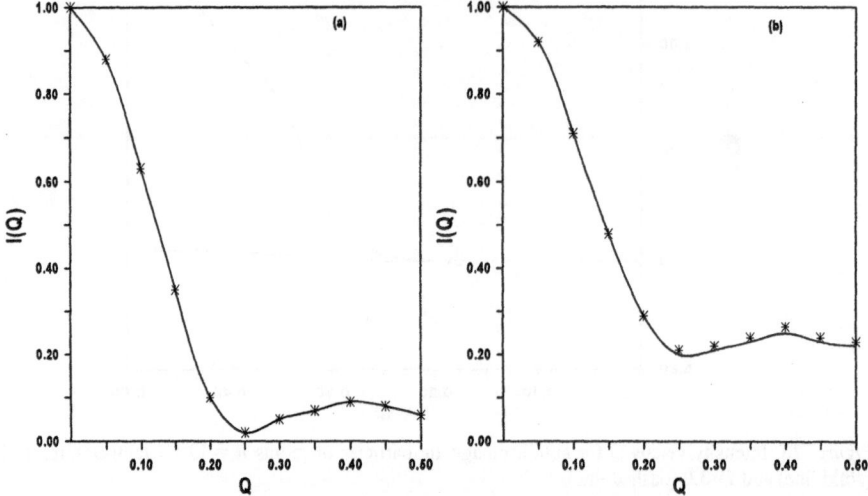

Figure 19. Intensity versus Q for antiferromagnetic particles for temperature (a)$T = 2.0$ and (b)$T=3.0$.
Solid line gives the functional form of Eq. (8). Data points give the Monte Carlo results.

6. Final Remarks

We have reviewed Monte Carlo simulation results on the magnetic behavior of
nanoparticles with core/shell morphology and in particular ferromagnetic,
antiferromagnetic and ferromagnetic nanoparticles covered by an antiferromagnetic

shell. We have started with a set of parameters for these systems and we examined the effect of adding one by one the factors that are present in real systems.

A radial surface anisotropy causes reduction of the magnetization compared with that obtained with z-easy axis anisotropy for all cases except those of very high surface anisotropy and small number of uncompensated spins in antiferromagnetic particles, where we found an increase of the magnetization.

In the antiferromagnetic particles the coercivity scales with the ratio of the number of spins in the particle to the number of uncompensated spins. For the antiferromagnetic and the oxidized nanoparticles we conclude that the exchange interaction between the antiferromagnetic and the ferromagnetic layers plays a predominant role in the magnetic properties of particles with large shell to core ratio and at low temperatures. This interaction causes a reversal in the size dependence of the coercivity at temperatures well below the blocking temperature of the particles.

The simulation results are in qualitative agreement with the experimental observations and give physics insight to the properties of the magnetic nanoparticles. Experimentally the behavior of small particles is very dependent on preparation and the annealing treatment. Obviously, using particles of constant radius would make the analysis of results considerably easier. It may be that the biological systems will prove to be ideal samples.

There are still several interesting issues to be addressed in the field of magnetic nanoparticles. The influence of interparticle exchange interactions on the magnetic properties of isolated particles in dense nanoparticle assemblies is of major importance especially for the technological exploitation of magnetic nanoparticle assemblies. MC simulations provide an extremely valuable tool for the study of these systems.

Acknowledgments

We would like to acknowledge the contribution of Dr. J.A. Blackman to parts of the work presented in this chapter. This work was supported by the projects PENED96, TMR No ERB-CHRXT9313598 and Growth No G5RD-CT-2001-00478.

References

[1]   L. Néel, Ann. Geophys. **5**, 99 (1949).
[2]   S. Jacobs and C. P. Bean, G. T. Rado, and H. Suhl (eds) in *Magnetism III* (Academic, New York, 1963), pp. 271-290.
[3]   K. Binder, H. Rauch, and V. Wildpaner, J. Phys. Chem. Solids **31**, 391 (1970).
[4]   K. Binder, Physica **62**, 508 (1972).
[5]   V. Wildpaner, Z. Phys. **270**, 215 (1974).
[6]   D. P. Landau, Phys. Rev. B **13**, 2997 (1976); Phys. Rev. B **14**, 255 (1976).
[7]   C. P. Bean and J. D. Livingston, J. Appl. Phys. **30**, 1205 (1959).
[8]   A. Hahn, Phys. Rev. B **1**, 3133 (1970).
[9]   G. Bottoni, D. Candolfo, and A. Cecchetti, J. Magn. Magn. Mater. **155**, 297 (1996).
[10] L. Del Bianco, A. Hernando, M. Multigner, C. Prados, J. C. Sanchez-Lopez, A. Fernandez, C. F. Conde, and A. Conde, J. Appl. Phys. **84**, 2189-2192 (1998).
[11] S. Gangopadhyay, G.C. Hadjipanayis, C.M. Sorensen, K..J. Klabunde, in *NanophaseMaterials*. George C. Hadjipanayis, and Richard W. Siegel (eds) (Kluwer Academic Publishers, Dordrecht, 1994), pp 573-578.

[12] L. Néel, C. R. Seances Acad. Sci. (Paris): 252, 4075 (1961); 253, 9 (1961); 253, 203 (1961); 253, 1286 (1961).

[13] K. M. Creer, Geophys. J. 5, 16 (1961).

[14] K. G. Srivastava, C. R. Seances, Acad. Sci. (Paris), 253, 2887 (1961).

[15] T.G. Stpierre, D.H. Jones and D.P.E. Dickson, J. Magn. Magn. Mat. 69, 276(1987).

[16] X. Batlle and A. Labarta, J. Phys. D 35, R15(2002).

[17] D.V. Dimitrov, G.C. Hadjipanayis, V. Papaefthymiou, and A. Simopoulos, J. Magn. Magn. Mater. 188, 8(1998).

[18] Jianmin Zhao, Frank Huggins, Zhen Feng, and Gerald P. Huffman, Phys. Rev. B 54, 3403(1996).

[19] K.N. Trohidou and J.A.Blackman, Phys. Rev. B 41, 9345 (1990).

[20] K.N. Trohidou, X. Zianni, and J.A. Blackmann, J.Appl. Phys. 84, 2795 (1998).

[21] K.N. Trohidou, X. Zianni, and J.A. Blackmann, IEEE 34, 1121 (1999).

[22] K.N. Trohidou, X. Zianni, and J.A. Blackmann, Phys. Status Solidi A 189, 305(2002).

[23] K.N. Trohidou, C.M. Soukoulis, A. Kostikas, and G.C. Hadjipanayis, J. Magn. Magn. Mater. 104-107, 1587 (1992).

[24] X. Zianni and K.N. Trohidou, J.Phys. Condens. Matter 10, 7475(1998).

[25] W.H. Meiklejohn and C.P. Bean, Phys. Rev. 105, 904 (1957).

[26] Bi Hong, Li Shandong, Jiang Xiqun, Du Youwei, and Yang Changzheng, Phys. Lett. A 307, 69 (2003).

[27] C. Prados, M. Multigner, A. Hernando, J.C. Sanchez, A. Fernandez, C.F. Conde, and A. Conde, J. Appl. Phys. 85, 6118(1999).

[28] E. Tronc, A. Ezzir, R. Cherkaoui, C. ChaneHac, M. Nogues, H. Kachkachi, D. Fiorani, A.M. Testa, J.M. Greneche, and J.P. Jolivet. J. Magn. Magn. Mater. 221, 63(2000).

[29] Mingzhong Wu, Y. D. Zhang, S. Hui, T. D. Xiao, Shihui Ge, W. A. Hines, and J. I. Budnick. J. Appl. Phys. 92, 491(2002).

[30] S.T. Chui and Tian De-Cheng, J. Appl. Phys. 78, 3965(1995).

[31] T. Kaneyoshi, J Phys.: Condens. Matter 3, 4497 (1991).

[32] J.G. Gay and P. Richter, J. Appl. Phys. 61, 3362 (1987).

[33] K. Binder, Monte Carlo Methods in Statistical Physics (Springer, Berlin, 1979); Applications of Monte Carlo Methods in Statistical Physics (Springer, Berlin, 1984).

[34] K. Fukumura, A. Nakanishi, T. Fujii, and T. Kobayashi, J. Magn. Magn. Mater. 145, 175 (1995).

[35] S. Gangopadhyay, G.C. Hadjipanayis, B. Dale, C.M. Sorensen, K.J. Klaubunde, V. Papaefthymiou, and A. Kostikas, Phys. Rev. B 45, 9778 (1992).

[36] P.V. Hendriksen, S. Linderoth, and P.A. Lindgard, Phys. Rev. B 48, 7259 (1993).

[37] J. P. Bucher and L. A. Bloomfield, Int. J. Mod. Phys. B 7, 1097 (1993).

[38] J. P. Bucher, D. C. Douglas, and L. A. Bloomfield, Phys. Rev. Lett. 66, 3052 (1991).

[39] I. M.L. Billas, J.A. Becker, A. Chatelain, and A. de Heer Walt, Phys. Rev. Lett. 71, 4067(1993).

[40] S.N. Khanna and S. Linderoth, Phys. Rev. Lett. 67, 742 (1991).

[41] M. Respaud, J. M. Broto, H. Rakoto, A. R. Fert, L. Thomas, B. Barbara, M. Verelst, E. Snoeck, P. Lecante, A. Mosset, J. Osuna, T. Ould Ely, C. Amiens, and B. Chaudret, Phys. Rev. B 57, 2925 (1998).

[42] F. Bodker, S. Morup, and S. Linderoth, Phys. Rev. Lett. 72, 282 (1994).

[43] M. Jamet, W. Wernsdorfer, C. Thirion, D. Mailly, V. Dupuis, P. Melinon, and A. Perez, J. Magn. Magn. Mater. 226, 1833(2001).

[44] C. Thirion, W. Wernsdorfer, M. Jamet, V. Dupuis, P. Melinon, A. Perez, and D. Mailly, J. Appl. Phys. 91, 7062(2002).

[45] E. F. Kneller and F. E. Luborsky, J. Appl. Phys. 34,656 (1963).

[46] D.A. Dimitrov and G.M.Wysin, Phys. Rev. B 50, 3077(1994).

[47] D.A. Dimitrov and G.M.Wysin, Phys. Rev. B 51, 11947(1995).

[48] H. Kachkachi and M. Dimian, Phys. Rev. B 66, 174419 (2002).

[49] X. Zianni, K.N. Trohidou, and J.A. Blackman in Magnetic Hysteresis in Novel Magnetic Materials. George Hadjipanayis(ed.) (Kluwer Academic Publishers, Dordrecht, 1977), pp45-48.

[50] X. Zianni, K.N. Trohidou, and J. A. Blackman, J. Appl. Phys. 81, 4739 (1997).

[51] C. Bellouard, M. Hennion, and I. Mirebeau, J. Magn. Magn. Mater. 140-144, 357(1995).

[52] C. Bellouard, I. Mirebeau, and M. Hennion, J. Magn. Magn. Mater. 140-144, 431(1995).

[53] C. Bellouard, I. Mirebeau, and M. Hennion, Phys. Rev. B **53**, 5570 (1996).

[54] I. Bergenti, A. Deriu, I. Savini, E. Bonetti, F. Spizzo, and H. Hoell, J. Magn. Magn. Mater. **262**, 60(2003).

[55] H. Casalta, P. Schleger, C. Bellouard, M. Hennion, I. Mirebeau, G. Ehlers, B. Farago, J.L. Dormann, M. Kelsch, M. Linde, and F. Phillippe, Phys. Rev. Lett. **82**, 1301(1999).

[56] A. Hoell, A. Wiedenmann, U. Lembke, and R. Kranold, Physica B **276-278**, 886(2000).

[57] D. Lin, A.C. Nunes, C.F. Majkrzak, and A.E. Berkowitz, J. Magn. Magn. Mater. **145**, 343(1995).

[58] C. Nunes, J. Appl. Crystallogr. **21**, 129 (1988).

[59] F. Spizzo, E. Angeli, D. Bisero, A. Da Re, F. Ronconi, P. Vavassori, I. Bergenti, A. Deriu, A. Hoell, and H.J. Lauter, J. Magn. Magn. Mater. **262**,124(2003).

[60] N. W. Dalton and D. W. Wood, J. Math. Phys. **10**, 1271 (1969).

[61] D.W. Wood and N.W. Dalton, Phys. Rev. **159**, 384 (1967).

[62] E. De Biasi, C. A. Ramos, R. D. Zysler, and H. Romero, Phys. Rev. B **65**, 144416(2002).

[63] T. Sinohora, T. Sato, T. Taniyama, and I. Nakatani. J. Magn. Magn. Mater. **196**, 94(1999).

[64] J.P. Chen, C.M.Sorensen, K.J. Klabunde, and G.C.Hadjipanayis, Phys. Rev. B **51**, 11527(1995).

[65] D. A. Garanin and H. Kachkachi, Phys. Rev. Lett. **90**, 065504(2003).

[66] Lord Rayleigh, Proc. R. Soc. London Ser. A **90**, 219 (1914).

[67] W. Marshall and R. J. Elliott, Rev. Mod. Phys. **30**, 75 (1958).

[68] B.J. Jonsson, T. Turkki, V. Strom, M.S. El-Shall, and K.V. Rao, J. Appl. Phys. **79**, 5063 (1996).

[69] V. Skumryev, S. Stoyanov, Y. Zhang, G. Hadjipanayis, D. Givord, and J. Nogues, Nature **423**, 850(2003).

[70] L. Savini, E. Bonetti; L. Del Bianco; L. Pasquini; L. Signorini, M. Coisson, and V. Selvaggini, J. Magn. Magn. Mater. **262**, 56(2003).

[71] D.L. Peng, K.Sumiyama, T. Hihara, S. Yamamuro, and T.J. Konno, Phys. Rev. B **61**, 3103(2000).

[72] L. Del Bianco, D. Florani, A.M. Testa, E. Bonetti, L. Savini, and S. Signoretti, J. Magn. Magn. Mater. **262**, 128(2003).

[73] X. Lin, A.S. Murthy, G.C. Hadjipanayis, C. Swann, and S.I. Shah, J. Appl. Phys. **76**, 6543 (1994).

[74] S. Yamamuro, K. Sumiyama, T. Kamiyama, and K. Suzuki, J. Appl. Phys. **86**, 5726(1999).

[75] C.M. Hsu, H.M. Lin, K.R. Tsai, and P.Y. Lee, J. Appl. Phys. **76**, 4793 (1994).

[76] R.H. Kodama, J. Magn. Magn. Mater. **200**, 359(1999).

[77] R. H. Kodama and A. S. Edelstein, J. Appl. Phys. **85**, 4316 (1999).

[78] L. Del Bianco, D. Fiorani, A.M. Testa, E. Bonetti, L. Savini, and S. Signoretti, Phys. Rev. B **66**, 174418 (2002).

[79] J. Nogues and I. K. Schuller, J. Magn. Magn. Mater. **192**, 203(2002).

[80] A.E. Berkowitz and K. Takano, J. Magn. Magn. Mater. **200**, 552(1999).

[81] L.T. Kuhn, A. Bojesen, L. Timmermann, M.M. Nielsen, and S. Morup, J. Phys. Conden. Matter **14**, 13551(2002).

[82] Y. Du, M. Xu, Y. Shi, H. Lu, and R. Xue, J. Appl. Phys. **70**, 5903 (1991).

[83] Y.D. Yao, Y.Y. Chen, M.F. Tai, D.H. Wang, and H.M. Lin, Mater. Scie. Eng. **218**, 281(1996).

[84] Z. Turgut, N.T. Nuhfer, H.R. Piehler, and M.E. McHenry, J. Appl. Phys. **85**, 4406(1999).

[85] E. R. Callen and H. B. Callen, Phys. Rev. **129**, 578 (1963); J. Phys. Chem. Solids **27**, 1271 (1966).

[86] B D Cullity, *Introduction to Magnetic Materials*, (Addison-Wesley, 1972).

[87] J. Harris, J. Grimaldi, D. C. Gilles, P. Bonville, H. Rakoto, J.M. Broto, K.K.W. Wong, and S. Mann, J. Magn. Magn. Mater. **241**, 430(2002).

[88] D.D. Awschalom, A. Chiolero, and D. Loss, Phys. Rev. B **60**, 3453(1999).

[89] S. Gider, D.D. Awschalom, T. Douglas, S. Mann, and M. Chaparal, Science **268**, 77 (1995).

[90] S. Gider, D.D. Awschalom, T. Douglas, K. Wong, S. Mann, and G. Cain, J. Appl. Phys. **79**, 5324 (1996).

[91] F.T. Parker, M.W. Foster, D.T. Margulies, and A.E. Berkowitz, Phys. Rev. B **47**, 7885(1993).

[92] R. Zysler, D. Fiorani, J.L. Dormann, and A.M. Testa, J Magn. Magn. Mater. **133**, 71 (1994).

[93] R.H. Kodama, A.E. Berkowitz, E.J. McNiff, and S. Foner, Phys. Rev. Lett. **77**, 394 (1996).

[94] S. Sako and K. Ohshima, J. Phys. Soc. Jpn. **64**, 944(1995).

[95] S. Sako, Y. Umemura, K. Ohshima, M. Sakai, and S. Bandow, J. Phys. Soc. Jpn. **65**, 280(1995).

[96] Z.G. Zhao, P.F. de Chatel, F.R. de Boer, and K.H.J. Buschow, J. Appl. Phys. **79**, 4626 (1996).

[97] S.G. Marchetti, R.A. Borzi, E.D. Cabanillas, S.J. Stewart, and R.C. Mercader, Hyperfine Interactions **139-140**, 513(2002).

[98] A.E. Berkowitz, R.H. Kodama, S.A. Makhlouf, F.T. Parker, F.E. Spada, E.J. McNiff, Jr., and S. Foner, J. Magn. Magn. Mater. **196-197,** 591(1999).

[99] M.Vasquez-Mansilla, R. Zysler, C. Arciprete, M. Dimitrijewits, C. Saragovi, and J. Greneche, J. Magn. Magn. Mater. **204,** 29(1999).

[100]S.A. Makhlouf, F.T. Parker, and A.E. Berkowitz, Phys. Rev B **55,** R14717(1997).

[101]S.A. Makhlouf, J. Magn. Magn. Mater. **246,** 184(2002).

[102]R.H. Kodama and A.E. Berkowitz, Phys. Rev. B **59,** 6321(1999).

[103]R.H. Kodama, S. A. Makhlouf, and A.E. Berkowitz, Phys. Rev. Lett. **79,** 1393 (1997).

[104]M. Banobre-Lopez, C. Vazquez-Vazquez, J. Rivas, and M. A. Lopez-Quintela, Nanotechnology **14,** 318(2003).

[105]R.C. Plaza, S.A. Gomez-Lopera, and A.V. Delgado, J. Colloid Interface Sci. **240,** 48(2001).

[106]M.S. Seehra and A. Punnoose, Phys. Rev. B **64,** 1324410(2001).

[107]L. Soriano, M. Abbate, A. Fernandez, A.R. Gonzalez-Elipe, F. Sirotti, and J.M. Sanz, J. Phys. Chem. B **103,** 6676(1999).

[108]S. Morup, F. Bodker, P.V. Hendriksen, and S. Linderoth, Phys. Rev. B **52,** 287 (1995).

[109]S. Bocquet, R.J. Pollard, and J.D. Cashion, J. Appl. Phys. **77,** 2809 (1995).

[110]M. Chudnovsky, J. Magn. Magn. Mater. **140-144,** 1821(1995).

[111]E. Del Barco, M. Duran, J.M. Hernandez, J. Tejada, R.D. Zysler, M. Vasqueez-Mansilla, and D. Fiorani, Phys. Rev. B **65,** 052404 (2002).

[112]B. Barbara, G. Barbero, V. V. Dobrovitski, and A. K. Zvezdin, J. Appl. Phys. **81,** 5539(1997).

[113]D. P. Di Vinzenzo, Physica B **197,** 109 (1994).

[114]B.V. Reddy and S.N. Khanna, Phys. Rev. B **45,** 10103(1992).

[115]E.Vittala, J. Merikoski, M. Manninen, and J. Timonen, Phys. Rev. B **55,** 11541(1997).

[116]D. Fiorani,A.M. Testa, F. Lucari, F. D'Orazio, and H. Romero, Physica B **320,** 122(2002).

[117]A.R.B. De Castro and R.D.Zysler, J. Magn. Magn. Mater. **257,** 51(2003).

[118]Q.A. Pankhurst and R.J. Pollard, Phys. Rev. Lett. **67,** 248(1991).

[119]O. Iglesias, A. Labarta, and F. Ritort, J. Appl. Phys. **89,** 7597 (2001).

[120]O. Iglesias and A. Labarta, Phys. Rev. B **63,** art. no. 184416 (2001).

[121]H. Kachkachi, A. Ezzir , M. Nogues, and E. Tronc, Eur. Phys. J. B **14,** 681(2000).

[122]N. Amin and S. Arajs, Phys. Rev. B **35,** 4810(1987).

[123]S. Kilcoyne and R. Cywinski, J. Magn. Magn. Mater. **140-144,** 1466(1995).

[124]A.R.B. De Castro, R.D.Zysler, M. Vasquez-Mansilla,, C. Arciprete, and M. Dimitrijevits, J. Magn. Magn. Mater. **231,** 287(2001).

[125]S.N.Klausen, P.A. Lindgard, K. Lefmann, F. Bødker, and S. Mørup, Phys. Status Solidi A **189,** 1039(2002).

[126]M.F.Hansen, F. Bodker, S. Morup, K. Lefmann, K.N. Clausen, and P.A. Lindgaard, J. Magn. Magn. Mater. **221,** 10(2000).

MAGNETIC NANOPARTICLES AS MANY-SPIN SYSTEMS

H. Kachkachi

*Laboratoire de Magnétisme et d'Optique, Université de Versailles St. Quentin,
45 av. des Etats-Unis, 78035 Versailles, France*

D. A. Garanin

Institut für Physik, Johannes-Gutenberg-Universität, D-55099 Mainz, Germany

1 Introduction

Magnetic nanoparticles, or nanoscale magnetic systems, have generated continuous interest since late 1940s as the investigation of their properties turned out to be challenging from both scientific and technological point of view. In 1949, in a pioneering work [1], Néel set the pace towards understanding of the magnetic behavior of nanoparticles, leading to an important development of fundamental theories of magnetism and modeling of magnetic materials, as well as remarkable technological advances, e.g., in the area of information storage and data processing, fostering the development of magnetorecording media with increasingly higher densities. Nanoparticles, as compared to bulk materials, possess very important novel properties such as enhanced remanence and giant coercivity, as well as exponentially slow relaxation at low temperature due to anisotropy barriers, which ensures great stability of the information stored. However, nanoparticles become *superparamagnetic* [1] at finite temperatures for very small sizes, and this is an impairment to the information storage. On the other hand, discovery of superparamagnetism that results from thermally activated crossing of the anisotropy energy barrier by the magnetic moment of the particle opened a rich area to the application of nonequilibrium statistical mechanics. There are mainly two types of nanoparticle samples: i) assemblies of, e.g., cobalt, nickel or maghemite, nanoparticles with volume distribution and randomly oriented axes of magnetocrystalline anisotropy; ii) isolated single particles of cobalt or nickel that can be probed by the micro-SQUID technique [2]. While most of the experiments have been done on assemblies, isolated particles are more important both as units of information storage and as a physical system.

In the investigation of the static properties of magnetic nanoparticles a great deal of work up to date has been based on the Monte Carlo (MC) technique. In addition to numerous simulations of the Ising model, this technique has been used with the more adequate classical Heisenberg model to simulate idealized isotropic models with simple cubic (sc) lattice and spherical shape in [3]. Magnetic nanoparticles with realistic lattice structure were recently simulated in [4] taking into account the surface anisotropy (SA) and dipole-dipole interactions (DDI).

On the other hand, in most theoretical approaches to the dynamics of a small magnetic particle the latter is considered as a single magnetic moment. This is the *one-spin approximation* that is only valid for particles that are not too large and thus are single-domain, and not too small to be free from surface effects. Letting apart spin tunneling (see, e.g., [5]) that becomes important for extremely small sizes such as those of molecular magnets, the magnetic moment can overcome the anisotropy-energy barrier and thus reverse its direction, at least in two ways [1] : either under applied magnetic field that suppresses the barrier, or via thermal fluctuations. The former, at zero temperature, is well described for particles with the uniaxial anisotropy by the Stoner-Wohlfarth model [7]. Thermally activated crossing of the energy barrier is described by the Néel-Brown model [8] and its extensions (see, e.g., [9]). At elevated temperatures, rotation of the magnetization in materials with strong anisotropy is always accompanied by changing its magnitude. This results in a shrinking of the Stoner-Wohlfarth astroid as described by the modified Landau theory [10], and (qualitatively) confirmed by experiments [2].

Both Stoner-Wohlfarth and Néel-Brown models have been confirmed by experiments on individual cobalt particles [2]. However, for magnetic particles with strong surface anisotropy magnetization switching occurs as result of successive switching of individual (or clusters of) spins inside the particle [11]. Such deviations from the one-spin approximation have been observed in metallic particles [14], [16], and ferrite particles [12], [13]. Deviations from the one-spin approximation and temperature effects lead to the absence of magnetization saturation at high fields [14], [15], [16], shifted hysteresis loops after cooling in field, and field dependence of the magnetization at very low temperatures. The latter effect has been clearly identified in dilute assemblies of maghemite particles [17] 4 nm in diameter. In addition, aging effects have been observed in single particles of cobalt and have been attributed to the oxidation of the sample surface into antiferromagnetic CoO (see [2] and references therein). It was argued that the magnetization reversal of a ferromagnetic particle with antiferromagnetic shell is governed by two mechanisms that are supposed to

---

[1] It has quite recently been shown [6], experimentally and theoretically, that efficient magnetization switching can be triggered by transverse field pulses of a duration that is half the precession period.

result from the spin frustration at the core-shell interface of the particle. Some of the above-mentioned novel features are most likely due to magnetic disorder at the surface which induces a canting of spins inside the particle, or in other words, an inhomogeneous magnetic state. This effect was first observed with the help of Mössbauer spectroscopy by Coey [18] and by Morrish and Haneda (see [19] for a review), and later by Prené et al. [20]. To sum up, the picture of a single-domain magnetic particle with all spins pointing into the same direction is no longer valid when one considers the effect of misaligned spins on the surface, which makes up to 50% of the total volume in a particle 4 nm in diameter.

One of our goals is thus to understand the effect of surfaces on the thermodynamics and magnetization profiles in small systems, and subsequently on their dynamics. This requires a microscopic approach to account for the local environment inside the particle, microscopic interactions such as spin-spin exchange, DDI, and the magneto-crystalline bulk and surface anisotropy. As this task is difficult owing to the large number of degrees of freedom involved, one has to gain a sufficient understanding of static properties before proceeding to the dynamics. There are three main effects that distinguish magnetic particles from bulk magnets and that were investigated in a series of our recent publications:

(1) Finite-size effect in isotropic magnetic particles with idealized periodic boundary conditions. The spin-wave spectrum of such particles is discrete and there is the mode with $\mathbf{k} = 0$ that corresponds to the global rotation of the particle. As a result, the standard spin-wave theory fails and one has to distinguish between *induced* and *intrinsic* magnetizations.

(2) Boundary effect, i.e., pure effect of the free boundary conditions at the surface in the absence of the surface anisotropy. This effect leads to the decrease of the particle's intrinsic magnetization and it makes the latter inhomogeneous at $T \neq 0$.

(3) Effect of surface anisotropy that changes the ground state of the particle and makes the intrinsic magnetization inhomogeneous even at $T = 0$.

This paper is organized as follows: In Sec. 2 we describe the Hamiltonian and introduce the basic notions of the induced and the intrinsic magnetizations. In Sec. 3, we first study inhomogeneities in small magnetic particles of a box shape induced by pure boundary effects in the absence of surface anisotropy, i.e., the effect of free boundary conditions (fbc), and compare their influence with that of the finite-size effects in an idealized model with periodic boundary conditions (pbc). We consider spins as D-component classical vectors. At first we present analytical and numerical results in the whole range of temperatures in the limit $D \to \infty$ where the problem simplifies. Then for the classical Heisenberg model, $D = 3$, at low temperatures we formulate the modified spin-wave theory accounting for the global rotation of the particle's magnetization. The results are compared with those of our MC simulations. In Sec. 4 we consider round-shaped systems and include surface anisotropy. In the case of the SA much weaker than the exchange interaction we study the problem perturbatively in small deviations from the perfectly ordered, collinear state. Then we investigate the hysteretic properties and the behavior of the magnetization as a function of temperature and applied field by MC simulations. The last section summarizes the results and points out open problems.

2 Basic relations

2.1 The Hamiltonian

Within the classical approximation it is convenient to represent the atomic spin as the three-component spin vector s_i of unit length on the lattice site i. We will consider the Hamiltonian that in general includes the exchange interaction, magneto-crystalline anisotropic energy, Zeeman energy, and the energy of dipolar interactions (DDI)

$$\mathcal{H} = -\frac{1}{2}\sum_{ij} J_{ij}s_i \cdot s_j - \mu_0 \mathbf{H} \cdot \sum_i s_i + \mathcal{H}_{an} + \mathcal{H}_{DDI}, \qquad (1)$$

where $\mu_0 = g\mu_B S$ and S is the value of the atomic spin. For materials with uniaxial anisotropy $\mathcal{H}_{an}$ in Eq. (1) reads

$$\mathcal{H}_{an}^{(uni)} = -\sum_i K_i(s_i.e_i)^2, \qquad (2)$$

with easy axis e_i and constant $K_i > 0$. This anisotropy model can be used to describe the surface effect if one attributes the same easy axis and the same anisotropy constant K_c for all core spins and different easy axes and anisotropy constants to surface spins. Within the simplest transverse surface anisotropy (TSA) model all surface spins have the same anisotropy constant K_s, whereas their easy axes are perpendicular to the surface, see, e.g., [21–23] . More realistic is Néel's surface anisotropy (NSA) model [24] ,

$$\mathcal{H}_{an}^{(NSA)} = -L\sum_i \sum_{j=1}^{z_i} (s_i \cdot e_{ij})^2, \qquad e_{ij} \equiv r_{ij}/r_{ij}, \qquad r_{ij} \equiv r_i - r_j, \qquad (3)$$

where z_i is the coordination number of site i that for the surface atoms is smaller than the bulk value z and e_{ij} is the unit vector connecting site i to its nearest neighbors j. One can check that for the simple cubic (sc) lattice the contributions from the bulk spins in (3) $\sim \mathbf{s}_i^2 = 1$ are irrelevant constants.

For materials with magneto-crystalline cubic anisotropy $\mathcal{H}_{an}$ reads

$$\mathcal{H}_{an}^{(4)} = -K^{(4)} \sum_i \left(s_{ix}^4 + s_{iy}^4 + s_{iz}^4 \right). \tag{4}$$

For $K^{(4)} > 0$, the energy $\mathcal{H}_{an}^{(4)}$ has minima for six orientations of type $[100]$ and maxima for eight orientations of type $[111]$.

If one discards $\mathcal{H}_{an}^{(4)}$ and $\mathcal{H}_{DDI}$, the Hamiltonian (1) can be generalized for D-component spin vectors. This is useful as in the limit $D \to \infty$ the problem simplifies while retaining important physics, see below.

2.2 Magnetization of finite systems

Magnetic particles of *finite size* do not show magnetic ordering at nonzero temperatures at $H = 0$ as the global magnetization of the particle can assume all possible directions (superparamagnetism). It is thus convenient to define two magnetizations, m and M, the first being the magnetization induced by the magnetic field and the second being a measure of the short-range order in the particle. Omitting the factor μ_0 that can be restored later, we first define the magnetization of a microscopic spin configuration

$$\mathbf{M} = \frac{1}{\mathcal{N}} \sum_i \mathbf{s}_i, \tag{5}$$

where $\mathcal{N}$ is the number of magnetic atoms in the system. The thermodynamic average of $\mathbf{M}$ yields what we call the *induced magnetization*

$$\mathbf{m} = \langle \mathbf{M} \rangle = \frac{1}{\mathcal{N}} \sum_i \langle \mathbf{s}_i \rangle. \tag{6}$$

The *intrinsic magnetization* is related to the spin correlation function:

$$M = \sqrt{\langle \mathbf{M}^2 \rangle} = \sqrt{\left\langle \left(\frac{1}{\mathcal{N}} \sum_i \mathbf{s}_i \right)^2 \right\rangle} = \frac{1}{\mathcal{N}} \sqrt{\sum_{ij} \langle \mathbf{s}_i \cdot \mathbf{s}_j \rangle}. \tag{7}$$

If the temperature is low and there is no surface anisotropy, all spins in the particle are bound together by the exchange interaction and $\mathbf{M}$ behaves as a rigid "giant spin", $|\mathbf{M}| \cong M \cong 1$. If a magnetic field $\mathbf{H}$ is applied, $\mathbf{M}$ exhibits an average in the direction of **H**, which leads to a nonzero value of the induced magnetization **m**. For isotropic D-component vector models (including the Ising model, $D = 1$) [25] the latter is given by the well known formula

$$m = MB_D(Mx), \qquad x \equiv \mathcal{N}H/T, \tag{8}$$

where $B_D(x)$ is the Langevin function [$B_3(x) = \coth x - 1/x$ for the isotropic Heisenberg model and $B_1(x) = \tanh x$ for the Ising model]. An important question is whether Eq. (8) remains valid at elevated temperatures where $M = M(T, H)$. We have shown that the *superparamagnetic relation*, Eq. (8), becomes exact for $D \to \infty$ but otherwise it contradicts the exact relation

$$M^2 = m^2 + \frac{dm}{dx} + \frac{(D-1)m}{x}. \tag{9}$$

One also can introduce the local intrinsic magnetization M_i according to

$$M_i = \frac{1}{M}\left\langle \mathbf{s}_i \cdot \frac{1}{\mathcal{N}}\sum_j \mathbf{s}_j \right\rangle, \qquad \frac{1}{\mathcal{N}}\sum_i M_i = M. \tag{10}$$

This quantity is smaller near the boundaries of the particle than in the core for the model with fbc because of boundary effects at $T > 0$.

3   Nanoparticle as a multi-spin system: finite-size vs boundary effects

Finite-size magnetic systems with free boundary conditions (fbc) present a spatially inhomogeneous many-body problem. In this section we shall only deal with finite-size versus boundary effects, leaving the more profound effects of the surface anisotropy for the next section. One of the interesting problems here is the interplay between boundary effects due to fbc and the "pure" finite-size effects. In systems of hypercubic shape, the latter can be singled out by using artificial periodic boundary conditions (pbc). The standard mean-field approximation (MFA) and spin-wave theory (SWT) are inappropriate for finite systems because of the Goldstone mode associated with the global rotation of the magnetic moment in zero field. Appropriate improvements include the so-called $D \to \infty$ model (see Sec. 3.1) operating at all temperatures and the modified spin-wave theory for finite magnets at low temperatures (see Sec. 3.2). Also the MC routine should incorporate global rotations of spins, in addition to the Metropolis algorithm of individual spin rotations.

3.1 $D \to \infty$ model

3.1.1   The model

One can improve upon the MFA by taking into account correlations in a wide temperature range for bulk and finite magnets by replacing 3-component spin vectors in (1) by D-component ones and taking the limit $D \to \infty$. This model was introduced by Stanley [25] who showed that in the bulk its partition function coincides with that of the exactly solvable spherical model (SM) [26]. On the other hand, for spatially inhomogeneous and anisotropic systems the $D \to \infty$ model is the only physically acceptable model of both (see, e.g., [27]). So far, the $D \to \infty$ model was only applied to spatially inhomogeneous systems in the plane geometry [27]. In [10] we extended it to *finite* box-shaped magnetic systems with free and periodic boundary conditions. In the MFA the Curie temperature of the D-component model is $T_c^{\mathrm{MFA}} = J_0/D$, where J_0 is the zero Fourier component of J_{ij}. It is convenient to use T_c^{MFA} as the energy scale and introduce the dimensionless variables

$$\theta \equiv T/T_c^{\mathrm{MFA}}, \qquad \mathbf{h} \equiv \mathbf{H}/J_0, \qquad \lambda_{ij} \equiv J_{ij}/J_0. \qquad (11)$$

For the nearest-neighbor (nn) interaction J_{ij} with z neighbors, λ_{ij} is equal to $1/z$ if sites i and j are nearest neighbors and zero otherwise. In the bulk the $D \to \infty$ model is described by two coupled nonlinear equations for the magnetization m and the so-called gap parameter G:

$$m = \frac{hG}{1-G}, \qquad m^2 + \theta G P(G) = 1, \qquad P(G) = \int \frac{d^3k}{(2\pi)^3} \frac{1}{1 - G\lambda_{\mathbf{k}}}, \qquad (12)$$

where $P(G)$ is the lattice Green function and $\lambda_{\mathbf{k}} = \left(\cos k_x + \cos k_y + \cos k_z\right)/3$ is the Fourier transform of λ_{ij}. The Curie temperature is defined by $G = 1$ and is $T_c = T_c^{\mathrm{MFA}}/W$, i.e., $\theta_c = 1/W$, where $W \equiv P(0)$ is the Watson integral ($W = 1.51639$ for the sc lattice). The system of equations describing the inhomogeneous $D \to \infty$ model can be obtained using the diagram technique for classical spin systems [28,29] in the limit $D \to \infty$ and generalising the results of Ref. [27] for spatially inhomogeneous systems to include the magnetic field $\mathbf{h} = h\mathbf{e}_z$. This is a system of equations for the average magnetization $m_i \equiv \langle s_{zi} \rangle$ directed along the field, gap parameter G_i, and correlation functions for the remaining "transverse" spin components labeled by $\alpha \geq 2$, i.e., $s_{ij} \equiv D \langle s_{\alpha i} s_{\alpha j} \rangle$ (all transverse correlation functions are the same). This system of equations has the form

$$\sum_j \mathcal{D}_{ij} m_j = h, \qquad \sum_j \mathcal{D}_{ij} s_{jl} = \theta \delta_{il}, \qquad s_{ii} + m_i^2 = 1, \qquad (13)$$

where $\mathcal{D}_{ij} \equiv G_i^{-1}\delta_{ij} - \lambda_{ij}$ is the Dyson matrix and δ_{il} is the Kronecker symbol. Solving this system of equations consists in determining m_i and s_{ij} as functions of G_i from the first two *linear* equations and inserting the solutions into the third nonlinear equation (the constraint equation) that leads to a system of nonlinear equations for all G_i that is in general subject to numerical solution. Combining Eqs. (13) results in $m^2 + \theta m/(\mathcal{N}h) - M^2 = 0$ that yields

$$m = M\frac{2\mathcal{N}Mh/\theta}{1 + \sqrt{1 + (2\mathcal{N}Mh/\theta)^2}} = MB_\infty(\mathcal{N}MH/T), \tag{14}$$

where

$$M = \sqrt{m^2 + \frac{1}{\mathcal{N}^2}\sum_{ij} s_{ij}}, \qquad B_\infty(\xi) = \frac{2\xi/D}{1 + \sqrt{1 + (2\xi/D)^2}} \tag{15}$$

and $B_\infty(\xi)$ is the Langevin function for $D \gg 1$. Alternatively Eq. (14) can be derived from Eq. (9) replacing $D - 1 \Rightarrow D$ and neglecting dm/dx in the limit $D \to \infty$ and then solving the resulting algebraic equation for m. One can find in the literature formulae of the type $m = M_s B(\mathcal{N}M_s H/T)$, where the saturation magnetization M_s is usually associated with the bulk magnetization at a given temperature (see, e.g., [30]). In our case, Eq. (14) is exact and $M = M(T, H)$ is defined by Eq. (15). For large sizes $\mathcal{N}$, Eq. (14) describes two distinct field ranges separated at $H \sim H_V$ where

$$H_V \equiv \frac{TD}{\mathcal{N}M}. \tag{16}$$

In the range $H \lesssim H_V$ the total magnetic moment of the system is disoriented by thermal fluctuations, $m < M$. In the range $H \gtrsim H_V$, the total magnetic moment is oriented by the field, m approaches M, and both further increase with the field towards saturation ($m = M = 1$) due to the suppression of spin waves in the system. This scenario is inherent to all $O(D)$ models [30].

Having established the superparamagnetic relation, Eq. (14) we are left with the problem of calculating $M(T, H)$. For the pbc the solution becomes homogeneous and one obtains (12) where in $P(G)$ the integral is replaced by a sum over discrete wave vectors [10] , whereas $M = \sqrt{m^2 + \theta G/[\mathcal{N}(1 - G)]}$. For the model with fbc analytical solution is only possible at low and high temperatures. At $\theta \ll 1$ small deviations from the collinear state with $M = 1$ can be described by the modified SWT for arbitrary D, see Sec. 3.2.

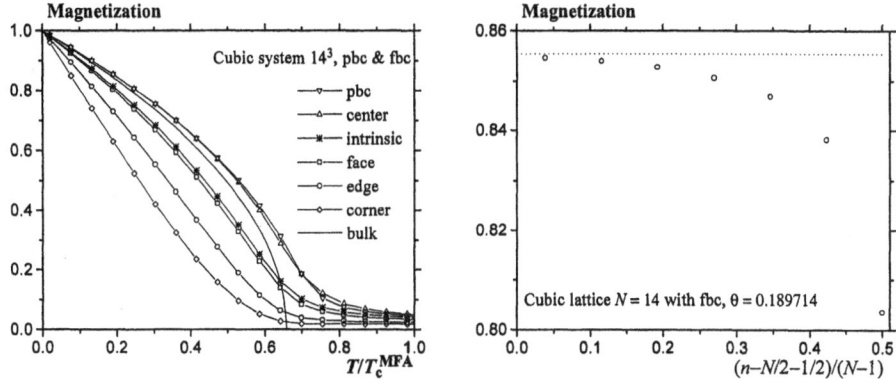

Fig. 1. Left: Temperature dependence of the intrinsic magnetization M, Eq. (15), and local magnetizations M_i, Eq. (10) in zero field. Right: Long-range magnetization profile in the direction from the center of the cube to the center of a face at temperature $\theta \equiv T/T_c^{\mathrm{MFA}} = 0.189714$.

3.1.2 Numerical results and discussion

The method for solving the $D \to \infty$ model consists in obtaining the correlation functions s_{ij} and magnetization m_i from the first two linear equations in Eq. (13), substituting them into the third equation of Eq. (13), and solving the resulting system of nonlinear equations for the gap parameter G_i numerically. Figure 1 (left) shows the temperature dependence of the intrinsic magnetization M, Eq. (15), and local magnetizations M_i, Eq. (10), of the 14^3 cubic system with free and periodic boundary conditions in zero field. For periodic boundary conditions, M exceeds the bulk magnetization at all temperatures. In particular, at low temperatures this is in accord with the positive sign of the finite-size correction to the magnetization, Eqs. (24) and (26). Local magnetizations at the center of the faces and edges and those at the corners decrease with temperature much faster than the magnetization at the center. One can see that below the bulk critical temperature M is smaller than the bulk magnetization. This means that the boundary effects suppressing M are stronger than the finite-size effects that increase M. This is also seen from the low-temperature expression of M given in Eq. (24), see also Figure 3. Figure 1 (right) shows the magnetization profile in the direction from the center of the cube to the center of a face. It is seen that perturbations due to the free boundaries extend deep into the particle, whereas the MFA predicts, on the contrary, a fast approach to a constant magnetization when moving away from the boundary [3]. This is a consequence of the Goldstone mode which renders the correlation length of an isotropic bulk magnet infinite below T_c. MC simulations of the classical Heisenberg model [3], [4] yield a similar result (see Figure 14 right).

We have shown that the critical indices for the magnetization at the faces,

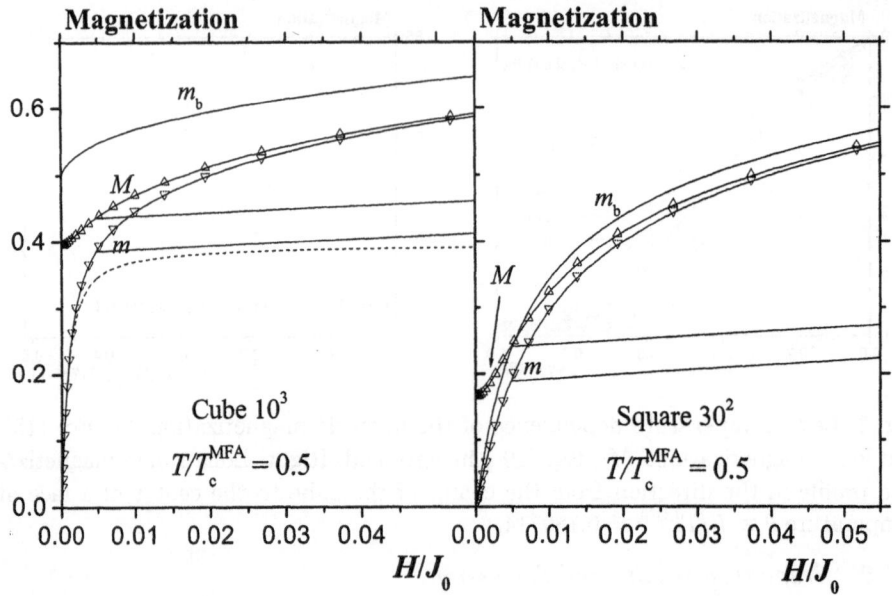

Fig. 2. Field dependence of the intrinsic magnetization M and induced magnetization m for hypercubic lattices with fbc in three and two dimensions. Dashed line is a plot of Eq. (14) in which $M(H,T)$ is replaced by its zero-field value. Bulk magnetization m_b in two and three dimensions is shown by solid lines.

edges, and corners are higher than the bulk critical index $\beta = 1/2$ for the present $D = \infty$ model. The critical index at the face β_1 is the most studied surface critical index (see, for a review, [31]). The exact solution of Bray and Moore [32] for the correlation functions at criticality in the $D = \infty$ model and application of the scaling arguments yield the value $\beta_1 = 1$ (see Table II in [31]). Exact values of the edge and corner magnetization indices, β_2 and β_3, seem to be unknown for $D = \infty$. Cardy [33] used the first-order ε-expansion to obtain $\beta_2(\alpha)$ for the edge with an arbitrary angle α. For $\alpha = \pi/2$ and $D = \infty$ in three dimensions the result for the edge critical exponent reads $\beta_2 = 13/8 + O(\varepsilon^2) = 1.625 + O(\varepsilon^2)$. To estimate the magnetization critical indices in our model we have performed a finite-size-scaling analysis (see, for a review, [34]) assuming the scaling form $M = N^{-\beta/\nu} F_M(\tau N^{1/\nu})$ and plotting the magnetization times $N^{\beta/\nu}$ vs $\tau N^{1/\nu}$. Here $\nu = 1$ is the critical index for the correlation length in the bulk and $\tau \equiv T/T_c - 1$, where $T_c = T_c^{\mathrm{MFA}}/W$ is the bulk Curie temperature. Our results for the system with $N = 10$ and $N = 14$ merge into single "master curves" for $\beta_1 = 0.86$, $\beta_2 = 1.33$, and $\beta_3 = 1.79$, which have been obtained by fitting $M \propto N^{-\beta/\nu}$ at $T = T_c$, i.e., $\theta = \theta_c = 1/W$. Note that our value 0.86 for the surface magnetization critical index β_1 is substantially lower than the value $\beta_1 = 1$ following from scaling arguments. This disagreement is probably due to corrections to scaling which could be pronounced for our small linear sizes $N = 10$ and 14. A more efficient way to obtain an accurate value of β_1 is to perform a similar analysis for the semi-infinite model. The latter was considered analytically and numerically for $T \geq T_c$ and $H = 0$ in [35]. We also mention the Monte Carlo simulations of the Ising model [36] which yield $\beta_1 = 0.80$, $\beta_2 = 1.28$, and $\beta_3 = 1.77$.

The field dependence of M and m at fixed temperature, as obtained from the numerical solution of Eqs. (13) for cubic and square systems, is shown in Figure 2. Naturally the numerical results for m confirm Eq. (14) which describes both the effect of orientation of the system's magnetization by the field and the increase of M in field. Using the zero-field value of M in Eq. (14) leads to a poor result for m as shown by the dashed curve for the 10^3 system in Figure 2. The field dependence of the particle's magnetization similar to that shown in Figure 2 for the cubic system was experimentally obtained for ultrafine cobalt particles in [16], as well as in a number of previous experiments. The curves for the square system in Figure 2 illustrate the fact that in two dimensions thermal fluctuations are much stronger than in $3d$, which leads to lower values of both M and m at a given temperature. The bulk magnetization m_b in two dimensions vanishes at zero field and it thus goes below the intrinsic magnetization M in the low-field region.

3.2 Modified spin-wave theory: Low-temperature properties and the super-paramagnetic relation

As briefly discussed in Sec. 2, and in more detail in the previous section, it is important to investigate the precise relation between the induced magnetization $\mathbf{m}$ of Eq. (6) and intrinsic magnetization M of Eq. (7) for the more realistic Heisenberg model, $D = 3$. To do so, we have developed a finite-size spin-wave theory that yields analytical results at low temperatures. We also performed simulations with the improved Monte Carlo technique [37].

3.2.1 Spin-wave theory for finite-size magnetic particles

In the absence of SA, at low temperatures all spins in the particle are strongly correlated and form a "giant spin" $\mathbf{M}$ defined in Eq. (5) which behaves superparamagnetically. In addition, there are internal spin-wave excitations in the particle that are responsible for $M(H, T) < 1$ at $T > 0$. These excitations can be described perturbatively in small deviations of individual spins $\mathbf{s}_i$ from the global direction of $\mathbf{M}$. Thus we use $\mathbf{M} = \mathcal{M}\mathbf{n}$ with $|\mathbf{n}| = 1$ and insert an additional integration over $d\mathbf{M} = \mathcal{M}^{D-1}d\mathcal{M}d\mathbf{n}$ in the partition function,

$$\mathcal{Z} = \int \mathcal{M}^{D-1}d\mathcal{M}d\mathbf{n} \prod_i d\mathbf{s}_i \delta \left(\mathbf{M} - \frac{1}{\mathcal{N}} \sum_i \mathbf{s}_i \right) e^{-\mathcal{H}/T}, \qquad (17)$$

and first integrate over the magnetization magnitude $\mathcal{M}$. Thus we reexpress the vector argument of the δ-function in the coordinate system specified by the direction of the central spin $\mathbf{n}$:

$$\delta\left(\mathbf{M} - \frac{1}{\mathcal{N}}\sum_i \mathbf{s}_i\right) = \delta\left(\mathcal{M} - \frac{1}{\mathcal{N}}\sum_i (\mathbf{n}\cdot\mathbf{s}_i)\right)\delta\left(\frac{1}{\mathcal{N}}\sum_i [\mathbf{s}_i - \mathbf{n}(\mathbf{n}\cdot\mathbf{s}_i)]\right).(18)$$

Then after integration over $\mathcal{M}$ one obtains

$$\mathcal{Z} = \int d\mathbf{n}\,\mathcal{Z}_\mathbf{n}, \qquad \mathcal{Z}_\mathbf{n} = \int \prod_i d\mathbf{s}_i\,\delta\left(\frac{1}{\mathcal{N}}\sum_i [\mathbf{s}_i - \mathbf{n}(\mathbf{n}\cdot\mathbf{s}_i)]\right) e^{-\mathcal{H}_{\text{eff}}/T}, \quad (19)$$

where $\mathcal{Z}_\mathbf{n}$ is the partition function for the fixed direction $\mathbf{n}$ and

$$\mathcal{H}_{\text{eff}} = -(\mathbf{n}\cdot\mathbf{H})\sum_i (\mathbf{n}\cdot\mathbf{s}_i) - \frac{1}{2}\sum_{ij} J_{ij}\mathbf{s}_i\cdot\mathbf{s}_j - (D-1)T\ln\left[\frac{1}{\mathcal{N}}\sum_i \mathbf{n}\cdot\mathbf{s}_i\right].(20)$$

In Eq. (19), the δ-function says that the sum of all spins does not have a component perpendicular to $\mathbf{M}$. This will lead to the absence of the $\mathbf{k} = 0$ component of the transverse spin fluctuations. That is, the global-rotation Goldstone mode that is troublesome in the standard spin-wave theory for finite systems, has been transformed into the integration over the global direction $\mathbf{n}$ in Eq. (19). $\mathcal{Z}_\mathbf{n}$ is computed at low temperature by expanding $\mathcal{H}_{\text{eff}}$ up to bilinear terms in the transverse spin components $\mathbf{\Pi}_i = \mathbf{s}_i - \mathbf{n}(\mathbf{n}\cdot\mathbf{s}_i)$:

$$\mathcal{H}_{\text{eff}} \cong E_0 - \mathcal{N}\mathbf{n}\cdot\mathbf{H} + \frac{1}{2}\sum_{ij} A_{ij}\mathbf{\Pi}_i\cdot\mathbf{\Pi}_j,$$

$$A_{ij} \equiv \left[(D-1)T/\mathcal{N} + \mathbf{n}\cdot\mathbf{H} + \sum_l J_{il}\right]\delta_{ij} - J_{ij}, \qquad (21)$$

where $E_0 = -(1/2)\sum_{ij} J_{ij}$ is the zero-field ground-state energy. Next, upon computing the resulting Gaussian integrals over π_i^α, one obtains

$$\mathcal{Z}_\mathbf{n} \cong \exp\left(\frac{-E_0 + \mathcal{N}\mathbf{n}\cdot\mathbf{H}}{T}\right)\mathcal{N}^{D-1}\left[\frac{(2\pi T)^{\mathcal{N}-1}}{\prod_k{}' \frac{A_k}{\mathcal{N}}}\right]^{(D-1)/2}, \qquad (22)$$

where for particles of cubic shape

$$A_\mathbf{k} = A_0 + J_\mathbf{k} - J_0, \qquad A_0 = \sum_i A_{ij} = (D-1)T/\mathcal{N} + \mathbf{n}\cdot\mathbf{H}. \qquad (23)$$

The prime on the product in (22) means omitting the $k = 0$ mode. Results for particles of arbitrary shape can be found in [37]. In Eq. (22) the exponential factor corresponds to rigid spins whereas the second factor describes spin-wave corrections. The latter makes the angular dependence of $\mathcal{Z}_\mathbf{n}$ more complicated. Differentiating $\mathcal{Z}$ with respect to H yields the induced magnetization m, then the intrinsic magnetization M can be obtained from the exact relation (9), and the validity of the superparamagnetic relation (8) can be checked.

3.2.2 Induced and intrinsic magnetizations, and superparamagnetic relation

Now we consider particles of cubic shape ($\mathcal{N} = N^3$) at low temperatures. For both pbc and fbc, we have obtained the following correction to the magnetization in zero field [10,37]

$$M \cong 1 - t, \qquad t \equiv \frac{D-1}{2}\frac{W_N T}{J_0} = \frac{D-1}{2D}W_N\theta, \qquad W_N = \frac{1}{\mathcal{N}}{\sum_\mathbf{k}}'\frac{1}{1-\lambda_\mathbf{k}} \quad (24)$$

where W_N is the sum without the $\mathbf{k} = 0$ term. The results for pbc and fbc differ only by the values of the discrete wave vectors in Eq. (24) [10]:

$$k_\alpha = \begin{cases} 2\pi n_\alpha/N, \ \text{pbc} \\ \pi n_\alpha/N, \ \text{fbc} \end{cases}, \qquad n_\alpha = 0, 1, ..., N-1 \quad (25)$$

where $\alpha = x, y, z$. This subtle difference is responsible for much stronger thermal fluctuations in the fbc model due to boundary effects. The limit $N \to \infty$ of W_N is the so-called Watson's integral $W = P(0)$ of Eq. (12). The difference between W_N and W has different signs for pbc and fbc models [10]

$$\Delta_N \equiv \frac{W_N - W}{W} \cong \begin{cases} -\dfrac{0.90}{N}, & \text{pbc} \\ \dfrac{9\ln(1.17N)}{2\pi W N}, & \text{fbc.} \end{cases} \quad (26)$$

Therefore, Figure 3 shows that the coefficient in the linear-θ term in Eq. (24) is smaller than in the bulk for the pbc system and greater for the fbc system. That is, boundary effects suppress the intrinsic magnetization at low temperatures while finite-size effects lead to its increase. Figure 3 shows that boundary effects render a larger contribution than the finite-size effects, making the net magnetization well below that of the bulk.

The field dependence of m and M is defined by the expansion of the lattice Green function for small gaps, $1 - G \ll 1$

$$\tilde{P}_N(G) = \frac{1}{\mathcal{N}}{\sum_\mathbf{k}}'\frac{1}{1-G\lambda_\mathbf{k}} \cong W_N - \begin{cases} c_N N(1-G), & N^2(1-G) \ll 1 \\ c_0\sqrt{1-G}, & N^2(1-G) \gg 1, \end{cases} \quad (27)$$

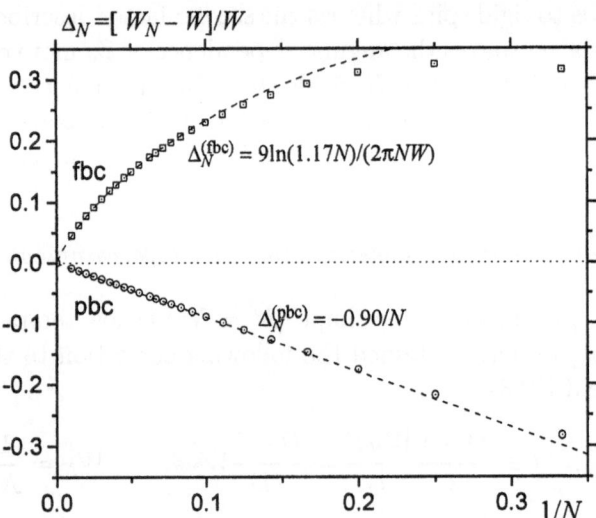

Fig. 3. Lattice sums W_N for cubic systems with free and periodic boundary conditions. $W = 1.51639$ is the bulk value for the sc lattice.

where for the sc lattice $c_0 = (2/\pi)(3/2)^{3/2}$ and the numerical results for c_N can, for $N \gg 1$, be fitted as

$$
c_N \cong \begin{cases} 0.384 - 1.05/N, \text{ pbc} \\ 1.90 - 1.17/N, \text{ fbc.} \end{cases} \tag{28}
$$

The spin-wave gap $1 - G$ depends on the temperature and field; for $H \ll H_V$ [see Eq. (16)] the gap approaches its zero-field value while for $H \gg H_V$ one has $1 - G \cong h \equiv H/J_0$. Thus Eq. (27) defines one more crossover in field, the crossover between its first and second lines at

$$
H_S \sim \frac{J_0}{N^{2/3}} = \frac{J_0}{N^2} \gg H_V = \frac{TD}{NM} = \frac{\theta J_0}{NM}. \tag{29}
$$

In the field range $H \gg H_S$ the discreteness of the lattice can be neglected and the bulk result $\Delta M \sim \sqrt{H}$ is reproduced. In the most interesting region $H \ll H_S$ one obtains

$$
M \cong 1 - t + 2\alpha x B(x), \qquad \alpha \equiv \frac{(D-1)c_N}{4N^2} \left(\frac{T}{J_0}\right)^2. \tag{30}
$$

On the other hand, Eq. (30) describes a crossover from the quadratic field dependence of M at low field, $x \ll 1$, to the linear dependence at $x \gg 1$. Note that for $x \gg 1$, where $m \cong M$ and a rigid magnetic moment would saturate, m continues to increase linearly as $m \cong 1 - t + 2\alpha x$. This is due to the field dependence of the intrinsic magnetization M. At higher fields there is another crossover to the standard spin-wave theory expression for M. Approximate expressions for M in the different field ranges are

$$M \cong 1 - t + \begin{cases} \dfrac{D-1}{2D} c_N \left(\dfrac{HN^2}{J_0} \right)^2 , & H \ll H_V \\[2ex] \dfrac{D-1}{2} c_N \dfrac{NHT}{J_0^2} , & H_V \ll H \ll H_S \\[2ex] \dfrac{D-1}{2} c_0 \dfrac{T}{J_0} \left(\dfrac{H}{J_0} \right)^{1/2} , & H_S \ll H \ll J_0. \end{cases} \tag{31}$$

A simple analysis shows [37] that the superparamagnetic relation (8) is a very good approximation for not too small systems, $\mathcal{N} \gg 1$ in the whole range below T_c. The deviation from Eq. (8) is controlled by the small parameter α of Eq. (30). Above T_c, however, deviations from Eq. (8) are large, except for the model with $D \to \infty$. In the close vicinity of T_c, there is a crossover to the high-temperature form of Eq. (8) given by the function $B_\infty(x)$ of Eq. (15).

The modified SWT developed above can be applied to study inhomogeneities in the fbc model. The local *intrinsic* magnetization defined by Eq. (10) shows stronger temperature dependence near the boundary than the averaged $M \cong 1 - t$ of Eq. (24). The biggest effect of the surface is naturally attained at the corners of the cube where $M_i \cong 1 - 8t$, at $H = 0$ [37].

3.2.3   MC simulations

Here we apply our Monte Carlo technique that accounts for the global-rotation Goldstone mode to compute the induced and intrinsic magnetizations, and to investigate the superparamagnetic relation between them, for the Heisenberg model with fbc. Our results for Ising model can be found in [37].

First of all, in Figure 4 we compare theoretical predictions of our analytical calculations within spin-wave theory for the Heisenberg model with our MC results at $T = T_c/4$, where $T_c = 0.722 T_c^{\mathrm{MFA}}$ is the actual bulk Curie temperature. For the small size $\mathcal{N} = 5^3$ the square-root field dependence of the magnetization [third line of Eq. (31)] does not arise and finite-size corrections are very important. For M one should use Eq. (30) with t given by Eq. (24) with numerically computed $W_N = 1.99$ and $c_N = 1.66$ for the fbc model [cf. Eqs. (26) and (28)]. This yields $t \simeq 0.119$ and $\alpha \simeq 1.20 \times 10^{-4}$. The theoretical dependence $M(H)$ is practically a straight line which goes slightly above the MC points. This small discrepancy can be explained by the fact that the applicability criterion for our analytical method, $t \ll 1$, is not fully satisfied

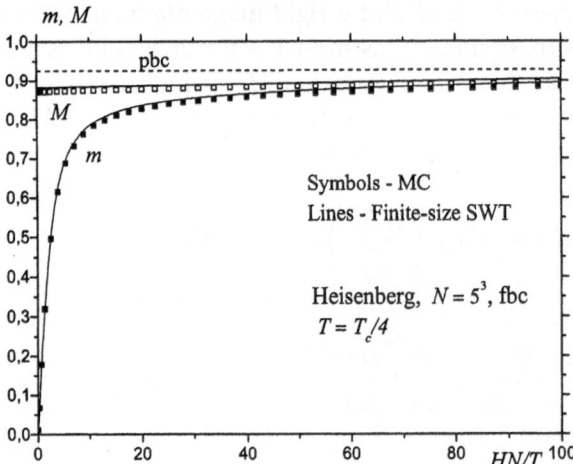

Fig. 4. Comparison of the theoretical and MC results for the field dependences of the magnetizations M and m for the Heisenberg model at $T = T_c/4$.

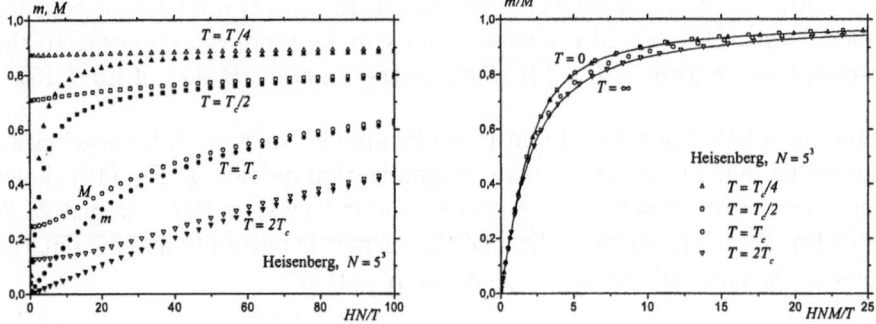

Fig. 5. Left: Field dependence of the intrinsic magnetization M and the induced magnetization m of the Heisenberg model on the sc lattice with fbc for different temperatures. Right: Scaled graph for m/M. Theoretical curves $B_3(x) = \coth x - 1/x$ for $T \ll T_c$ and $B_\infty(x)$ for $T \gg T_c$ are shown by solid lines.

at $T = T_c/4$, and a better agreement is achieved at lower temperatures. For comparison we also plot the theoretical $M(H)$ for the model with periodic boundary conditions. Here one has $W_N = 1.25$ and $c_N = 0.20$, thus $t \simeq 0.075$ and $\alpha \simeq 1.45 \times 10^{-5}$, so $M(H)$ goes noticeably higher and with a much smaller slope. The quadratic field dependence of M in the region $x \lesssim 1$ is not seen at this low temperature since the value of α is very small and thus much more accurate MC simulations would be needed. We also plot in Figure 4 the analytical result for the field dependence of m [37] which favorably compares with our MC data. Figure 5 (left) shows the intrinsic magnetization M and induced magnetization m versus the scaled field $x \equiv \mathcal{N}H/T$ for different temperatures. We see that the particle's magnetic moment is aligned and thus $m \sim M$ for $x \gtrsim 1$, if $T \ll T_c$. At $T \gg T_c$ the field aligns individual spins, and this requires $H \gtrsim T$, i.e., $x \gtrsim \mathcal{N}$. The quadratic dependence of $M(H)$ at small fields manifests itself strongly at elevated temperatures.

The results of Figure 5 (right) show that the superparamagnetic relation of Eq. (8) with $M = M(T, H)$ is a very good approximation everywhere below T_c, for the Heisenberg model. This also holds for the Ising model [37]. On the other hand, above T_c Eq. (8) with the function $B_\infty(x)$ of Eq. (15) is obeyed. The difference between these limiting expressions decreases with increasing number D of spin components and disappears in the spherical limit $(D \to \infty)$.

4   A nanoparticle as a multi-spin system: Effect of surface anisotropy

Surface anisotropy causes large deviations from the bulk behavior that are much stronger than just the effect of free boundaries. Also SA exerts influence upon the coercive field. Here we first consider magnetic structures and hysteresis loops at zero temperature induced by a strong transverse surface anisotropy (TSA). Then for the more physically plausible Néel's surface anisotropy (NSA) we investigate analytically and numerically its contribution to the effective anisotropic energy of the particle in the case when the NSA is weak in comparison to the exchange. After that we study the effect of the NSA at finite temperatures using the MC technique. At the end of this section we investigate the thermal and spatial dependence of the magnetization of a maghemite particle.

4.1   Magnetic structure and hysteresis at $T = 0$: Transverse surface anisotropy

Here we study the effect of TSA on the hysteresis loop and the angular dependence of the switching field. We construct an effective Stoner-Wohlfarth (SW) astroid for a single-domain spherical particle with free surfaces, a simple cubic (sc) crystal structure, ferromagnetic exchange J, uniaxial anisotropy in the core K_c, and TSA of strength K_s. Using the Hamiltonian (1) without the DDI at $T = 0$, we solve the coupled Landau-Lifshitz equations (LLE) for each spin in the particle [11] until a stationary state is attained. In [38] the same method was used for studying hysteresis loops in nanoparticles with a random bulk or surface-only anisotropy. Here we address the question of whether one can still use the simple SW model for a nanoparticle endowed with strong surface effects. We show that it is so as long as $K_s \lesssim J$. Otherwise switching of the particle's magnetization occurs via the reversal of clusters of spins, invalidating the simple SW model.

We consider K_s and exchange coupling on the surface as free parameters since there are so far no definite experimental estimations of them, whereas the core parameters are taken as for the bulk system. In the sequel, we use the reduced parameters, $j \equiv J/K_c, k_s \equiv K_s/K_c$.

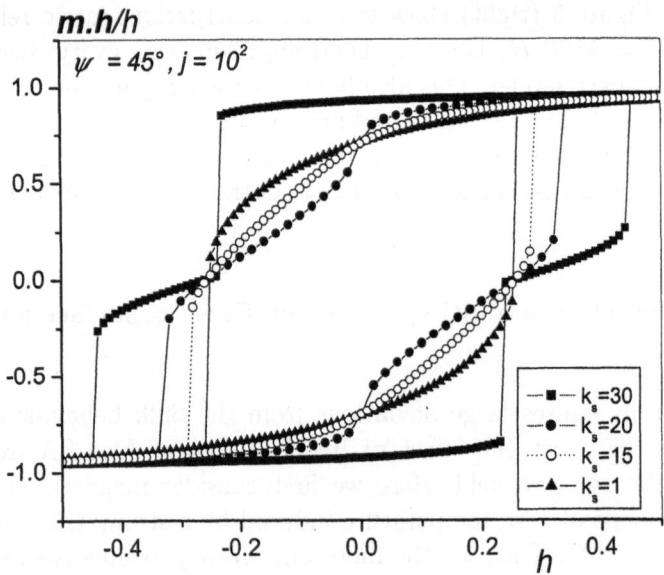

Fig. 6. Hysteresis for $\psi = \pi/4$, $j = 10^2$, $N = 10$ ($\mathcal{N} = 360$) and different k_s.

Figure 6 shows that when k_s becomes comparable with j, the hysteresis loop exhibits multiple jumps, which can be attributed to the switching of different spin clusters containing surface spins whose easy axes make the same angle with the field direction. The hysteresis loop is characterized by two field values: one that marks the limit of metastability, called the *critical field* or the *saturation field*, and the other that marks the magnetization switching, and is called the *switching field* or the coercive field. This progressive switching of spins is illustrated in Figure 7 for simplicity for the non-interacting case.

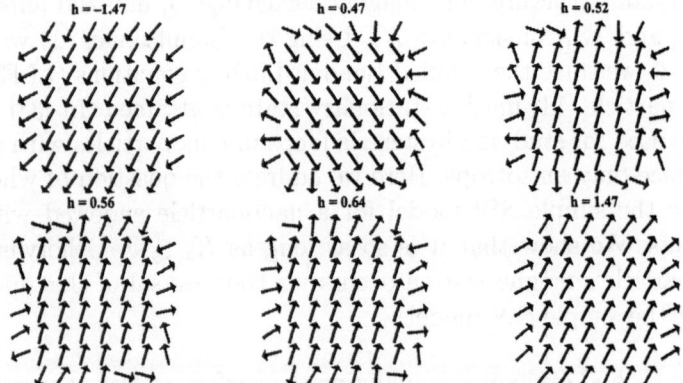

Fig. 7. Magnetic structures in the middle plane of the particle with $\mathcal{N} = 360$ in the TSA model for $j = 0$, $k_s = 1$, $\psi = \pi/4$ and different fields.

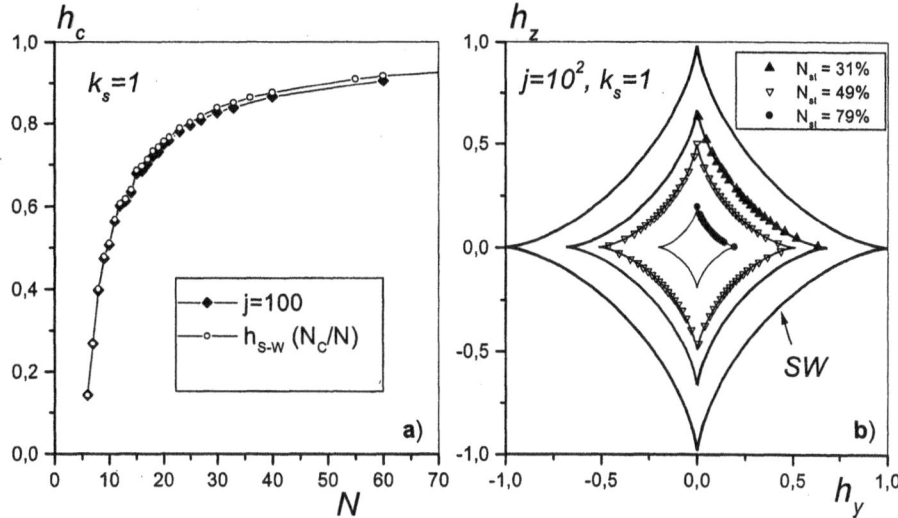

Fig. 8. a) (◇) Switching field for $k_s = 1$, $j = 10^2$ versus the particle's diameter N. (○) SW switching field multiplied by N_c/N. b) Astroid for $k_s = 1$, $j = 10^2$ for different values of the surface-to-volume ratio $N_{st} \equiv N_s/N$.

For small values of $k_s/j \sim 0.01$ the TSA model renders hysteresis loops and limit-of-metastability curves that scale with the SW results for all values of the angle ψ between the core easy axis and the applied field, the scaling constant being $N_c/N < 1$, see in Figure 8.

For larger values of k_s/j, but $k_s/j \lesssim 0.2$, we still have the same kind of scaling but the corresponding constant now depends on the angle ψ between the core easy axis and field direction. This is reflected by a deformation of the limit-of-metastability curve (see Figure 9a). More precisely, the latter is depressed in the core easy direction and enhanced in the perpendicular direction. However, there is still only one jump in the hysteresis loop implying that the magnetization reversal can be considered uniform. For $k_s/j \gtrsim 1$, there appear multiple steps in the hysteresis loop associated with the switching of spin clusters. It makes the hysteresis loop both qualitatively and quantitatively different from those of the SW model, as the magnetization reversal can no longer be considered uniform. In addition, in the present case, there are two more new features: the values of the switching field are much higher than in SW model, and more importantly, its behavior as a function of the particle's size is opposite to that of the previous cases (compare Figures 8a and 9b). More precisely, in this case one finds that this field increases when the particle's size is lowered. This is in agreement with the experimental observations in nanoparticles (see, e.g., [14] for cobalt particles). The whole situation is summarized in Figure 10 where we plot the critical field h_c as a function of $\tilde{k}_s \equiv k_s/j$ for different values of the surface-to-core ratio of the exchange coupling. For large values of k_s, surface spins are aligned along their radial easy axes, and because of strong exchange

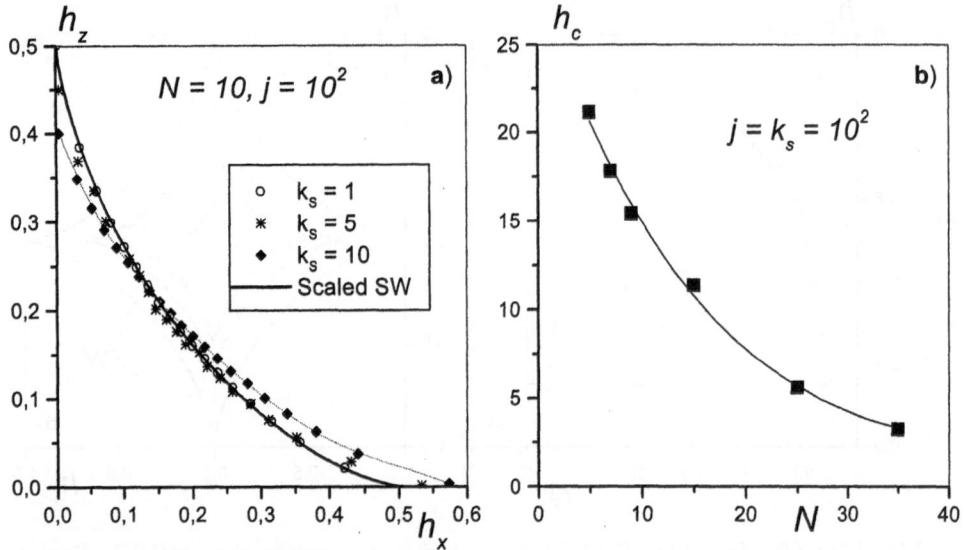

Fig. 9. a) Astroid for $j = 10^2$, $\mathcal{N} = 360$ and different values of surface anisotropy constant k_s. The full dark line is the SW astroid scaled with $N_c/\mathcal{N}$, but the dotted line is only a guide for the eye. b) Switching field versus the particle's diameter N for $\psi = 0$, $j = k_s = 10^2$.

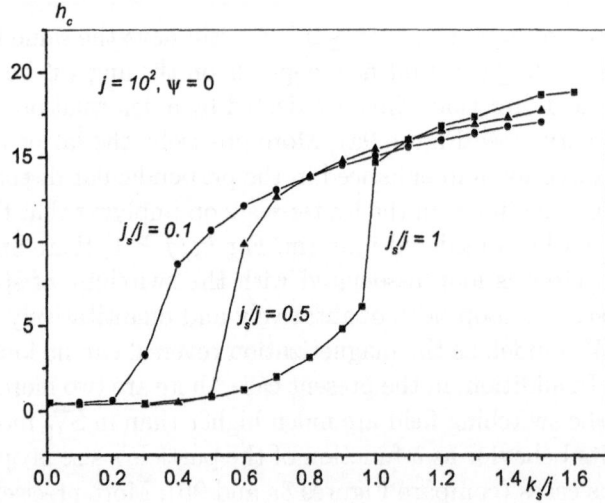

Fig. 10. Critical field versus the surface anisotropy constant for $\psi = 0$, and different values of surface-to-core ratio of exchange couplings; $N = 10$.

coupling they also drive core spins in their switching process, which requires a very strong field to be completed. The value of k_s where h_c jumps up (e.g., $= 1$ for $J_s/J = 1$) marks the passage from a regime where scaling with the SW results is possible (either with a ψ-dependent or independent coefficient) to the second regime where this scaling is no longer possible because of completely different switching processes.

To estimate K_s and the critical field, consider a 4 nm cobalt particle of fcc crystal structure, for which the lattice spacing is $a = 0.3554$ nm, and there are 4 cobalt atoms per unit cell. The (bulk) magneto-crystalline anisotropy is $K_c \simeq 3 \times 10^{-17}$ erg/spin or 2.7×10^6 erg/cm^3, and the saturation magnetization is $M_s \simeq 1422$ emu/cm^3. The switching field is given by $H_c = (2K_c/M_s)h_c$. For $\psi = 0$, $\tilde{k}_s^c = 1$ and $h_c = 15$, so $H_c \simeq 6$ T. On the other hand, $\tilde{k}_s^c = 1$ means that the effective exchange field experienced by a spin on the surface is of the order of the anisotropy field, i.e. $zSJ/2 \sim 2K_s$. Then using $J \simeq 8$ meV we get $K_s \simeq 5.22 \times 10^{-14}$ erg/spin, or using the area per surface spin (approximately $a^2/8$), $K_s \simeq 5$ erg/cm^2. For the case of $\psi = \pi/4$, $\tilde{k}_s^c \simeq 0.2$ and $h_c \simeq 0.3$, which leads to $H_c \simeq 0.1$ T and $K_s \simeq 1.2 \times 10^{-14}$ erg/spin or 1.2 erg/cm^2.

4.2 Surface contribution to the energy of magnetic nanoparticles: NSA

We calculate the contribution of the NSA [24] to the effective anisotropy of magnetic nanoparticles of spherical shape cut out of a simple cubic lattice. The effective anisotropy arises because of deviations of atomic magnetic moments from collinearity and dependence of the energy on the orientation of the global magnetization with respect to crystallographic directions. We show [39] that the result is second order in the NSA constant, scales with the particle's volume, and has cubic symmetry with preferred directions $[\pm 1, \pm 1, \pm 1]$.

As shown many times before, as the size of the magnetic particle decreases, surface effects become more and more pronounced. In many cases surface atoms yield a contribution to the anisotropy energy that scales with particle's surface, i.e., to the *effective* volume anisotropy decreasing with the particle's linear size R as $K_{V,\text{eff}} = K_V + K_S/R$, as was observed in a number of experiments (see, e.g., [16], [40]). The $1/R$ surface contribution to $K_{V,\text{eff}}$ is in accord with the picture of all magnetic atoms tightly bound by the exchange interaction while only surface atoms feel the surface anisotropy. This is definitely true for magnetic films where a huge surface contribution to the effective anisotropy has been observed. The same holds for cobalt nanoclusters of the form of truncated octahedrons [41] where contributions from different faces, edges, and apexes compete, resulting in a nonzero, although significantly reduced, surface contribution to $K_{V,\text{eff}}$.

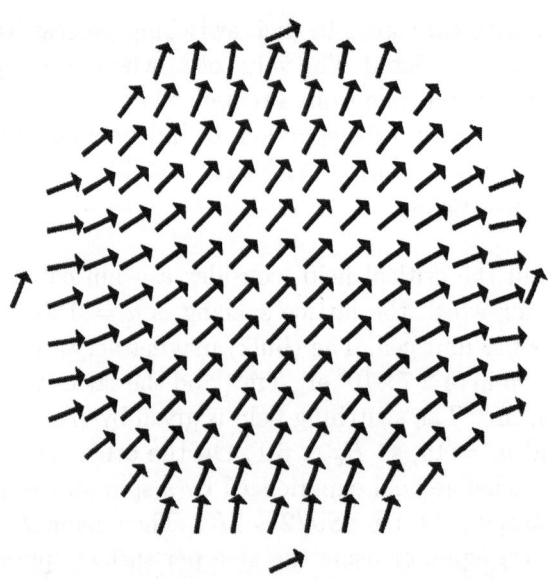

Fig. 11. Magnetic structure in the plane $z = 0$ of a spherical nanoparticle of linear size $N = 15$ with $L/J = 2$ for the global magnetization directed along [110].

However, for symmetric particle's shapes such as cubes or spheres, the symmetry leads to vanishing of this first-order contribution. In this case one has to take into account deviations from the collinearity of atomic spins that result from the competition of the surface anisotropy and exchange interaction J. The resulting magnetic structures (for the simplified radial SA model) can be found in [38], [42], [11] (see also Figure 11 for the NSA). In the case $L \gtrsim J$ deviations from collinearity are very strong, and it is difficult, if not impossible, to characterize the particle by a global magnetization suitable for the definition of the effective anisotropy. For $L \ll J$ the magnetic structure is nearly collinear with small deviations that can be computed perturbatively in $L/J \ll 1$. The global magnetization vector $\mathbf{m}_0$ [this is the same as the vector defined in Eq. (6)] can be used to define the anisotropic energy of the whole particle. The key point is that deviations from collinearity, and thereby the particle's energy, depend on the orientation of $\mathbf{m}_0$, even for a particle of a spherical shape, due to the crystal lattice. To illustrate this idea, we neglect the bulk anisotropy and the DDI in the Hamiltonian (1). For sc lattice Eq. (3) reduces to

$$\mathcal{H}_{an}^{(NSA)} = \sum_i \mathcal{H}_{an,i}^{(NSA)}, \qquad \mathcal{H}_{an,i}^{(NSA)} = \frac{L}{2} \sum_{\alpha=x,y,z} z_{i\alpha} s_{i\alpha}^2, \qquad (32)$$

where $z_{i\alpha} = 0, 1, 2$ are the numbers of available nearest neighbors of atom i along the $\alpha-$axis. One can see that the NSA is in general biaxial. For $L > 0$ and $z_{i\alpha} = 0 < z_{i\beta} = 1 < z_{i\gamma} = 2$ the α-axis is the easy axis and the $\gamma-$axis is the hard axis. If the local magnetic moments $\mathbf{s}_i$ are all directed along one of the crystallographic axes α, then the anisotropy fields $\mathbf{H}_{Ai} = -\partial \mathcal{H}_{Ai}/\partial \mathbf{s}_i$ are

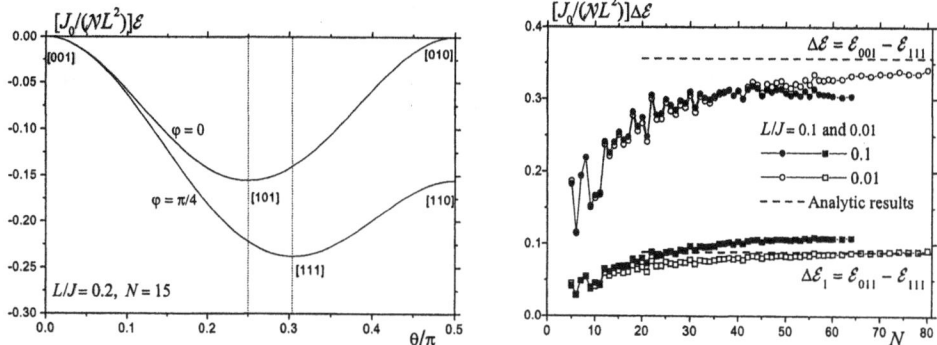

Fig. 12. Left: Effective anisotropy energy of the particle for different orientations of its global magnetization showing cubic symmetry of Eq. (41). Right: Differences of particle's energies between main orientations of the global magnetization vs the particle size in the scaled form for $L/J = 0.1$ and 0.01. The scaling is valid for $N \lesssim J/L$, and its violation for $L/J = 0.1$ is seen in the right part of the figure.

also directed along α and are thus collinear with s_i. Hence, at least for $L \ll J$, there are no deviations from collinearity if the global magnetization m_0 is directed along one of the crystallographic axes. For other orientations of m_0, the vectors s_i and H_{Ai} are not collinear, since then at least two components of s_i are nonzero, and the transverse component of H_{Ai} with respect to s_i causes a slight canting of s_i and thereby a deviation from the collinearity of magnetizations on different sites. This adjustment of the magnetization to the surface anisotropy lowers the energy. As we shall see, this effect is strongest for the $[\pm 1, \pm 1, \pm 1]$ orientations of m_0. For both signs of L these are easy orientations, whereas $[\pm 1, 0, 0]$, $[0, \pm 1, 0]$, and $[0, 0, \pm 1]$ are hard orientations.

To solve the problem numerically, we have to fix the global magnetization of the particle in a desired direction ν_0 ($|\nu_0| = 1$) by using the energy function with a vector Lagrange multiplier λ:

$$\mathcal{F} = \mathcal{H} - N\lambda \cdot (\nu - \nu_0), \qquad \nu \equiv \frac{\sum_i s_i}{|\sum_i s_i|}. \tag{33}$$

To minimize $\mathcal{F}$ we numerically integrate the evolution equations

$$\dot{s}_i = -[s_i \times [s_i \times F_i]], \qquad F_i \equiv -\partial \mathcal{F}/\partial s_i,$$
$$\dot{\lambda} = \partial \mathcal{F}/\partial \lambda = -N(\nu - \nu_0), \tag{34}$$

starting from $s_i = \nu_0 = m_0$ and $\lambda = 0$, until the stationary state is attained and an energy minimum is found. Our numerical results for the magnetic energy of spherical particles as a function of the orientation of the global magnetization are shown in Figure 12 (left). Figure 12 (right) shows differences between the basic directions [001], [011], and [111]. It is seen that $\Delta E/N \rightarrow$ const for

$N \to \infty$ limit, i.e., ΔE scales with the particle's volume $V \propto \mathcal{N}$.

To analytically solve the problem in the continuous limit, we replace in Eq. (32) the number of nearest neighbors of a surface atom by its average value

$$z_{i\alpha} \Rightarrow \overline{z}_{i\alpha} = 2 - |n_\alpha| / \max \{|n_x|, |n_y|, |n_z|\}. \tag{35}$$

Here n_α is the α-component of the normal to the surface $\mathbf{n}$. The surface-energy density can then be obtained by dropping the constant term and multiplying Eq. (32) by the surface atomic density $f(\mathbf{n}) = \max \{|n_x|, |n_y|, |n_z|\}$:

$$E_S(\mathbf{m}, \mathbf{n}) = -\frac{L}{2} \left[|n_x| m_x^2 + |n_y| m_y^2 + |n_z| m_z^2 \right]. \tag{36}$$

At equilibrium the Landau-Lifshitz equation reads

$$\mathbf{m} \times \mathbf{H}_{\text{eff}} = 0, \qquad \mathbf{H}_{\text{eff}} = \mathbf{H}_A + J\Delta\mathbf{m}. \tag{37}$$

For small deviations from collinearity one can seek its solution in the form

$$\mathbf{m}(\mathbf{r}) \cong \mathbf{m}_0 + \boldsymbol{\psi}(\mathbf{r}, \mathbf{m}_0), \qquad \psi \equiv |\boldsymbol{\psi}| \ll 1, \tag{38}$$

where $\boldsymbol{\psi}$ is the solution of the internal Neumann boundary problem

$$\Delta\boldsymbol{\psi} = 0, \qquad \left.\frac{\partial\boldsymbol{\psi}}{\partial r}\right|_{r=R} = \mathbf{f}(\mathbf{m}, \mathbf{n})$$

$$\mathbf{f} = -\frac{1}{J} \left[\frac{dE_S(\mathbf{m}, \mathbf{n})}{d\mathbf{m}} - \left(\frac{dE_S(\mathbf{m}, \mathbf{n})}{d\mathbf{m}} \cdot \mathbf{m} \right) \mathbf{m} \right]. \tag{39}$$

This equation can be solved with the help of the Green function $G(\mathbf{r}, \mathbf{r}')$ (see [39]), and the final result for the second-order energy contribution is

$$\mathcal{E}_2 \cong \frac{1}{2\pi J} \int\int_S d^2r\, d^2r'\, G(\mathbf{r}, \mathbf{r}') E_S(\mathbf{m}, \mathbf{n}) E_S(\mathbf{m}, \mathbf{n}'). \tag{40}$$

Taking into account the cubic symmetry and computing numerically a double surface integral one can write the result of Eq. (40) as

$$\mathcal{E}_2 \cong \kappa \frac{L^2 \mathcal{N}}{J_0} \left(m_x^4 + m_y^4 + m_z^4 \right), \qquad \kappa = 0.53465, \tag{41}$$

where $J_0 = zJ = 6J$. This defines the large-N asymptotes in Figure 12 (right) shown by the horizontal lines.

The analytical results above are valid for particle sizes N in the range

$$1 \ll N \ll J/L. \tag{42}$$

The lower boundary is the applicability condition of the continuous approximation. Since the surface of a nanoparticle is made of atomic terraces separated by atomic steps, each terrace and each step with its own form of NSA [see Eq. (32)], the variation of the local NSA along the surface is very strong. Approximating this variation by a continuous function according to Eq. (35) requires pretty large particle sizes N. This is manifested by a slow convergence to the large-N results in Figure 12 (right). The upper boundary in Eq. (42) is the applicability condition of the linear approximation in ψ. For $N \gtrsim J/L$ deviations from the collinear state are strong, and the effective anisotropy of a magnetic nanoparticle cannot be introduced.

As we have seen in Eq. (41), the contribution of the surface anisotropy to the overall anisotropy of a magnetic particle scales with its volume $V \propto N^3 \sim \mathcal{N}$. This surprising result is due to the penetration of perturbations from the surface deeply into the bulk. If a uniaxial bulk anisotropy K_c was present in the system, perturbations from the surface would be screened at the bulk correlation length (or the domain-wall width) $\delta \sim \sqrt{J/K_c}$. Then for $N \gtrsim \delta$ the contribution of the surface anisotropy to the overall anisotropy would scale as the surface: $\mathcal{E}_2 \sim (L^2/J) N^2 \delta$. For not too large particles, $N \lesssim \delta$, contributions of both anisotropies to the anisotropic energy are additive and scale as the volume. If the bulk anisotropy is cubic, both contributions have the same cubic symmetry [cf. Eqs. (4) and (41)], and the experiment should provide a value of the effective cubic anisotropy different form the bulk value [41]. For the uniaxial bulk anisotropy, the two contributions have different functional forms. Even if the bulk anisotropy is dominant so that the energy minima are realized for $\mathbf{m} \| \mathbf{e}_z$, the surface anisotropy makes the energy dependent on the azimuthal angle φ. This modifies particle's energy barrier by creating saddle points and strongly influences the process of thermal activation [43].

For small deviations from the cubic or spherical shape, i.e., for weakly elliptic or weakly rectangular particles, there should emerge a corresponding weak first-order contribution $\mathcal{E}_1$ that would add up with our second-order contribution. For an ellipsoid with axes a and $b = a(1 + \epsilon), \epsilon \ll 1$, the anisotropy energy scales with the surface, $\mathcal{E}_1 \sim L N^{2/3} \epsilon m_z^2$ [cf. Eq. (41)], so that

$$\frac{\mathcal{E}_2}{\mathcal{E}_1} \sim \frac{L}{J} \frac{N}{\epsilon} \tag{43}$$

can be large even for $L/J \ll 1$.

The Néel constant L is in most cases poorly known. However, for metallic Co

[44] quotes the value of surface anisotropy -1.5×10^8 erg/cm^3, i.e., $L \sim -10$ K. This is much smaller than $J \sim 10^3$ K, which makes our theory valid for particle sizes up to $N \sim J/L \sim 100$, according to Eq. (42). For this limiting size one has $\mathcal{E}_2/\mathcal{E}_1 \sim 1/\epsilon$ that is large for nearly spherical particles, $\epsilon \ll 1$.

4.3 Surface effects on the magnetization of a nanoparticle at $T > 0$: MC

In this section, we consider a more realistic model of round-shaped (spherical or ellipsoidal) nanoparticles of simple cubic or spinel crystalline structure, uniaxial or cubic anisotropy in the core, and transverse or Néel anisotropy on the surface, described by the Hamiltonian defined by Eqs. (1), (2), (4), and (3). Using various techniques explained above, we compute the magnetization, induced and intrinsic, as a function of temperature and applied magnetic field, for different values of the surface anisotropy constant and exchange coupling [4], [45]. We shall mainly focus on novel features stemming from the combination of anisotropy, field and temperature effects on the magnetization.

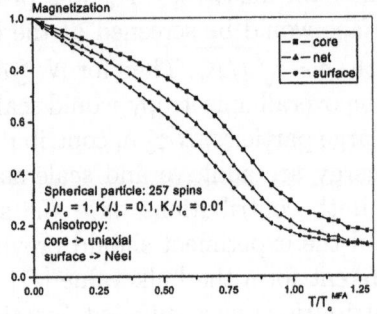
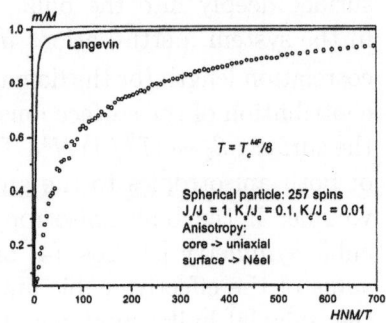

Fig. 13. Left: Core, surface and net magnetizations as functions of temperature. Right: Circles: Scaled graph for the induced magnetization m as a function of HNM/T, where M is the intrinsic magnetization. Full line: Langevin function $L(x) = \coth x - 1/x$.

4.3.1 Ferromagnetic particles with Néel's surface anisotropy

In our simulations we consider the core ferromagnetic coupling J_c and uniaxial anisotropy $K_c = 0.01 J_c$. On the surface we adopt $J_s = J_c$, while the anisotropy is given by the Néel expression (3) with constant $K_s = 0.1 J_c$. We ignore the DDI for simplicity. In particular, we are interested in how anisotropy affects the superparamagnetic relation (8) that has been shown to hold at all temperatures below T_c for isotropic systems. As no analytical calculations are possible here, we resort to the Monte Carlo technique with global spin rotations. In Figure 13 (left) we plot the core, surface and net magnetizations of a nanoparticle of 257 spins, as functions of temperature in zero magnetic field. These results do confirm what was obtained from the spherical model [see Figure 1 (left)] for isotropic box-shaped systems, namely that boundary effects suppress the magnetization, and here we see that this effect is enhanced by the SA. A drastic effect of the SA is clearly seen in the field dependence of the scaled induced magnetization m as shown in Figure 13 (right) at $T = T_c^{MFA}/8$. In the presence of a strong SA the superparamagnetic relation (8) is no longer valid, and m/M strongly deviates from the Langevin function.

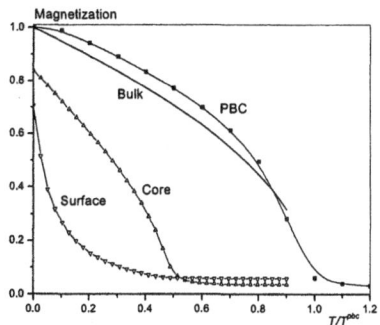 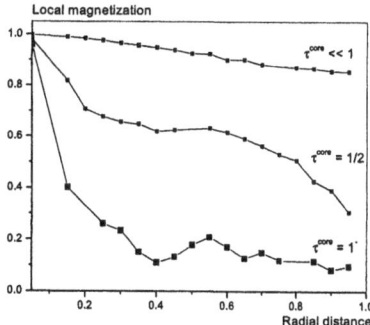

Fig. 14. Left: Temperature dependence of the surface and core magnetizations for $\mathcal{N} = 3766$, magnetization of the bulk system, and that of the cube with the spinel structure and periodic boundary conditions (pbc) with $\mathcal{N} = 40^3$. Right: Spatial variation of the net magnetization of a spherical nanoparticle of 3140 spins, as a function of the normalized particle radius, for $\tau^{core} \equiv T/T_c^{core} \ll 1$, and $\tau^{core} = 0.5$, $\tau^{core} \simeq 1^-$.

4.3.2   Maghemite (γ-Fe$_2$O$_3$) nanoparticles

In this section we deal with the ferrimagnetic maghemite nanoparticles (γ-Fe$_2$O$_3$) having spinel crystalline structure, summarizing the results of [4], [45]. This time we include the DDI in the Hamiltonian of Eq. (1). The bulk anisotropy in such materials is cubic, still we are using a uniaxial anisotropy to simplify the study of effects that are of more interest to us here. We consider maghemite particles of various sizes ($\mathcal{N} \simeq 10^3 - 10^5$ that correspond to a radius of 2-3.5 nm) and with the physical properties in the core (spinel crystal structure with lacuna, exchange and dipolar interactions, anisotropy constant, etc.) as those of the bulk, except that the anisotropy is taken as uniaxial. We use the TSA model with $K_s = 0.06$ erg/cm^2. All spins in the core and on the surface are identical but interact via different couplings depending on their locus in the lattice. We assume that the exchange interactions between the core and surface spins are the same as those inside the core. Although we treat only the crystallographically "ideal" surface, we do allow for a scatter in the exchange constants on the surface. In contrast, in [12] it was assumed that all exchange interactions are the same but there was postulated the existence of a fraction of missing bonds on the surface.

In Figure 14 (left), we see that the surface magnetization decreases more rapidly than the core magnetization as the temperature increases. Moreover, it is seen that even the (normalized) core magnetization per site does not reach its saturation value of 1 at very low temperatures, which shows that the magnetic order in the core is disturbed by the surface. This may also be due to lacuna in the spinel structure. In Figure 14 (right) we plot the spatial evolution of the local magnetization from the center to the border of the particle, at different temperatures. At all temperatures it decreases with distance from the center. At high temperatures, the local magnetization exhibits a temperature-dependent jump, and then continues to decrease. This indicates that there is a radius within which the magnetization assumes relatively large values. This result agrees with that of [3] where this radius was called the *magnetic radius*.

5   Conclusion

We have demonstrated by different analytical and numerical methods the importance of accounting for the magnetization inhomogeneities in magnetic nanoparticles, especially in the presence of surface anisotropy. The latter makes the magnetization inhomogeneous even at $T = 0$ and in general modifies the relation between the intrinsic and induced magnetizations. It also changes the magnetization switching mechanism, since for strong surface anisotropy the particle's spins switch cluster-wise. For weak surface anisotropy we have been able to calculate the spin canting in the particle analytically and to obtain a novel second-order contribution to the particle's overall anisotropy. It remains to generalize this result for nonzero bulk anisotropy. Another important task is to study dynamical implications of the many-body effects in magnetic nanoparticles.

References

[1] Néel L., C. R. Acad. Sci. Paris **228**, 664 (1949); Ann. Geophys. **5**, 99 (1949).

[2] Wernsdorfer W., Adv. Chem. Phys. **118**, 99 (2001).

[3] Wildpaner V., Z. Phys. B **270**, 215 (1974).

[4] Kachkachi H. et al., Eur. Phys. J. B **14**, 681 (2000).

[5] Chudnovsky E. M. and Tejada J. (1998) *Macroscopic quantum tunneling of the magnetic moment*, Cambridge University Press, Cambridge.

[6] Schumacher H. W. et al., Phys. Rev. Lett. **90**, 17201; 17204 (2003).

[7] Stoner E. C. and Wohlfarth E. P., Philos. Trans. R. Soc. London Ser. A **240**, 599 (1948); IEEE Trans. Magn. **27**, 3475 (1991).

[8] Brown W. F., Phys. Rev. **130**, 1677 (1963); IEEEM **15**, 1196 (1979).

[9] Coffey W. T. et al., Adv. Chem. Phys. **117**, 483 (2001).

[10] Kachkachi H. and Garanin D. A., Physica A **300**, 487 (2001).

[11] Kachkachi H. and Dimian M. , Phys. Rev. B **66** , 174419 (2002); Dimian M. and Kachkachi H., J. Appl. Phys. **91**, 7625 (2002).

[12] Kodama R. H. et al., Phys. Rev. Lett. **77**, 394 (1996); Kodama R.H. and Berkovitz A.E., Phys. Rev. B **59**, 6321 (1999).

[13] Richardson J. T. et al., J. Appl. Phys. **70**, 6977 (1991).

[14] Chen J. P. et al., Phys. Rev. B **51**, 11527 (1995).

[15] Ezzir A., *Propriétés Magnétiques d'une assemblée de nanoparticules: modélisation de l'aimantation*, thesis, Université Paris-Sud, Orsay 1998.

[16] Respaud M. et al., Phys. Rev. B **57**, 2925 (1998).

[17] Tronc E. et al., J. Magn. Magn. Mat. **221**, 110 (2000).

[18] Coffey W. T. et al., Adv. Chem. Phys. **117**, 483 (2001).

[19] Haneda K., Can. J. Phys. **65**, 1233 (1987).

[20] Prené P. et al., Hyperfine Interactions **93**, 1049 (1994).

[21] Aharoni A. (1996) *Introduction to the theory of ferromagnetism* Oxford Science Pubs., Oxford .

[22] Brown W. F., Jr. (1963) *Micromagnetics* Interscience, New York.

[23] Shilov V., *Effects of surface anisotropies of the ferromagnetic resonance in ferrite nanoparticles*, Ph.D. thesis, University Paris VII, 1999.

[24] Néel L., J. Phys. Radium **15**, 225 (1954).

[25] Stanley H. E., Phys. Rev. Lett. **20**, 589 (1968); Phys. Rev. **176**, 718 (1968).

[26] Berlin T. N. and Kac M., Phys. Rep. **86**, 821 (1952).

[27] Garanin D. A., Z. Phys. B **102**, 283 (1997); J. Phys. A **29**, L257 (1996); **29**, 2349 (1996); **32**, 4323 (1999); Phys. Rev. E **58**, 254 (1998).

[28] Garanin D. A. and Lutovinov V. S., Solid State Commun. **50**, 219 (1984).

[29] Garanin D. A., J. Stat. Phys. **74**, 275 (1994); Phys. Rev. B **53**, 11593 (1996).

[30] Fisher M. E. and Privman V., Phys. Rev. B **32**, 447 (1985); Commun. Math. Phys. **103**, 1986 (1986).

[31] Binder K. (1983), in C. Domb and J. L. Lebowitz (eds.), *Phase transitions and critical phenomena*, Academic Press, New York, vol.8, pp.75–267; Diehl H. W., ibid 1986, vol. 10, pp. 75–267.

[32] Bray A. J. and Moore M. A., Phys. Rev. Lett. **38**, 735 (1977).

[33] Cardy J. L., J. Phys. A **16**, 3617 (1983).

[34] Binder K. (1992), in H. Gausterer and C. B. Lang (eds.), *Computational methods in field theory*, Springer, Berlin.

[35] Garanin D. A., Phys. Rev. E **58**, 254 (1998).

[36] Pleimling M. and Selke W., Eur. Phys. J. B **5**, 805 (1998).

[37] Kachkachi H. and Garanin D. A., Eur. Phys. J. B **22**, 291 (2001).

[38] Dimitrov D. A. and Wysin G. M., Phys. Rev. B **50**, 3077 (1994).

[39] Garanin D. A. and Kachkachi H., Phys. Rev. Lett. **90**, 65504 (2003).

[40] Chen C. et al., J. Appl. Phys. **86**, 2161 (1999).

[41] Jamet M. et al., Phys. Rev. Lett. **86**, 4676 (2001).

[42] Labaye Y. et al., J. Appl. Phys. **91**, 8715 (2002).

[43] Garanin D. A. et al, Phys. Rev. E **60**, 6499 (1999).

[44] Chuang D. S. et al., Phys. Rev. B **49**, 15084 (1994).

[45] Kachkachi H. et al., J. Magn. Magn. Mat. **221**, 158 (2000).

FROM FINITE SIZE AND SURFACE EFFECTS TO GLASSY BEHAVIOUR IN FERRIMAGNETIC NANOPARTICLES

Amílcar Labarta, Xavier Batlle and Òscar Iglesias

Departament de Física Fonamental Universitat de Barcelona, Diagonal 647, 08028 Barcelona, Spain
`amilcar@ffn.ub.es, xavier@ffn.ub.es, oscar@ffn.ub.es`

This chapter is aimed at studying the anomalous magnetic properties (glassy behaviour) observed at low temperatures in nanoparticles of ferrimagnetic oxides. This topic is discussed both from numerical results and experimental data. Ferrimagnetic fine particles show most of the features of glassy systems due to the random distribution of anisotropy axis, interparticle interactions and surface effects. Experiments have shown that the hysteresis loops display high closure fields with high values of the differential susceptibility. Low magnetisation as compared to bulk, shifted loops after field-cooling, high-field irreversibilities between zero-field and field cooling processes and ageing phenomena in the time-dependence of the magnetisation, are also observed. This phenomenology indicates the existence of some kind of freezing phenomenon arising from a complex hierarchy of the energy levels, whose origin is currently under discussion. Two models have been proposed to account for it: i) the existence of a spin-glass state at the surface of the particle which is coupled to the particle core through an exchange field; and ii) the collective behaviour induced by interparticle interactions. In real systems, both contributions simultaneously occur, being difficult to distinguish their effects. In contrast, numerical simulations allow us to build a model just containing the essential ingredients to study solely one of two phenomena.

1 Frustration in ferrimagnetic oxides

Ferrimagnetic oxides are one kind of magnetic systems in which there exist at least two inequivalent sublattices for the magnetic ions. The antiparallel alignment between these sublattices (ferrimagnetic ordering) may occur provided the intersublattice exchange interactions are antiferromagnetic (AF) and some requirements concerning the signs and strengths of the intrasublattice interactions are fulfilled. Since usually in ferrimagnetic oxides the magnetic cations are surrounded by bigger oxygen anions (almost excluding the direct overlap between cation orbitals) magnetic interactions occur via indirect superexchange mediated by the p oxygen orbitals. It is well-known that the sign of these superexchange interactions depends both on the electronic structure of the cations and their geometrical arrangement [1]. In most ferrimagnetic oxides, the crystallographic and electronic structure give rise to antiferromagnetic inter- and intrasublattice competing interactions. Consequently, in these cases not all magnetic interactions can be fulfilled, and in spite of the collinear alignment of the spins, some degree of magnetic frustration exists.

The most common crystallographic structures for ferrimagnetic oxides are hexagonal ferrites, garnets, and spinels, all of them having intrinsic geometrical frustration when interactions are all antiferromagnetic [2]. A typical ferrimagnetic oxide with spinel structure, having a variety of technological applications, is γ-Fe_2O_3 (maghemite), in which the magnetic Fe^{3+} ions with spin $S = 5/2$ are disposed in two sublattices with different coordination with the O^{2-} ions. Each unit cell (see Fig. 1) has 8 tetrahedric (T), 16 octahedric (O) sites, and one sixth of the O sites have randomly distributed vacancies to achieve neutrality charge. The T sublattice has larger coordination than O, thus, while the spins in the T sublattice have $N_{TT} = 4$ nearest neighbours in T sites and $N_{TO} = 12$ in O sites, the spins in the O sublattice have $N_{OO} = 6$ nearest neighbours in O and $N_{OT} = 6$ in T. The values of the nearest-neighbour exchange constants are [3, 4, 5]: $J_{TT} = -21$ K, $J_{OO} = -8.6$ K, $J_{TO} = -28.1$ K. So, the local magnetic energy balance favours ferrimagnetic order, with spins in each sublattice ferromagnetically aligned and antiparallel intrasublattice alignment.

The substitution of magnetic ions by vacancies or the presence of broken bonds between both sublattices destabilises the collinear arrangement of the spins inducing spin canting in the sublattice with less degree of substitution, a fact that has been observed in many ferrimagnetic oxides with substitution by non-magnetic cations [6, 7, 8, 9, 10, 11]. It is worth noting that when the ferrimagnetic oxide is in the form of small particles, the structural modifications at the boundaries of the particle (vacancies, broken bonds and modified exchange interactions) may induce enough frustration so as to destabilise the ferrimagnetic order at the surface layer giving rise to different canted magnetic structures.

In this chapter, we will discuss how these canted spins at the surface of the particle may freeze giving rise to a glassy state below a certain temperature.

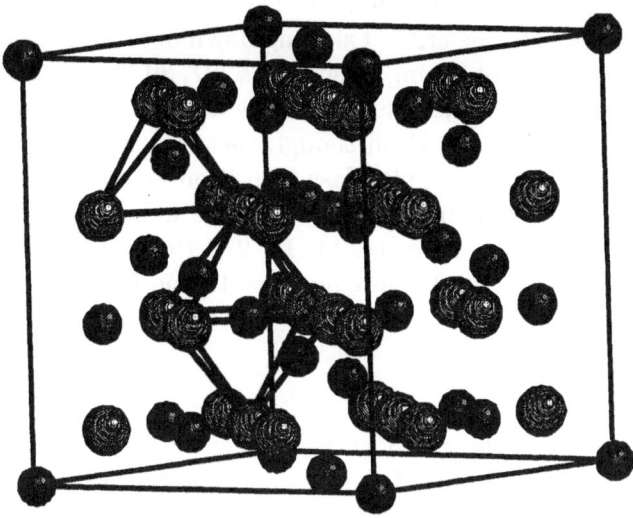

Fig. 1. Unit cell of maghemite. The magnetic Fe^{3+} ions occupying the two sublattices, in different coordination with the O^{2-} ions (light-grey colour), are coloured in black (T sublattice, tetrahedric coordination) and in dark grey (O sublattice, octahedric coordination).

2 Glassy behaviour in ferrimagnetic nanoparticles

The assemblies of fine magnetic particles with large packing fractions and/or nanometric sizes show most of the features which are characteristic of glassy systems (for a recent review see Ref. [12]). This glassy behaviour results from a complex interplay between surface and finite-size effects, interparticle interactions and the random distribution of anisotropy axis throughout the system. In many cases, these contributions are mixed and in competition, and their effects are thus generally blurred. However, several works propose that glassy behaviour in fine particle systems is mostly attributed to the frustration induced by strong magnetic interactions among particles with a random distribution of anisotropy axes [13, 14, 15], although some experimental and numerical results suggest the dominant role of the spin disorder at the surface layer in ferrimagnetic oxides [16, 17, 18, 19, 20, 21].

2.1 Anomalous magnetic properties at low temperature

The most important features characterising the glassy state in fine-particle systems include the flattening of the field-cooling (FC) susceptibility [22], the existence of high-field irreversibility in the magnetisation curves and between the field and zero-field-cooling (ZFC) susceptibilities [3, 21, 23, 24], the occurrence of shifted loops after FC [24, 25, 23, 26], the increase in the magnetic viscosity [27], the critical slowing down observed by ac susceptibility [28], the occurrence of ageing phenomena [27, 29, 28, 30, 31, 32] and the rapid increase of the non-linear susceptibility as the blocking temperature is approached from above [27]. Even though these features are not associated with the occurrence of a true spin-glass transition in interacting fine-particle systems, some authors claim that they may be indicative of the existence of some kind of collective state [13, 14, 15, 27]. Here we discuss an example of a complete magnetic study in a nanocrystalline barium ferrite which is representative of this glassy behaviour.

Sample characterisation

The phenomenology of the glassy state in strong interacting fine particles is illustrated through the study of the magnetic properties of nanocrystalline $BaFe_{10.4}Co_{0.8}Ti_{0.8}O_{19}$ [11]. M-type barium ferrites have been studied for a long time because of their technological applications [33, 34, 35, 36], such as microwave devices, permanent magnets, and high-density magnetic and magneto-optic recording media, as well as their large pure research interest [37, 38]. The compounds obtained by cationic substitution of the pure $BaFe_{12}O_{19}$ ferrite display a large variety of magnetic properties and structures, which go from collinear ferrimagnetism to spin-glass-like behaviour [37, 39, 38], depending on the degree of frustration introduced by cationic substitution. In particular, the $BaFe_{10.4}Co_{0.8}Ti_{0.8}O_{19}$ compound seems to be ideal for perpendicular magnetic recording [40], since the Co^{2+}-Ti^{4+} doping scheme reduces sharply the high values of the coercive field of the pure compound, which precludes their technological applications. For this composition the magnetic structure is still ferrimagnetic, the Curie temperature is well above room temperature and the coercive field is reduced to usual values for magnetic recording technologies.

Nanocrystalline $BaFe_{10.4}Co_{0.8}Ti_{0.8}O_{19}$ samples were prepared by the glass crystallisation method (GCM). The GCM is characterised by the presence of homogenised melt fluxes of Fe_2O_3, BaO, B_2O_3 and the corresponding oxides of the doping cations (Ti_2O and CoO for the Co-Ti substitution) at about 1300 °C, which are amorphised by rapid quenching in a two-roller equipment. Annealing the glass flakes above 550 °C leads to the nucleation and growth of the borate and hexaferrite phases. Barium ferrite particles crystallise to suitable sizes during this treatment and may be isolated by dissolving the matrix in dilute acetic acid in an ultrasonic field. After centrifugation, washing and drying, a fine powder of M-type doped barium ferrite particles with the desired stoichiometry is obtained. A more detailed description of this method can be found in Refs. [40, 41].

In order to maximise interparticle interactions the sample was studied in powder form. X-ray diffraction data showed very broad peaks and the fitting

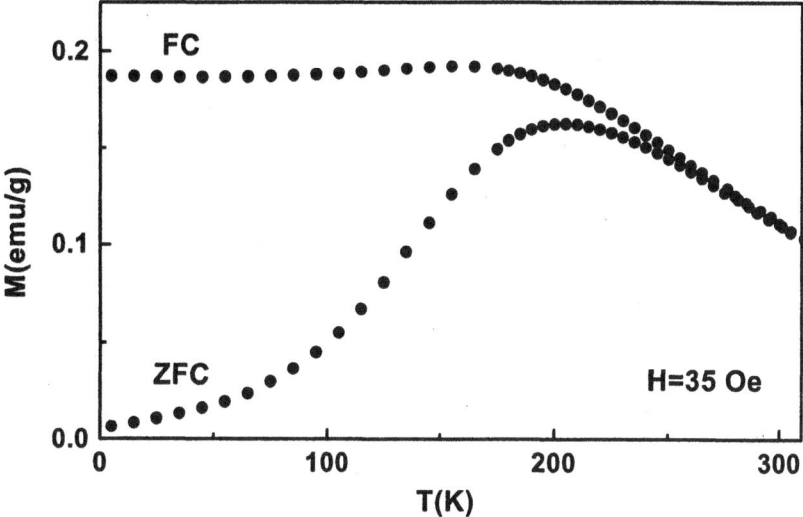

Fig. 2. Zero-field-cooling and field-cooling magnetisation as a function of temperature measured at a field of 35 Oe for a nanocrystalline $BaFe_{10.4}Co_{0.8}Ti_{0.8}O_{19}$ sample.

of the whole spectra to the M-type structure demonstrated the platelet-like morphology of the particles. The particle size distribution as determined from transmission electron microscopy (TEM) and X-ray diffraction was log normal with a mean platelet diameter of 10.2 nm, $\sigma = 0.48$ and an aspect ratio of 4 (mean volume of 10^5 nm^3). TEM also confirmed the platelet-like shape and showed a certain tendency of the particles to pile up producing stacks along the perpendicular direction to the (001) face of the platelet, which corresponds to the easy axis. TEM data showed the existence of large agglomerates composed of stacks, particle aggregates and quasispherical conglomerates, as expected from the value of the aspect ratio of these particles. The powder was mixed with a glue in order to avoid particle rotation towards the applied field axis during magnetic measurements.

Magnetic properties

Magnetisation measurements were carried out with a SQUID magnetometer under magnetic fields up to 55 kOe. High-field magnetisation measurements (up to 240 kOe) were performed at the Grenoble High Field Laboratory, using a water-cooled Bitter magnet with an extraction magnetometer. The temperature dependence of the AC magnetic susceptibility was recorded at frequencies between 10 Hz and 1 kHz, applying an AC magnetic field of 1 Oe after cooling the sample at zero field from room temperature.

    The zero-field-cooling and field-cooling curves measured at 35 Oe displayed all the typical features of an assembly of small magnetic particles with a distribution of energy barriers (see Fig. 2). The ZFC curve showed a broad maximum located at ca. $T_M = 205 \pm 5$ K, which originated from

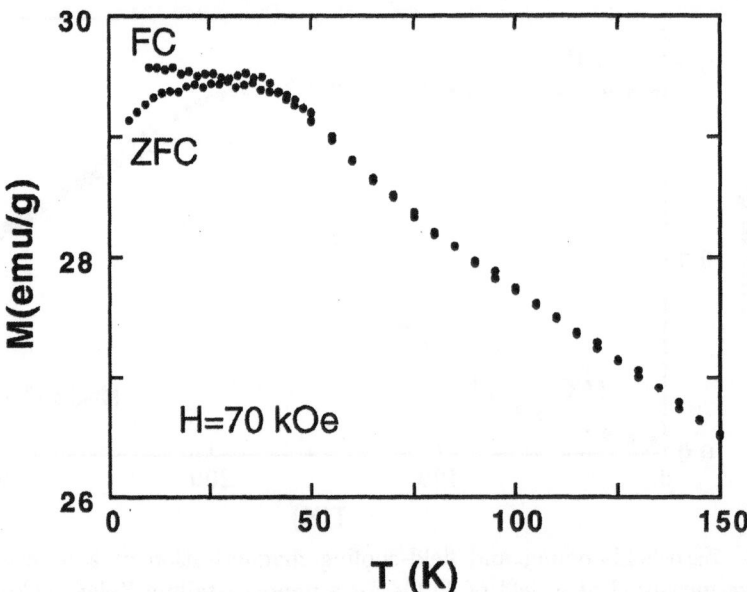

Fig. 3. Zero-field-cooling and field-cooling magnetisation as a function of temperature measured at a field of 70 kOe for a nanocrystalline $BaFe_{10.4}Co_{0.8}Ti_{0.8}O_{19}$ sample.

both blocking and freezing processes, the latter due to magnetic frustration induced by interparticle interactions. The FC was very flat below T_M, in comparison with the monotonously increasing behaviour characteristic of noninteracting systems, which reinforced the existence of strong interactions among particles. Moreover, in the superparamagnetic (SPM) regime, the low-field susceptibility of an assembly of interacting particles is expected to be of the form $\chi \approx \frac{\bar{\mu}^2}{3k_B(T-T_0)}$, where $\bar{\mu}$ is the mean magnetic moment per particle and T_0 is an effective temperature arising from interparticle interactions. In accordance with this equation, the reciprocal of the FC susceptibility showed a linear behaviour well above T_M when it was multiplied by $M_S^2(T)/M_S^2(0)$ in order to correct the temperature dependence of $\bar{\mu}$. The value $T_0 = -170 \pm 30$ K was obtained by extrapolating this linear behaviour to the temperature axis, suggesting also the existence of strong interparticle interactions with demagnetising character. It is worth noting that irreversibility between the ZFC and FC susceptibilities was still present up to fields as high as 70 kOe (see Fig. 3), indicating that high energy barriers were freezing the system into a glassy state. From all these results, it can be concluded that the broad maximum at T_M in the ZFC susceptibility and the irreversibility between ZFC and FC curves originate from the blocking of the relaxation of the particle magnetisation and the freezing associated with frustrated interactions among particles. Both processes run parallel in the system giving rise to a collective state at low temperature which shares with spin-glass-like systems in bulk form most of the characteristic features of a glassy state.

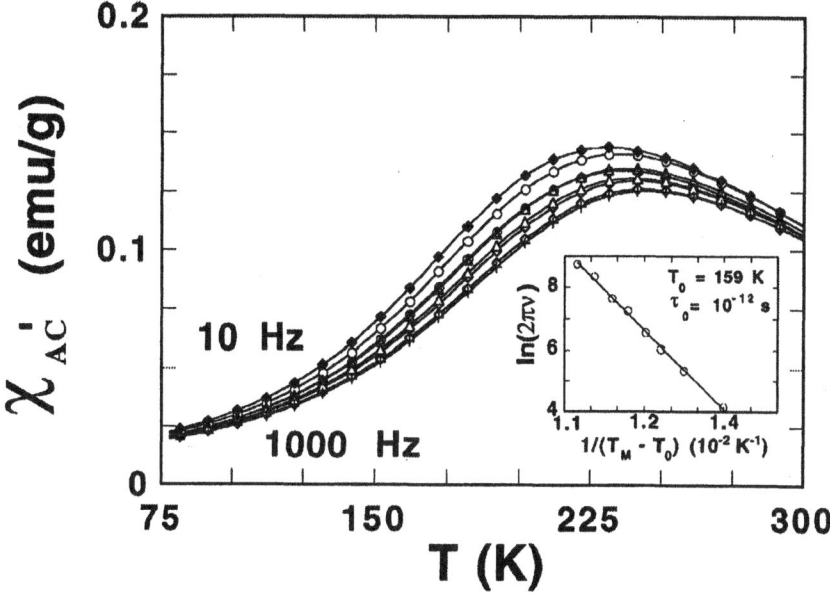

Fig. 4. Temperature dependence of the in-phase component of the AC susceptibility (χ'_{AC}) of a nanocrystalline $BaFe_{10.4}Co_{0.8}Ti_{0.8}O_{19}$ sample measured at different frequencies in the range between 10 Hz and 1 kHz. Inset: logarithm of the measuring frequency as a function of the reciprocal of the difference between the temperature of the peak and T_0.

Further insight into the dynamics of the processes responsible for the glassy state in this system can be gained by analysing the in-phase component of the AC susceptibility (χ'_{AC}) measured at different frequencies from 75 to 300 K [42]. The measured curves (see Fig.4) behave as the DC low-field susceptibility, but the temperature at which the peak locates increases with the measuring frequency as expected for a blocking/freezing process. The Vogel-Fulcher law [43] describes the slowing down of a system composed of magnetically interacting clusters as the temperature is reduced and can be expressed in the form

$$\nu^{-1} = \tau_0 \ \exp\left[\langle E \rangle / k_B (T_M - T_0)\right] , \qquad (1)$$

where T_0 is an effective temperature with a similar origin to that used to reproduce the DC susceptibility in the SPM regime and T_M is a characteristic temperature signalling the onset of the blocking process (e.g. the temperature of the peak position in the AC susceptibility). Experimental data can be fitted to Eq. 1 as shown in the inset of Fig. 4, giving the following values for the parameters: $\tau_0 = 10^{-12}$ s, $\langle E \rangle = 2.45 \times 10^{-13}$ erg and $T_0 = 159$ K, a value which is very close to the effective temperature deduced from the reciprocal of the FC susceptibility.

The good agreement of the experimental data with the Vogel-Fulcher law evidences that the phenomenon taking place at T_M is related to the blocking of an assembly of interacting particles rather than a collective freezing as that

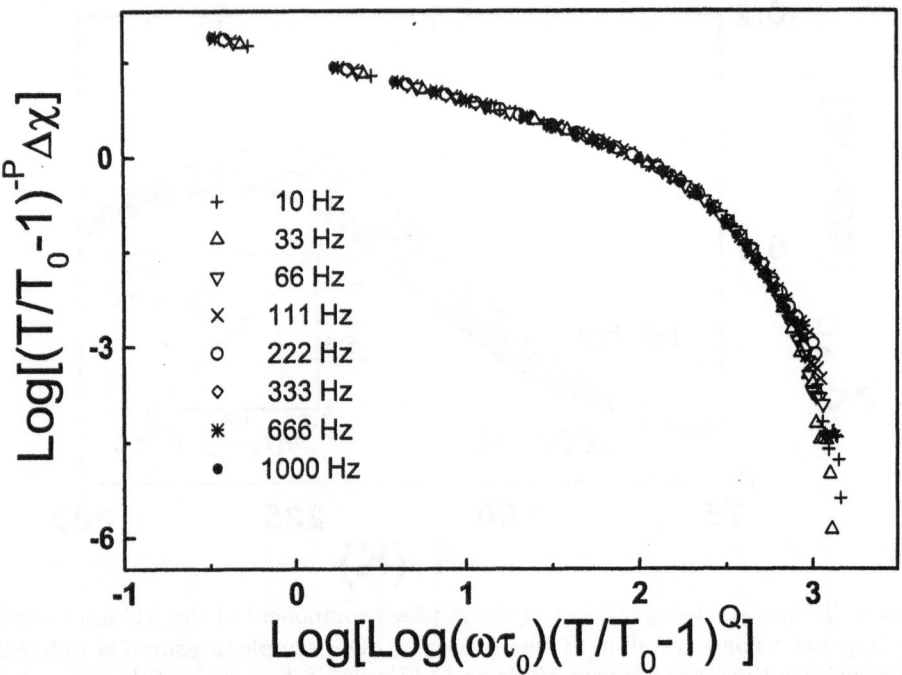

Fig. 5. Activated dynamic scaling for the in-phase AC susceptibility shown in Fig. 4.

occurring in a canonical spin-glass system. As a consequence, the temperature dependence of the in-phase AC susceptibility at different frequencies can be scaled within the framework of a model based on the activated dynamics [44, 45], in which it is supposed that $\Delta\chi = t^P G\left[-t^Q \log(\omega\tau_0)\right]$, with $\Delta\chi = (\chi_0 - \chi'_{AC})/\chi_0$, where χ_0 is the equilibrium susceptibility at zero frequency, P and Q are the critical exponents, $t = T/T_0 - 1$ is the reduced temperature, and G is the scaling function. The quality of the achieved scaling is shown in Fig. 5, in which the values of the parameters are as follows: $P = 0.45 \pm 0.2$, $Q = 0.7 \pm 0.1$, $T_0 = 150 \pm 15$ K and $\tau_0 = 10^{-12}$ s. On the contrary, AC susceptibility cannot be scaled assuming a critical slowing down with the form of a power law [46], fact which almost discards the occurrence of a true spin-glass transition in this assembly of interacting nanoparticles.

More evidences of the glassy behaviour are obtained from the study of the isothermal magnetisation as a function of the magnetic field. The hysteresis loop at 5 K in the first quadrant with a maximum applied field of 200 kOe is shown in Fig. 6, displaying the typical features of fine-particle systems: the saturation magnetisation is about half of the bulk value, the high-field differential susceptibility is about double and the coercive field is about four times larger. Moreover, the onset of irreversibility takes place at 125 kOe, which is much larger than the typical values for bulk barium ferrites (thousands of Oe), and the hysteresis loop recorded after FC the sample at 70 kOe is shifted

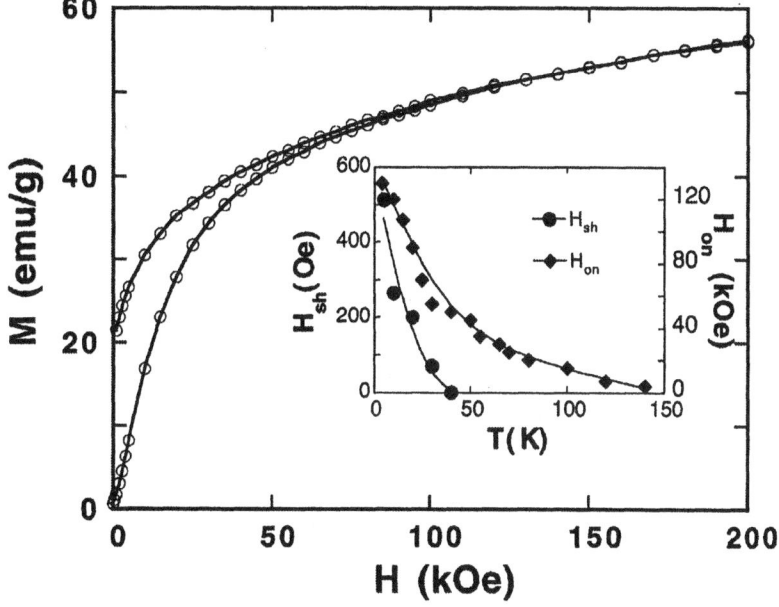

Fig. 6. Magnetisation vs magnetic field at 5 K for a nanocrystalline $BaFe_{10.4}Co_{0.8}Ti_{0.8}O_{19}$ sample. The inset shows the dependence on the temperature of the onset of the irreversibility (H_{on}) and the shifting of the hysteresis (H_{sh}) loops after an FC process at 70 kOe.

ca. 500 Oe in the opposite direction to the cooling field (magnetic training effect). The thermal dependence of the onset of the irreversibility and the shifting of the hysteresis loops are shown in the inset of Fig. 6. The latter decays rapidly to zero as the temperature is raised and glassy behaviour disappears, while the former decreases slowly towards a value larger than the intrinsic anisotropy field because of the surface anisotropy contribution.

Ageing is also commonly considered a characteristic feature of systems with enough magnetic frustration as to give rise to a multivalley energy structure at low temperature. In particular, it has been observed in fine-particle systems with strong interactions [29, 28, 30] and in many glassy systems in bulk form, e.g. canonical spin glasses and dilute antiferromagnets [47, 48]. The phenomenon consists in the slow evolution of the magnetic order at a microscopic scale while an external magnetic field is applied, without changing significantly the net magnetisation of the system. The waiting time elapsed before the removal of the applied magnetic field is a relevant parameter determining the shape of the subsequent relaxation curves. The system studied in this work also showed ageing below about the temperature of the maximum of the ZFC curve. Fig. 7 shows the relaxation curves at 150 K after an FC process from room temperature at 200 Oe, for waiting times of 5×10^2 and

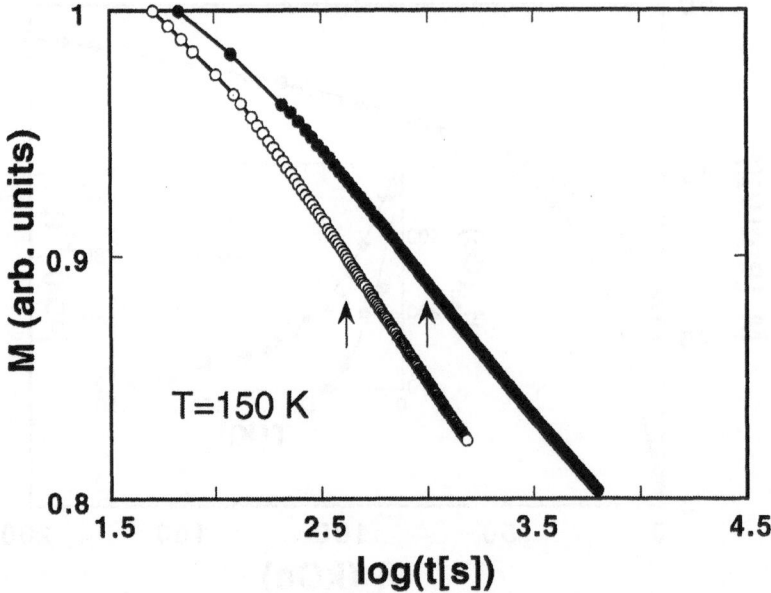

Fig. 7. Relaxation curves of a nanocrystalline $BaFe_{10.4}Co_{0.8}Ti_{0.8}O_{19}$ sample at 150 K after an FC process at 200 Oe for waiting times of 5×10^2 s (open circles) and 10^3 s (solid circles). Arrows indicate the position of the inflection point of the curves, which is located at about the waiting time.

10^3 s. The characteristic trends of ageing were present in these curves proving the existence of a glassy state: (i) the relaxation rate decreased as the waiting time increased, and (ii) the relaxation curves plotted on a logarithmic time scale showed an inflection point at an elapsed time which was roughly the waiting time.

All these phenomena are common to many ferrimagnetic and antiferromagnetic oxides in the form of fine particles [3, 17, 21, 23, 24, 25, 26, 27, 49], suggesting that they are part of the characteristic fingerprints of the glassy behaviour associated with the collective state existing at low temperature for strong interacting particles and/or of the surface effects appearing in the nanoscale limit.

2.2 Surface to core exchange coupling and interparticle interactions

Surface effects dominate the magnetic response of fine particles since for diameters smaller than ca. 2 nm more than one third of the total spins are located at the surface. Consequently, the ideal picture of a superspin formed by the collinear arrangement of all the spins of the particle is no longer valid, and the misalignment of the surface spins yields strong deviations from the bulk behaviour. This is also true for particles of many ferrimagnetic oxides with strong exchange interactions, in which magnetically competing sublattices usually exist. In these cases, broken bonds and defects at the surface layer destabilise magnetic order giving rise to magnetic frustration, which is enhanced with the strength of the magnetic interactions. As a consequence, the profile of the magnetisation is not uniform across the particle, the surface layer being more demagnetised than the core spins. In early models, this fact was modelled postulating the existence of a dead magnetic layer giving no contribution to the magnetisation of the particle [50]. Coey proposed the existence of a random canting of the surface spins due to the competing antiferromagnetic interactions between sublattices to account for the reduction of the saturation magnetisation observed in γ-Fe_2O_3 ferrimagnetic nanoparticles [18]. He found that a magnetic field of 5 T was not enough to achieve full alignment of the spins in the field direction for particles 6 nm in size. A further verification of the existence of spin canting in nanoparticles of different ferrimagnetic oxides (γ-Fe_2O_3, $NiFe_2O_4$, $CoFe_2O_4$, $CuFe_2O_4$) was gained by using Mössbauer spectroscopy [18, 51], polarised [52] and inelastic neutron scattering [53], and ferromagnetic resonance [54]. However, in contrast with the original suggestion by Coey that spin canting occurs chiefly at a surface layer due to magnetic frustration, some works based on Mössbauer spectroscopy support the idea that it is a finite-size effect which is uniform throughout the particle [16, 19, 55, 56, 24]. In fact, the origin of the spin misalignment observed in ferrimagnetic nanoparticles is still under discussion and there is not a complete understanding of the phenomenon. Nevertheless, spin canting has not been found in metallic ferromagnetic particles, which reinforces the hypothesis that magnetic frustration originated from antiferromagnetically competing sublattices is a necessary ingredient to explain the non-collinear arrangement of the particle spins.

Based on all the foregoing results and on the glassy properties observed at low temperatures, a model has been proposed [24, 21] suggesting the existence of a magnetically ordered core surrounded by a surface layer of canted spins, which undergoes a spin-glass-like transition to a frozen state below a certain temperature, T_f. The glassy state of the surface below T_f creates an exchange field acting over the ordered core of the particle, which could be responsible for the shifting of the hysteresis loops after an FC process by a mechanism similar to that giving rise to exchange bias in layered structures [57, 58]. Besides, the magnetic frustration at the surface may increase the effective anisotropy energy giving rise to both the high-field irreversibility observed in the ZFC-FC curves and the high values of the closure field in the hysteresis loops. In fact, these anomalous and enhanced properties vanish or are strongly reduced above T_f, a fact that indicates that they are related to some kind of frozen state occurring below T_f.

The existence of a frozen state at the particle surface has been experimentally established by different techniques [17, 23, 3, 21, 59, 60]. For instance, electron paramagnetic resonance measurements in iron-oxide nanoparticles 2.5 nm in diameter diluted in a polyethylene matrix [60], showed a sharp line broadening and a shifting of the resonance field on sample cooling below T_f, which is about the temperature at which an anomaly in the FC susceptibility was also observed. Furthermore, the study of the field dependence of T_f in γ-Fe_2O_3 nanoparticles of 10 nm average size demonstrated that this magnitude follows the Almeida-Thouless line [21] which is a characteristic behaviour found in many magnetic glassy phases. This field dependence of T_f was not affected by diluting the magnetic particles with a nanometric SiO_2 powder [21], which excludes that the glassy state in these systems could originate from interparticle interactions only.

A micromagnetic model at atomic scale proposed by Kodama and Berkowitz [17, 3] and several numerical simulations of a single ferrimagnetic particle with a variety of assumptions but with the common condition of free boundaries at the surface [4, 61, 62, 63, 64, 65, 66, 67, 68, 69, 70], have evidenced the non-uniformity of the magnetisation profile across the ferrimagnetic particle, with a fast decreasing towards the surface. These results demonstrate that the non-uniform profile of the particle magnetisation is merely a surface effect. However, enhanced anisotropy (normal to the surface) [71], vacancies and broken bonds at the surface have to be included in these particle models in order to induce enough magnetic frustration so as to freeze the disordered surface layer giving rise to a glassy state [3, 4]. Therefore, surface and finite-size effects seem not to be enough as to produce the glassy layer, even in the case of ferrimagnetic particles with competing antiferromagnetic sublattices, surface anisotropy and disorder being necessary additional ingredients.

In highly concentrated samples (e.g. powder samples) with a random distribution of easy axis, interparticle interactions are a supplementary source of magnetic frustration which may lead to a frozen collective state of the particle spins at low temperature, apart from the effects of the surface-to-core exchange coupling discussed above. In fact, both processes may occur in parallel contributing simultaneously to the glassy phenomenology in ferrimagnetic fine particles. The main types of magnetic interactions that can be found in fine-particle assemblies are dipole-dipole interaction, which always exists, and exchange interactions through the surface of the particles being in close contact [72]. Taking into account the anisotropic character of dipolar interactions, which may favour parallel or antiparallel arrangements of the spins depending on the geometry, and the random distribution of local easy axis, concentrated samples have the required elements of magnetic frustration to give rise to a glassy state. The complex interplay between both sources of magnetic disorder determines the state of the system and its dynamical properties. In particular, the effective distribution of energy barriers, which block the inversion of particle spins, is further enhanced by magnetic frustration leading to high field irreversibilites.

In the limit of strong dipolar interactions, individual energy barriers coming from intrinsic anisotropy of the particles can no longer be considered, the total energy of the assembly being the only relevant magnitude. In this limit, relaxation is governed by the evolution through an energy landscape similar to that of a spin glass with a complex hierarchy of energy minima. The inversion of one particle moment may also modify the energy barriers of the whole assembly. Consequently, the energy barrier distribution may evolve as the total magnetisation of the system relaxes [73, 74, 75, 76] and non-equilibrium dynamics may appear, for instance, showing ageing effects as observed in several fine-particle systems with strong interactions [28, 27, 32, 29, 31, 30]. In this way, the occurrence of ageing phenomena in fine particles could be considered as a clear indication of a strong interacting scenario with a phenomenology which largely mimics that corresponding to spin glasses, including memory effects in which magnetic relaxation depends on heating or cooling rates and/or on the cycles followed by the temperature [30, 32]. However, three main differences between ageing in bulk spin-glasses and fine particles may be established: i) the dependence of the relaxation on the waiting time is weaker (see, for instance, Fig. 7); ii) in the collective state, the relaxation times are widely distributed and strongly dependent on temperature; and iii) the moment of the largest particles is blocked during long time scales, acting as a random field throughout the system.

Finally, one of the facts that complicates the study of these systems is the coexistence of the freezing associated with magnetic frustration and the intrinsic blocking of the particles. Consequently, depending on the time window of the experiment, one or both phenomena are observed. For example, blocking processes usually determine the results of Mössbauer spectroscopy, since the measured blocking temperature decreases with increasing interactions [77], while freezing phenomena determine the temperature location of the cusp of the real part of the AC susceptibility for concentrated samples, which moves to higher temperatures with increasing interactions [31, 78, 27].

2.3 Effective distribution of energy barriers from relaxation measurements

One way to obtain a further insight into the nature of the glassy phenomenology described in the preceding sections consists in analysing the time dependence of the magnetisation of the system in terms of the so-called $T \ln(t/\tau_0)$ scaling [79, 80, 81], since this method allows one to calculate the effective distribution of energy barriers [82, 83, 84]. In this scaling procedure, the value of τ_0 (characteristic attempt time) is chosen so as to make all the relaxation curves, measured at different temperatures, scale onto a single master curve, which stands for the relaxation curve at the lowest measuring temperature

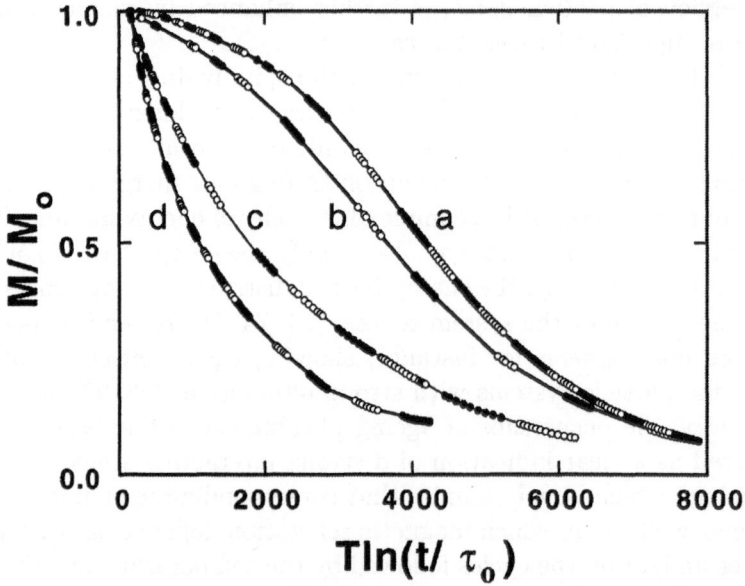

Fig. 8. $T\ln(t/\tau_0)$ scaling of the magnetic relaxation measured in the temperature range 5-230 K after an FC process at 200 Oe (a), 500 Oe (b), 10 kOe (c), 50 kOe (d). Alternated open and full circlesb correspond to consecutive different temperatures.

extended to very long times. The effective distribution of energy barriers is then obtained by performing the derivative of the experimental master curve with respect to the $T\ln(t/\tau_0)$ variable [82, 83, 84]. However, we could wonder whether this method is applicable in the strong interacting regime, for which it is not possible to identify individual energy barriers blocking the reversal of the particle moments as pointed out before. Monte Carlo simulations and experimental results [75, 76, 82, 85] have demonstrated that a good $T\ln(t/\tau_0)$ scaling is also achievable in this case, from which a static effective distribution of energy barriers (non-time-evolving) may also be derived. This energy distribution is similar to that observed in non-interacting systems, even though it is not only a direct consequence of the distribution of particle anisotropies, being an average of the contributions due to volume, surface and shape anisotropies, and interparticle interactions. Through this simple method, it is possible to ascertain the differences between fine-particle systems with glassy phenomenology, for which the existence of a large number of quasidegenerate energy levels makes the dynamics complex, since their actual energy barrier distribution is time-dependent.

As an example of this procedure applied to a strong interacting system, in Fig. 8, we show the scaling of the relaxation data measured after FC of the sample at different fields for the $BaFe_{10.4}Co_{0.8}Ti_{0.8}O_{19}$ particles discussed in

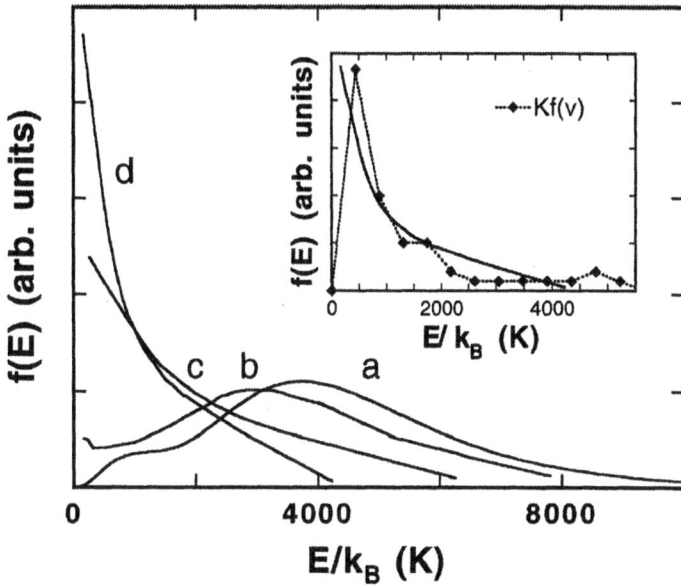

Fig. 9. Effective distribution of energy barriers obtained from the scaling of Fig.8. The labels of the curves are as in Fig.8. Inset: energy distribution obtained after an FC process at 50 kOe [curve (d)] compared with $Kf(v)$ (symbols and dashed line respectively). K stands for the anisotropy constant and $f(v)$ is the volume distribution.

Section 2.1. The value of τ_0 used in this scaling is 10^{-12} s, which is consistent with those values deduced from the frequency dependence of the maximum of the real part of the AC susceptibility (Vogel-Fulcher law) and the activated dynamic scaling of these data (see Figs. 4, 5). It is evident from Fig. 8 that the cooling field drastically modifies the relaxation curves, which demonstrates that the initial arrangement of the particle moments (FC state) determines the time evolution of the magnetisation when interparticle interactions are strong [86, 87, 88]. This is in contrast with the noninteracting case, for which the observed results are independent of the cooling field, at least for fields lower than that at which the lowest energy barriers start to be destroyed [89, 90].

2.4 Effects of the magnetic field on the glassy state

The effective distribution of energy barriers, $f(E)$, characterising the glassy state of the assembly as a function of the cooling field, can be obtained from the scaling curves in Fig. 8 by the procedure detailed in Refs. [75, 79, 84, 89, 90], the results being shown in Fig. 9. At low cooling fields, $f(E)$ extends to extremely high energies, and the energy corresponding to the maximum of the

A. Labarta et al.

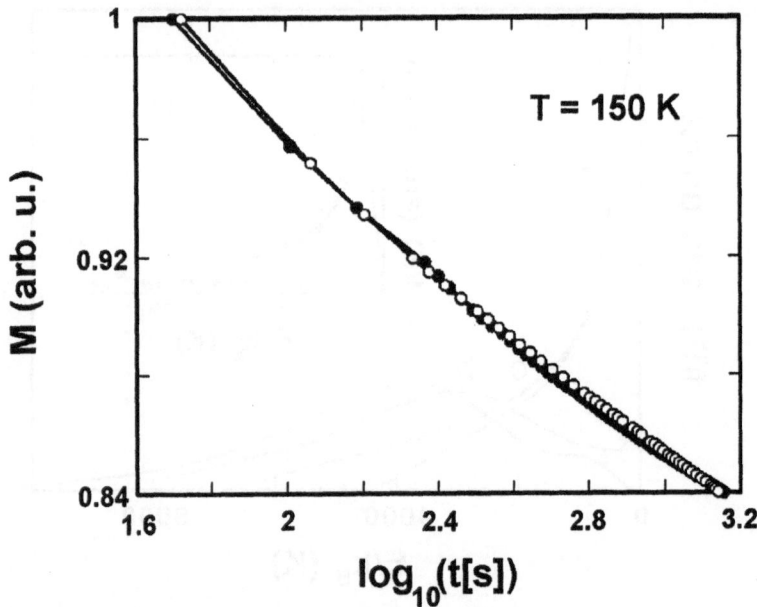

Fig. 10. Relaxation curves at 150 K after an FC process at 10 kOe for waiting times of 5×10^2 s (open circles) and 10^3 s (solid circles).

distribution is one order of magnitude higher than that expected from bulk anisotropy, $Kf(V)$, where K is the bulk anisotropy constant [39] and $f(V)$ is the volume distribution derived from TEM [91]. In contrast, $f(E)$ progressively resembles $Kf(V)$ as the cooling field increases (see inset of Fig. 9). This field dependence of $f(E)$ may be interpreted as follows. The component of the energy barrier distribution centred at high energies, which is dominant at low cooling fields, is then attributed to the collective behaviour associated with the glassy state, since particle magnetisation is mostly randomly distributed in the FC state. Nevertheless, $f(E)$ at high cooling fields, which is centred at much lower energies, corresponds to the intrinsic anisotropy of the individual particles. In the high-field-cooled state, the particle magnetisation is mostly aligned parallel to the field. Therefore, the overall dipolar interactions are demagnetising and their effect can be considered through a mean field which reduces the height of the energy barrier associated with the intrinsic anisotropy of the particles [75, 89]. Consequently, a slight shift to lower energies is observed in $f(E)$ with respect to $Kf(V)$ (see inset of Fig. 9). At intermediate fields, a bimodal $f(E)$ arising from both contributions is observed, their relative importance being determined by the strength of the cooling field (see $f(E)$ obtained after FC processes at 200 and 500 Oe in Fig. 9). This interpretation is also confirmed by repeating the ageing relaxation experiments shown in Fig. 7 with a cooling field of 10 kOe instead of 200 Oe.

In this case, no significant differences in the relaxation curves are detected as the waiting time is varied (see Fig. 10). Accordingly, the characteristic ageing associated with the glassy state can be detected only at low enough cooling fields. Therefore, the glassy state in particle assemblies originated from strong interparticle interactions can be erased by field-cooling the sample at moderate fields, which precludes the occurrence of a true spin-glass transition at the freezing temperature. Even though interacting particle systems seems to exhibit an effective irreversibility line similar to that found in other cluster and spin-glass systems in the bulk form, the magnetic fields at which the disordered state is erased are very low compared, for example, to either the anisotropy field of the particles or from those values corresponding to archetypal spin glasses [92].

All in all, cooling fields monitor the dynamics of interacting fine magnetic particles through determining the initial state of the magnetic moment arrangement. Consequently, at high cooling fields the dynamics of the system is mostly dominated by the intrinsic energy barriers of the individual particles, while at low cooling fields, the energy states arising from collective glassy behaviour play the dominant role. Thus, care should be taken when comparing relaxation data from isothermal remanent magnetisation and thermoremanent magnetisation, since the initial magnetic state may be very different in both kinds of experiments depending on the field strength. Similar results have been obtained by Jönson et al. [93] in strongly interacting FeC nanoparticles, for which the collective glassy dynamics can be destroyed by the application of moderate fields and also in the Monte Carlo simulations by Ulrich et al. [88]. Fine-particle systems are thus relevant because, although they display an important degree of magnetic frustration, the collective state may be destroyed by the application of a moderate field, which precludes a true spin-glass behaviour.

3 Monte Carlo simulations

Numerical modeling and, in particular, Monte Carlo (MC) simulations, are well suited to discern between the contributions to the glassy behavior due to inter-particle interactions and single-particle effects, such as surface and finite-size effects. By considering the particles as dipolar interacting superspins, MC simulations reproduce the enhancement of the magnetic viscosity and the increase in the energy barriers [75, 76] and also the non-equilibrium dynamics showing ageing and memory effects [94]. On the contrary, some relaxation simulations starting from different states discard the occurrence of cooperative freezing since all the curves converge to the equilibrium state [86, 87] whereas on similar simulations performed in the absence of a magnetic field the system seem to approach some glassy ferromagnetic state [88]. Alternatively, by taking a single particle as the simulation unit, numerical models can also be used to study intrinsic effects related to the finite volume and boundary limits of the particle, disregarding collective behaviour due to inter-particle interactions. In this approach, the atomic spins and the underlying lattice are considered in detail focusing the attention on the magnetic characterisation in terms of the microscopic structure. In what follows, we summarise the chief previous studies of single particle models.

The first atomic-scale model of the magnetic behaviour of individual ferrimagnetic nanoparticles is due to Kodama and Berkowitz [3]. The authors presented results of calculations of a micromagnetic model of maghemite particles which were based on an energy minimisation procedure, instead of the Monte Carlo method. They used Heisenberg spins with enhanced anisotropy at the surface with respect to the core and included vacancies and broken bonds at the surface, arguing that these are indeed necessary to obtain hysteresis loops with enhanced coercivity and high-field irreversibility. Later, Kachkachi et al. [4, 95, 96] performed MC simulations of a maghemite particle described by a Heisenberg model, including exchange and dipolar interactions and using surface exchange and anisotropy constants different to those of the bulk. Their study was mainly focused on the thermal variation of the surface (for them consisting of a shell of constant thickness) and core magnetisation, concluding that surface anisotropy is responsible for the non-saturation of the magnetisation at low temperatures. More recently [63, 64], they studied the influence of surface anisotropy in the zero-temperature hysteretic properties of a ferromagnetic particle by means of numerical evaluation of Landau-Lifschitz equations.

Other computer simulations studying finite-size and surface effects on ferro- and antiferromagnetic cubic lattices have also been published. Bucher and Bloomfield [97] presented a quantum mechanical calculation and performed MC simulations of a Heisenberg model to explain the measured reduction in the magnetic moment of small free Co clusters. Trohidou et al. [66, 70] performed MC simulations of AF small spherical clusters. By using an Ising model on a cubic lattice [66], they computed the thermal and magnetic field dependence of the magnetisation and structure factor, concluding that the particle behaved as a hollow magnetic shell. By means of a Heisenberg model [70] with enhanced surface anisotropy, they studied the influence of different kinds of surface anisotropy on the magnetisation reversal mechanisms and on the temperature dependence of the switching field. Dimitrov and Wysin [61, 62] studied the hysteresis phenomena of very small spherical and cubic ferromagnetic (FM) fcc clusters of Heisenberg spins by solving the Landau-Lifshitz equations. They observed an increase of the coercivity with decreasing cluster size and steps in the loops due to the reversal of surface spins at different fields. However, they did not consider the finite temperature effects. Also Altbir and co-workers [98] investigated the evolution of nanosize Co cluster under different thermalisation processes by MC simulations.

In what follows, we will present the results of extensive MC simulations [5, 99, 100] which aim at clarifying what is the specific role of the finite size and surface on the magnetic properties of the particle, disregarding the inter-particle interactions. In particular, we study the magnetic properties under a magnetic field and at finite temperature, thus extending other simulation works. In choosing the model, we have tried to capture the main features of real particles with the minimum ingredients allowing us to interpret the results without any other blurring effects.

3.1 One-particle model for γ-Fe$_2$O$_3$

In maghemite γ-Fe$_2$O$_3$ with the spinel structure described in Sec. 1, the spins interact via antiferromagnetic (AF) exchange interactions with the nearest neighbours on both sublattices and with an external magnetic field H, the corresponding Hamiltonian of the model being

$$\mathcal{H}/k_B = - \sum_{\alpha,\beta=T,O} \sum_{i,n=1} J_{\alpha\beta} S_i^\alpha S_{i+n}^\beta - h \sum_{\alpha=T,O} \sum_{i=1}^{N_\alpha} S_i^\alpha , \qquad (2)$$

where we have defined the field in temperature units as $h = \frac{\mu H}{k_B}$, with bb S and μ the spin value and magnetic moment of the Fe^{3+} ion, respectively. In our model, the Fe^{3+} magnetic ions are represented by Ising spins $S_i^\alpha = \pm 1$, which allows us to reproduce a case with strong uniaxial anisotropy while keeping computational efforts within reasonable limits. The maghemite values of the nearest neighbour exchange constants given in Sec. 1 are considered.

We have assumed two kinds of boundary conditions on the lattice defined above with the aim to study the changes induced by the finite size of the nanoparticle in the magnetic properties with respect to the bulk material. By using periodic boundary (PB) conditions for systems with large enough linear size N (linear sizes are measured in multiples of the unit cell length), we will simulate the bulk behaviour. When studying a finite-sized particle, we will cut the lattice in the shape of a sphere of diameter D (measured in multiples of the unit cell length) with free boundary (FB) conditions. Moreover, in the last case, we distinguish two regions in the particle: the surface formed by the outermost unit cells, and a core formed by a sphere of spins with diameter D_{Core} unit cells. The particle sizes considered range from $D = 3$ to $D = 10$, corresponding to real diameters from 25 to 83 Å. The different measured magnetisations M_{Unc} are given in normalised units with respect to the number of uncompensated spins, $M_{Unc} = (N_O - N_T)/N_{Total}$ the ratio of the difference of O and T spins to the total number of spins N_{Total}, that for an infinite lattice with PB conditions is 1/3.

3.2 Equilibrium properties: surface and core magnetisations

Let us begin by studying the finite-size and surface effects on the equilibrium magnetic properties. The simulation protocol uses the standard Metropolis algorithm and proceeds in the following way. Starting from a disordered

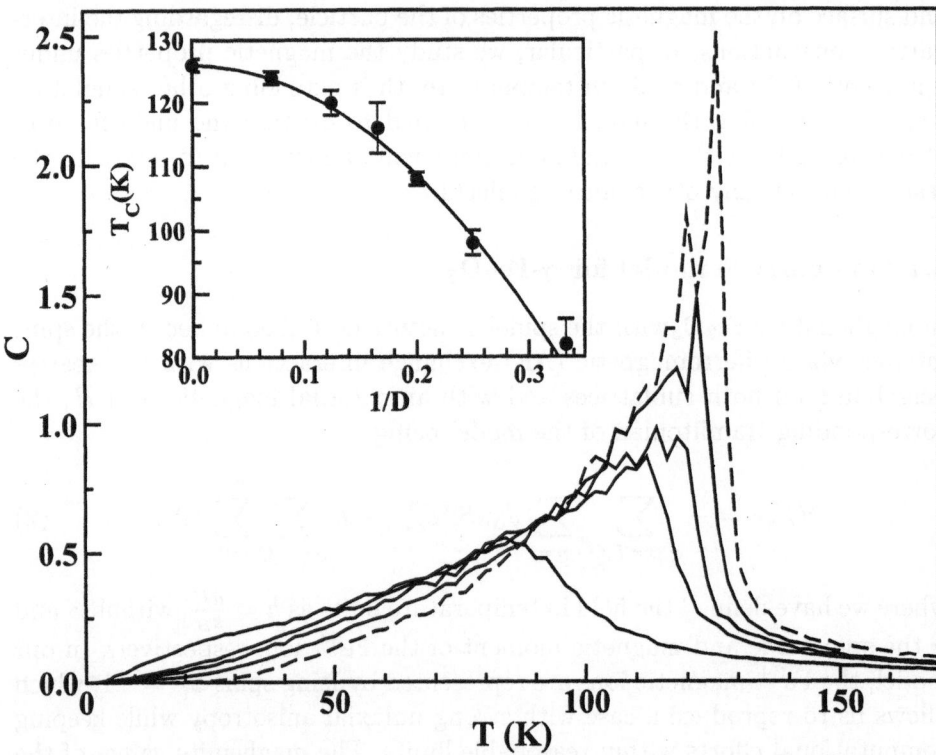

Fig. 11. Thermal dependence of the specific heat C for different diameters $D =$ 3, 6, 8, 14 (from the uppermost curve) and periodic boundary conditions for $N = 14$ (dashed curve). Inset: Particle size dependence of the transition temperature T_c from paramagnetic to ferrimagnetic phases for spherical particles with FB. The displayed values have been obtained from the maximum in the specific heat. The continuous line is a fit to Eq. 3.

configuration of spins at a high temperature ($T = 200$ K), the temperature is progressively reduced in steps $\delta T = -2$K. At each temperature, thermal averages of the thermodynamic quantities were performed during 10000 to 50000 MC steps after discarding the first 1000 MC steps for thermalisation.The quantities monitored during the simulation are the energy, specific heat, susceptibility, and different magnetisations: sublattice magnetisations (M_O, M_T), surface, core and total magnetisation ($M_{Surf}, M_{Core}, M_{Total}$). Note that with the above-mentioned normalisation, M_{Total} is 1 for ferromagnetic order, 0 for a disordered system and 1/3 for ferrimagnetic order of the O and T sublattices.

   In Fig. 11, we present the thermal dependence of the specific heat, C, for different particle diameters and we compare it to the PB case for $N = 14$. In all the cases, the sharp peak in C at the critical ordering temperature $T_C(D)$ is a sign of a second order transition from paramagnetic to ferrimagnetic order. As we can see, finite-size effects are clearly noticeable in the FB case

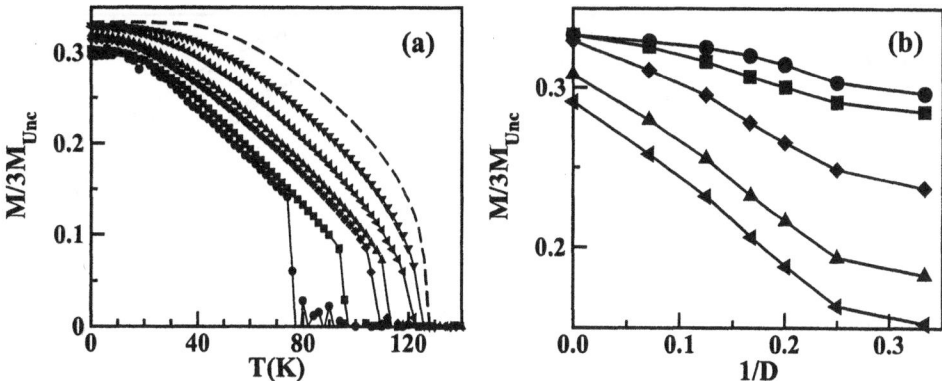

Fig. 12. (a) Thermal dependence of the magnetisation M. The results for particle diameters $D = 3, 4, 5, 6, 8, 14$ (from the lowermost curve in circles) and PB conditions $N = 14$ (dashed line) are shown. (b) Size dependence of the magnetisation of a spherical particle at different temperatures $T = 0, 20, 40, 60, 70$ K (from upper to lowermost curves). M_{Unc} is the ratio of the difference of O and T spins to the total number of spins.

even for D's as large as $D = 14$. In particular, $T_C(D)$ increases as the particle size is increased, approaching the infinite-size limit which can be estimated from the PB curve for $N = 14$ as $T_c(\infty) = 126 \pm 1$ K, as shown in the Inset of Fig. 11. Finite-size scaling theory [101, 102, 103] predicts the following scaling law for $T_c(D)$

$$\frac{T_c(\infty) - T_c(D)}{T_c(\infty)} = \left(\frac{D}{D_0}\right)^{-1/\nu} . \tag{3}$$

This expression fits nicely our MC data with $D_0 = 1.86 \pm 0.03$ being a microscopic length scale (in this case, it is roughly twice the cell parameter), and a critical exponent $\nu = 0.49 \pm 0.03$, which seems to indicate a mean field behaviour [104, 105]. This result can be ascribed to the high coordination of the O and T sublattices. The fitted curve is drawn in Fig. 11 where deviations from scaling are appreciable for the smallest diameters for which corrections to the finite-size scaling in Eq. 3 may be important [102]. Thus, these results discard any important surface effect on the ordering temperature and are consistent with spin-wave calculations [106] and old MC simulations [107]. Similar finite-size effects have been found in fine particles [108] of $MnFe_2O_4$, but with a surprising increase of $T_c(D)$ as D decreases, which has been attributed to surface effects due to the interactions with the particle coating.

In order to better understand how the finite-size effects influence the magnetic order it is also interesting to look at the thermal dependence of the magnetisation for different particle sizes. In Fig. 12a, we compare the results for spherical particles with different D to the bulk behaviour (system with $N = 14$ and PB). The most significant feature observed is the reduction in the total magnetisation with respect to the PB case due to the lower coor-

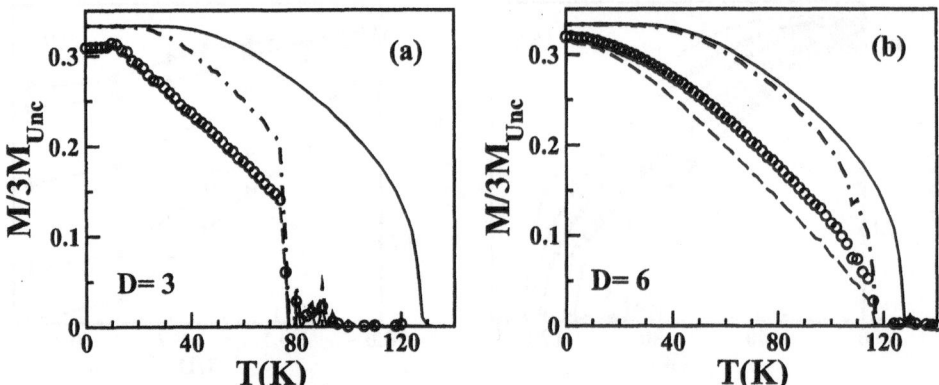

Fig. 13. (a) Thermal dependence of the magnetisation M. The results for two particle diameters are shown: $D = 3, 6$ (left and right-hand panels respectively). The contributions of the surface (dashed line, for $D = 3$ it cannot be distinguished from the total magnetisation) and core spins (dot-dashed line) have been distinguished from the total magnetisation (circles). The results for PB conditions, in a system of linear size $N = 14$, have also been included for comparison (continuous line).

dination at the surface. Moreover, by plotting separately M_{Core} and M_{Surf} (see Fig. 13), we can differentiate the roles played by the surface and core in establishing the magnetic order. On the one hand, the core (dash-dotted lines) tends to attain perfect ferrimagnetic alignment ($M = 1/3$) at low T independently of the particle size, whereas for the surface spins (dashed lines) a rapid thermal demagnetisation is observed that makes M_{Surf} depart from the bulk behaviour. The behaviour of M_{Total} is strongly dominated by the surface contribution for the smallest studied sizes ($D = 3, 4$) and progressively tends to the bulk behaviour as the particle size is increased (see Fig. 13b).

Notice also that for all the studied sizes there is a region of linear dependence of the demagnetisation curve at intermediate T that becomes wider as the particle size is decreased. In this region, the behaviour of the core is strongly correlated to that of the surface. This effect is indicative of a change from 3D to 2D effective behaviour of the surface layer of spin and has previously been observed in thin film systems [109, 110] and in simulations of rough FM surfaces [111].

In contrast to the size dependence of the ordering temperature, the spontaneous magnetisation M_{Total} follows, at any temperature, a quasi-linear behaviour with $1/D$ (see Fig. 12b) indicating that the reduction of M_{Total} is simply proportional to the ratio of surface-to-core spins, so it is mainly a surface effect. This is consistent with the existence of a surface layer of constant thickness Δr independent of D and with reduced magnetisation with respect to the core. With these assumptions, the size dependence of M can be expressed as

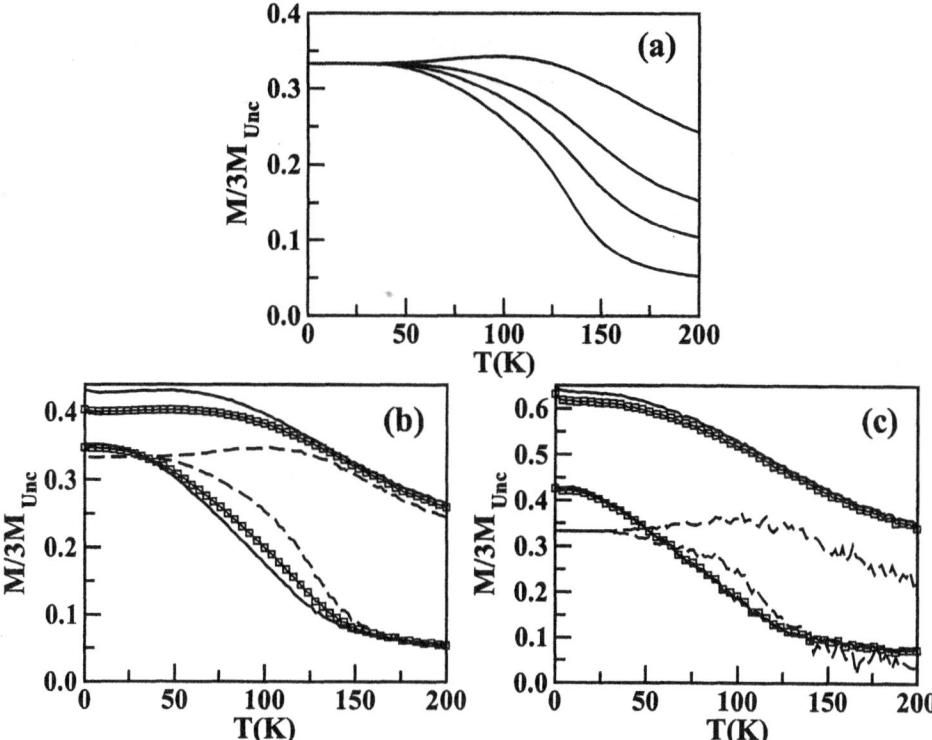

Fig. 14. Thermal dependence of M after cooling under h_{FC}. (a) Curves corresponding to a system with PB conditions and after cooling at $h_{FC} = 20, 40, 80, 100$ K
(from lower to uppermost curves). (b) Corresponding curves for a spherical particle with $D = 3$. The results for two cooling fields $h_{FC} = 20$ K (lower curve) and
$h_{FC} = 100$ K (upper curve) are shown. The contributions of the surface (thick
lines) and the core (dashed lines) to the total magnetisation (squares) have been
plotted separately. (c) The same than in (b) for $D = 6$.

$$M(D) = M_{Core} - \Delta M \frac{\Delta r S}{V} = M_{Core} - \Delta M \frac{6\Delta r}{D} , \qquad (4)$$

where S and V are the surface and volume of the particle, and $\Delta M =
M_{Core} - M_{Surface}$. Similar experimental behaviour has been found in γ-Fe_2O_3
[112], microcrystalline $BaFe_{10.4}Co_{0.8}Ti_{0.8}O_{19}$ [11] and the above-mentioned
$MnFe_2O_4$ system [108].

Deeper insight on the magnetic ordering of this system can be gained
by studying the thermal dependence of the equilibrium magnetisation in a
magnetic field. Several such curves obtained by the same cooling procedure
as described previously in the presence of different fields h_{FC} are presented
in Fig. 14. For particles of finite size, the curves at different fields do not
converge to the ferrimagnetic value at low T, reaching higher values of the
magnetisation at $T = 0$ the higher the cooling field (see lines with squares in
Fig. 14b,c). The total magnetisation for small particles is completely domi-

nated by the surface contribution and this is the reason why the ferrimagnetic order is less perfect at these small sizes and the magnetic field can easily magnetise the system. This is in contrast with the results for PB for which the system reaches perfect ferrimagnetic order (i.e. $M = 1/3$) even in cooling fields as high as 100 K (see Fig. 14a). Moreover, for this case, a maximum appears at high enough cooling fields $h_{FC} = 100$ K, which is due to the competition between the ferromagnetic alignment induced by the field and the spontaneous ferrimagnetic order (as the temperature is reduced, the strength of the field is not enough to reverse the spins into the field direction).

At low fields, the surface is in a more disordered state than the core since its magnetisation lies below M at temperatures for which the thermal energy dominates the Zeeman energy of the field. In contrast, a high field is able to magnetise the surface more easily than the core due to the fact that the broken links at the surface worsen the ferrimagnetic order while the core spins align towards the field direction in a more coherent way.

As a conclusion of the $M(T)$ dependence obtained in our simulations let us emphasize that, in spherical particles, there is a surface layer with much higher degree of magnetic disorder than the core, which is the Ising version of the random canting of surface spins occurring in several fine particles with spinel structure [17, 52, 18, 113, 114]. However, our model does not evidence the freezing of this disordered layer in a spin-glass-like state below any temperature and size, in contrast with the suggestions given by some authors [21, 24] and some experimental findings [21, 23, 53], for instance, the observation of shifted loops after an FC process (see Fig. 6 and Refs. [23, 115, 25, 26]). Furthermore, the surface layer, by partially breaking the ferrimagnetic correlations, diminishes the zero-field M_{Total} but, at the same time, enhances M_{Total} at moderate fields. Although the surface is easily thermally demagnetised and more easily magnetised by the field than the core is, it does not behave as a dead layer, since, at any T, it is magnetically coupled to the core. All these facts indicate that the surface has higher magnetic response than the core, excluding a spin-glass freezing. Moreover, we have not observed irreversibilities between field and zero-field cooled magnetisation curves in contrast to the experimental observations, which is a key signature that within the scope of our model, neither finite-size nor surface effects are enough to account for the spin-glass-like state.

3.3 Hysteresis loops

More insight into the magnetisation processes can be gained by studying the thermal and size dependence of hysteresis loops which, on the other hand, are the more common kind of magnetic measurements. In order to simulate the loops, the system has been initially prepared in a demagnetised state at $h = 0$ and the field subsequently increased in constant steps $\delta h = 1$ K up to a point well beyond the irreversibility field, from which the field is cycled. Measurements of the different magnetisation contributions are performed at

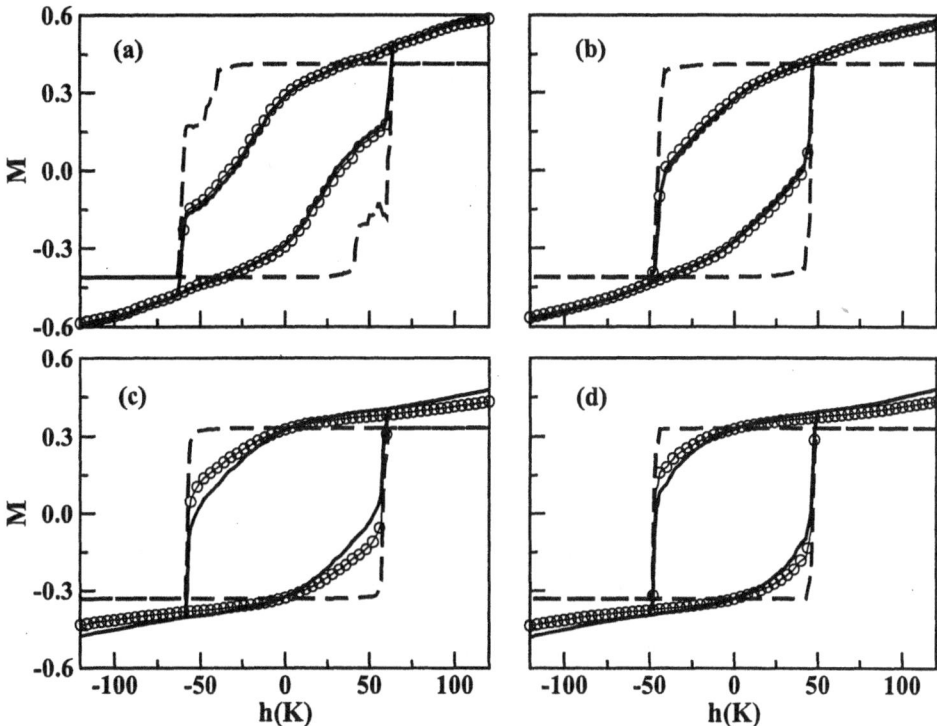

Fig. 15. Surface (continuous line), core (dashed line) and total (circles) contributions to the magnetisation of the hysteresis loops for particles of diameters $D = 3$, $T = 10$ K (a); $D = 3$, $T = 20$ K (b); $D = 6$, $T = 10$ K (c); $D = 6$, $T = 20$ K (d).

each step during 3000 MCS and the results averaged for several independent runs.

In Fig. 15, we show the core and surface contributions to the hysteresis loops for two particles sizes $D = 3, 6$ and temperatures $T = 10, 20$ K. First of all, let us notice that the shape of the loops for M_{Total} reproduces qualitatively the main features observed experimentally: high field susceptibilities and increasing saturation fields as the particle size is decreased. Both facts can be attributed to the progressive alignment of surface spins towards the field direction. The loops of the smallest particles resemble those found in ferrimagnetic nanoparticles [3, 17, 24] and other bulk systems with disorder [116, 117], increasing their squaredness (associated with the reversal of M as a whole) with the size.

Secondly, the panels shown in Fig. 15 allow us to distinguish between the very different roles played by the surface and core in the reversal process. On the one hand, the particle core presents an almost perfect squared loop (dashed lines) independently of the particle size, which indicates that the core spins reverse in a coherent fashion so that, in spite of the AF inter- and intra-lattice interactions, one can consider the interior of the ferrimagnetic particle behaving as a single superspin with magnitude equal to the number

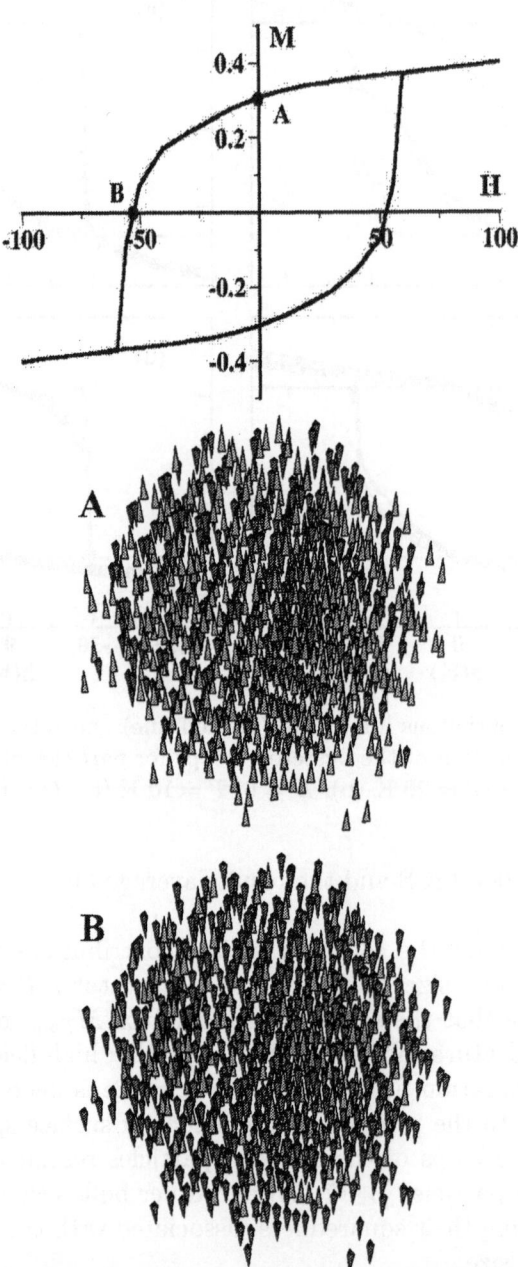

Fig. 16. Spin configurations of a particle of diameter $D = 4$ at different points of the hysteresis loop for $T = 20$ K. Configuration A corresponds to the remanent magnetisation state whereas B corresponds to the zero magnetisation state.

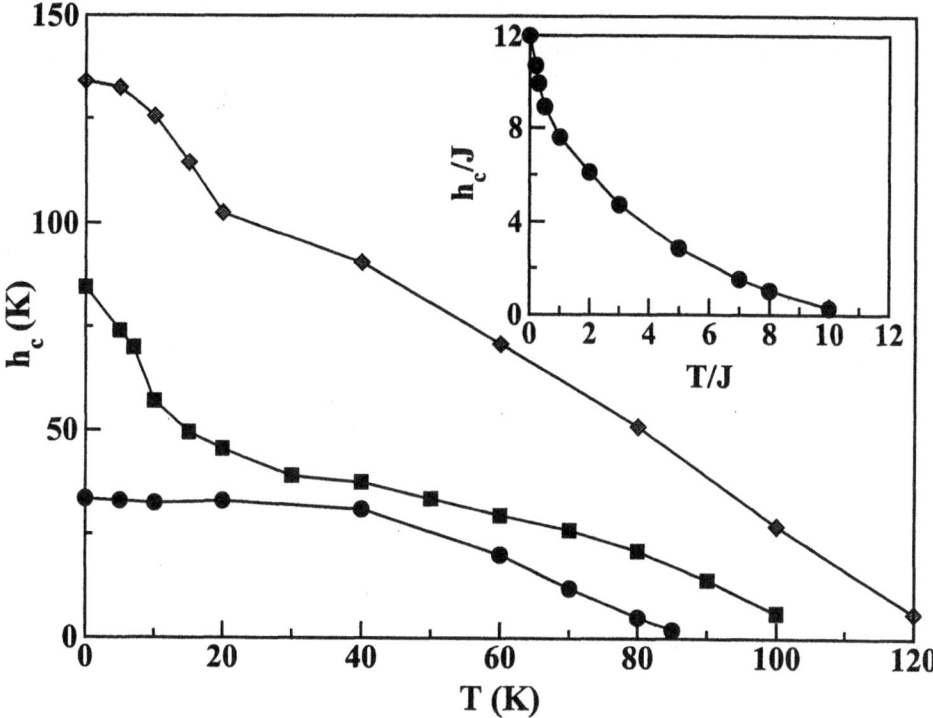

Fig. 17. Temperature dependence of the coercive field h_c for the real AF values of the exchange constants for maghemite for the case of FB spherical particles of diameters $D = 3$ (circles), $D = 6$ (squares), and for a system of linear size $N = 8$ with PB conditions (diamonds). Inset: Temperature dependence of the coercive field h_c for a system with the same structure as maghemite but ferromagnetic interactions ($J_{\alpha\beta} = J$), PB conditions and $N = 8$.

of uncompensated spins. On the contrary, the surface contribution dominates the reversal behaviour of the particle (compare the continuous curves with those in circles) and reveals a progressive reversal of M, which is a typical feature of disordered and frustrated systems [116, 117]. Nonetheless, for a wide range of temperatures and particle sizes, it is the reversal of the surface spins which triggers the reversal of the core. This is indicated by the fact that the coercive field of the core is slightly higher but very similar to the one of the surface. The surface spin disorder can be clearly observed in the snapshots of configurations taken along the hysteresis loop presented in Fig. 16, at the remanence point and at the coercive field (named A and B in the figure), where the different degree of disorder of the surface with respect to the core is more evident.

The thermal dependence of the coercive field $h_c(T)$ for all the studied particle sizes, at difference with ferromagnetic particles, shows a complex behaviour mainly related to the frustrating character of the antiferromagnetic interactions and the non-trivial geometry of the spinel spin lattice. Therefore,

whereas simulations for a system with PB conditions and with the same structure as maghemite but equal interspin FM interactions show a monotonous
decrease of h_c with increasing T (see the innset in Fig. 17), the results for
real maghemite display different thermal variations depending on the particle
size (see Fig. 17). The FM case, $h_c(T)$ presents a power law decay for high
enough temperatures $(T/J \gtrsim 1)$, that can be fitted to the expression

$$h_c(T) = h_c(0)[1 - (T/T_c)^{1/\alpha}] \, , \tag{5}$$

with $\alpha = 2.26 \pm 0.03$; close but different from what would be obtained by
a model of uniform reversal such as Stoner-Wohlfarth [118] ($\alpha = 2$). Thus,
we see that, even in this simple case for which M reverses as a whole, the
thermal variation of $h_c(T)$ cannot only be ascribed to the thermal activation
of a constant magnetisation vector over an energy barrier landscape, since
actually M is of course temperature dependent. In contrast with the FM case,
the $h_c(T)$ dependence for maghemite particles with PB or spherical shape,
presents opposite curvature and two regimes of variation. In the PB case at
high T, data can also be fitted to the power law of Eq. 5 for $T \gtrsim 20$ K with
$\alpha = 0.94 \pm 0.02$, $h_c(0) = 134 \pm 2$ K. Values of α close to 1 have been deduced
in the past for some models of domain wall motion [119, 120]. However, at low
T, a different regime is entered but tending to the same $h_c(0)$ value. In this
regime, the hysteresis loops become step-like and this change in behaviour is
associated with the wandering of the system through metastable states with
$M_{Total} \simeq 0$, which are induced by the frustration among AF interactions [5].

Finally, the thermal dependence for spherical maghemite particles strongly
depends on the particle size. Notice that h_c values are always smaller than
the bulk (PB) values independently of D because, for a finite particle, the
existence of surface spins with reduced coordination acts as a seed for the
reversal of the rest of the particle, which is not the case for PB, where all
equivalent spins have the same coordination. This fact explains why the h_c
values for PB are only recovered at low T in the limit of large particle size, although the decay of h_c is similar to that of the bulk for high T. In this regime,
h_c is dominated by the surface as indicated by the similarity of the core and
surface h_c already pointed out in Fig. 15. Instead, the low T ($T \lesssim 20$) regime
becomes dominated by the core contribution as $h_c^{Surf} < h_c^{Core}$ for any particle size (see Fig. 15). However, for the $D = 3$ particle, the prevalence of the
core is hindered due to the small ratio of core-to-surface spins in this case,
causing the saturation of h_c when lowering the temperature.

3.4 Effects of the disorder

In real maghemite particles, disorder and imperfections cause the system to
depart from perfect stoichiometry and distort the position of the atoms on the
lattice, their effect being more important at the surface [55]. There are several
ways to implement this disorder on the model. The simplest way to simulate

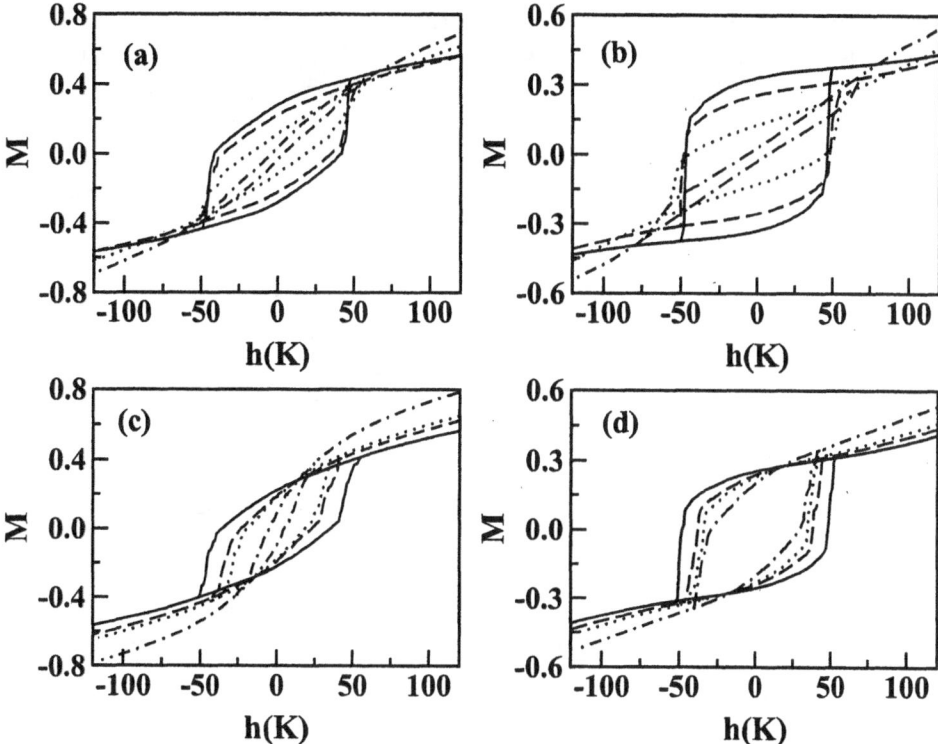

Fig. 18. Upper panels: hysteresis loops for systems with vacancy concentrations $\rho_v=$ 0.0, 0.166, 0.4, 0.6 (from outer to innermost) on the O sublattice at $T = 20$ K. Particle diameters $D = 3$ (a), $D = 6$ (b). Results have been averaged over 10 disorder realisations. Lower panels: hysteresis loops for systems with vacancy densities on the surface of the O and T sublattices $\rho_{sv}=$ 0, 0.1, 0.2, 0.5, vacancy density $\rho_{sv} = 0.1666$ on the O sublattice, and $T = 20$ K. Particle diameters $D = 3$ (c), $D = 6$ (d). Results have been averaged over 10 disorder realisations.

the deviation of the O and T sublattice atoms from ideal stoichiometry is by random removal of magnetic ions on the O and T sublattices.

Lattice disorder

The ideal maghemite lattice presents one sixth of randomly distributed vacancies on the O sublattice in order to achieve charge neutrality that, however, in real samples, can vary depending on the conditions and method of preparation or the size of the particles [55, 121, 122]. In the model presented so far, we have only considered perfect lattices, now we will study the effect of vacancies on the magnetic properties. Given the dominance of AF intersublattice interactions in maghemite, the inclusion of vacancies on one of the sublattices destabilises the perfect FM intralattice order of the other and therefore may result in a system with a greater degree of magnetic disorder. Since $N_{TO} > N_{OT}$ (see Section 1), this effect will be stronger when vacencies

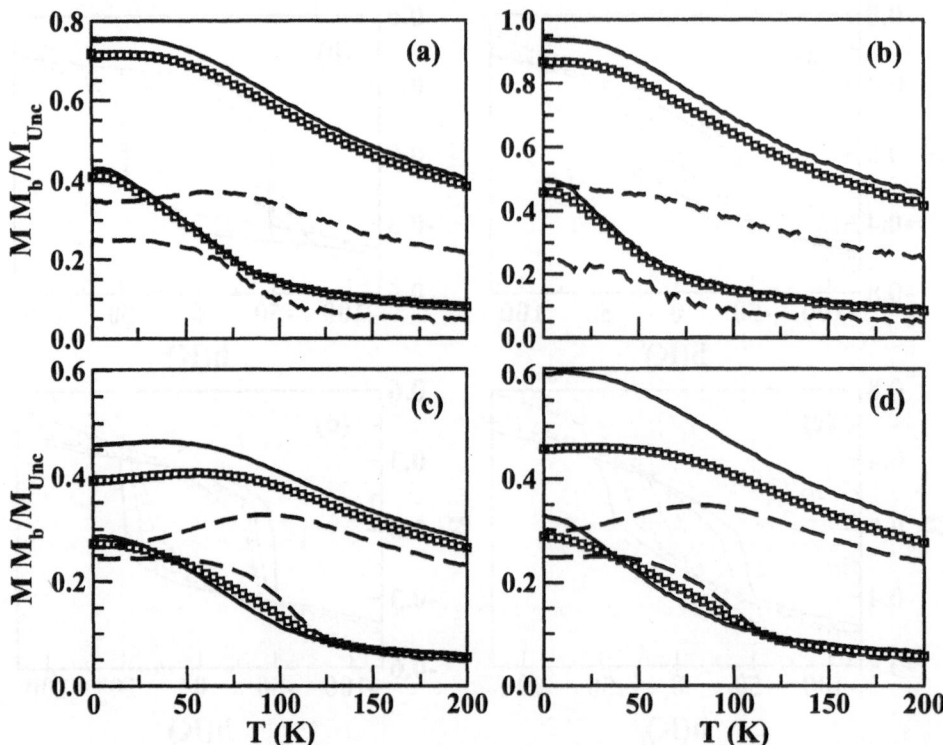

Fig. 19. Thermal dependence of M after cooling under a magnetic field for a spherical particle with $D = 3$ (upper panels) and $D = 6$ (lower panels), with vacancy densities on the surface of the O and T sublattices $\rho_{sv} = 0.2$ (a) and (c), $\rho_{sv} = 0.5$ (b) and (d), and $\rho_v = 0.166$ on the O sublattice. The results for two cooling fields $h_{FC} = 20, 100$ K (lower and upper curves respectively in each pannel) are shown. The contributions of the surface (thick lines) and the core (dashed lines) to the total magnetisation (squares) have been plotted separately. The magnetisation has been normalised to M_b, the magnetisation of a perfect ferrimagnetic configuration for a system of infinite size.

are introduced in the O sublattice. In the following, we will refer to ρ_v as the vacancy concentration on this sublattice.

To show the effect of these kinds of disorder, we have simulated the hysteresis loops for different ρ_v at two cooling fields $h_{FC} = 20, 100$ K. As can be seen in the upper panels of Fig. 18, the introduction of a low concentration of vacancies ($\rho_v = 1/6$ as in the real material) results in a reduction in the magnetisation and in an increase of the high field susceptibility, without any substantial change in the general shape of the loops. However, if ρ_v is increased beyond the actual value, the loops progressively close, loosing squareness and progressively resembling those for a disordered system [116, 117], with high values of the high field susceptibilities and much lower coercivity.

Surface disorder

Let us now consider the effects of the disorder at the surface of the particle, taking a $\rho_v = 1/6$ vacancy density on the O sublattice. Since the surface of the particles is not an ideal sphere, the outermost unit cells may have an increased number of vacancies on both sublattices with respect to those present in the core. Reduced coordination at the surface may also change the number of links between the surface atoms. We will denote by ρ_{sv} the concentration of surface vacancies in the outermost primitive cells.

As for the case with no vacancies, the thermal dependence of the magnetisation in the presence of a magnetic field h_{FC} helps to characterise the magnetic ordering of the system. Several such curves are shown in Fig. 19, in which the surface (continuous lines) and the core (dashed lines) contributions to the total magnetisation (open symbols) have been distinguished. The introduction of vacancies does not change the low field behaviour of the total magnetisation, which is still dominated by the surface for both $D = 3, 6$, although the smallest particles are easily magnetised by the field. However, at high fields, M_{Total} is lower than M_{Surf} and the surface progressively decouples from M_{total} with the introduction of vacancies in the surface, this effect being more remarkable for the biggest particle. With respect to the core, at difference with the non-disordered case ($\rho_v = \rho_{sv} = 0$), the low temperature plateau of M_{Core} tends to a higher value than that for perfect ferrimagnetic order, since the main effect of the disorder is to break ferrimagnetic correlations in the core; increasing the ferromagnetic order induced by the field. This is reflected in a progressive departure of the high and low field M_{Core} curves with increasing disorder (see the dashed lines). The maximum appearing at high h_{FC} is only slightly affected by disorder, shifting to lower T and eventually disappearing for $D = 3$ and $\rho_{sv} = 0.5$.

Hysteresis loops with surface disorder are given in Fig. 18c,d for two particle diameters. The introduction of surface vacancies facilitates the magnetisation reversal by progressive rotation, producing a rounding of the hysteresis loops when approaching h_c, in the same way as occurs when particle size is reduced. The same fact explains the increase of the high field susceptibility, since the vacancies act as nucleation centers of FM domains at the surface, which, from there on, extend the FM correlations to the inner shells of spins. Moreover, a considerable decrease of h_c is observed. All these facts yield a progressive elongation of the loops, giving loop shapes resembling those of disordered systems [116, 117]. The lower panels in Fig. 18, where the surface and core contributions are shown separately, clearly evidence that the increase of FM correlations at the surface, facilitated by the vacancies, induces FM order in the core. That is to say, M_{Core} follows the evolution of M_{Surf} at moderate fields above h_c, in contrast with the case with no surface vacancies, where the core keeps the ferrimagnetic order for the same field range.

4 Open questions and perspectives

We would like to conclude with an account of open questions and perspectives for future work. With respect to experiments, future work should be aimed at distinguishing which part of the effective energy barrier distribution in interacting fine-particle systems originates from the dipolar interactions and which one comes from frustration at the particle surface. In this respect, experiments in the spirit of those performed by Salling et al. [123], measuring the switching fields of individual ellipsoidal γ-Fe$_2$O$_3$ particles by Lorentz magnetometry, or by Wernsdorfer and co-workers [26, 124], measuring magnetisation processes of a single particle by using microSQUIDs, should be encouraged. This would clarify the surface contribution. The synthesis of diluted samples with well-isolated particles, such as fluid suspensions or a solid matrix, would allow the study of shifted loop effects and the contribution of the surface to the irreversibility and closure fields. In order to better characterise the intrinsic interaction effects on the formation of a collective state, the behavior of ferrimagnetic oxide systems could be compared to that of interacting metallic particles for which the surface glassy state is not observed.

From the point of view of simulations it would be interesting to extend the results presented here for Ising spins to a model with Heisenberg spins with finite anisotropy. This would allow one to account for the important role played by surface anisotropy on the magnetisation processes, a well-known phenomenon first described by Néel back in the 50's [71] and that arises from the breaking of the crystalline symmetry at the boundaries of the particle. This symmetry breaking changes also the effective exchange interactions at surface atoms with respect to the bulk values, a fact that could be easily incorporated in a model for MC simulation to study its possible influence on the appearance of shifted loops observed experimentally and that is not obtained within the scope of the Ising model.

5 acknowledgments

We are indebted to Dra. Montse García del Muro for her contribution to some parts of the work reported in this chapter. We also acknowledge CESCA and CEPBA under coordination of C^4 for the computer facilities. This work has been supported by SEEUID through project MAT2000-0858 and Catalan DURSI under project 2001SGR00066.

References

1. S. Krupicka and P. Novak, in *Ferromagnetic Materials*, edited by E. P. Wohlfarth (North-Holland, The Netherlands, 1982), Vol. 3, Chap. 4, pp. 189–304.
2. P. W. Anderson, Phys. Rev. **102**, 1008 (1956).

3. R. H. Kodama and A. E. Berkowitz, Phys. Rev. B **59**, 6321 (1999).
4. H. Kachkachi, A. Ezzir, M. Nogués, and E. Tronc, Eur. Phys. J. B **14**, 681 (2000).
5. O. Iglesias and A. Labarta, Phys. Rev. B **63**, 184416 (2001).
6. S. Geller, J. Appl. Phys. **37**, 1408 (1966).
7. A. Rosencwaig, Can. J. Phys. **48**, 2857 (1970).
8. C. E. Patton and Y. H. Liu, J. Phys. C: Solid State Phys. **16** , 5995 (1983).
9. G. Albanese, M. Carbucicchio, and G. Asti, Appl. Phys. **11**, 81 (1976).
10. X. Obradors, A. Isalgue, A. Collomb, A. Labarta, M. Pernet, J. A. Pereda, J. Tejada, and J. C. Joubert, J. Phys. C: Solid State Phys. **19**, 6605 (1986).
11. X. Batlle, X. Obradors, M. Medarde, J. Rodríguez-Carvajal, M. Pernet, and M. Vallet, J. Appl. Phys. **124**, 228 (1993).
12. X. Batlle and A. Labarta, J. Phys. D: Appl. Phys. **35**, R15 (2002).
13. H. Mamiya, I. Nakatani, and T. Furubayashi, Phys. Rev. Lett. **80**, 177 (1998).
14. S. Mørup, Europhys. Lett. **28**, 671 (1994).
15. S. Mørup, F. Bødker, P. V. Hendriksen, and S. Linderoth, Phys. Rev. B **52**, 287 (1995).
16. F. T. Parker, M. W. Foster, D. T. Margulies, and A. E. Berkowitz, Phys. Rev. B **47**, 7885 (1993).
17. R. H. Kodama, A. E. Berkowitz, E. J. McNiff, and S. Foner, Phys. Rev. Lett. **77**, 394 (1996).
18. J. M. D. Coey, Phys. Rev. Lett. **27**, 1140 (1971).
19. Q. A. Pankhurst and R. J. Pollard, Phys. Rev. Lett. **67**, 248 (1991).
20. M. Respaud, J. M. Broto, H. Rakoto, A. R. Fert, L. Thomas, B. Barbara, M. Verelst, E. Snoek, P. Lecante, A. Mosset, J. Osuna, T. O. Ely, C. Amiens, and B. Chaudret, Phys. Rev. B **57**, 2925 (1998).
21. B. Martínez, X. Obradors, L. Balcells, A. Rouanet, and C. Monty, Phys. Rev. Lett. **80**, 181 (1998).
22. P. Prene, E. Tronc, J.-P. Jolivet, J. Livage, R. Cherkaoui, M. Nogues, J.-L. Dormann, and D. Fiorani, IEEE Trans. Magn. **29**, 2658 (1993).
23. M. García del Muro, X. Batlle, and A. Labarta, Phys. Rev. B **59**, 13584 (1999).
24. R. H. Kodama, S. A. Makhlouf, and A. E. Berkowitz, Phys. Rev. Lett. **79**, 1393 (1997).
25. M. P. Morales, S. Veintesmillas-Verdaguer, M. I. Montero, C. J. Serna, A. Roig, L. Casas, B. Martínez, and F. Sandiumenge, Chem. Mater. **11**, 3058 (1999).
26. E. Tronc, D. Fiorani, M. Nogués, A. M. Testa, F. Lucari, F. D'Orazio, J. M. Grenèche, W. Wernsdorfer, N. Galvez, C. Chenéac, D. Mally, and J. Jolivet, J. Magn. Magn. Mat. **262**, 6 (2003).
27. J. Dormann, R. Cherkaoui, L. Spinu, M. Nogués, F. Lucari, F. D'Orazio, D. Fiorani, A. García, E. Tronc, and J. Jolivet, J. Magn. Magn. Mat. **187**, L139 (1998).
28. C. Djurberg, P. Svedlindh, P.Nordblad, M. F. Hansen, F. Bødker, and S. Mørup, Phys. Rev. Lett. **79**, 5154 (1997).
29. T. Jonsson, J. Mattsson, C. Djurberg, F. Khan, P. Nordblad, and P. Svedlindh, Phys. Rev. Lett. **75**, 4138 (1995).
30. H. Mamiya, I. Nakatani, and T. Furubayashi, Phys. Rev. Lett. **82**, 4332 (1999).
31. T. Jonsson, P. Nordblad, and P. Svedlindh, Phys. Rev. B **57**, 497 (1998).
32. P. Jonsson and P. Nordblad, Phys. Rev. B **62**, 1466 (2000).

33. M. P. Sharrock and L. Josephson, IEEE Trans. Magn. **22**, 723 (1986).
34. T. Fujiwara, IEEE Trans. Magn. **23**, 3125 (1987).
35. M. P. Sharrock and L. Josephson, IEEE Trans. Magn. **25**, 4374 (1989).
36. M. H. Kryder, J. Magn. Magn. Mat. **83**, 1 (1990).
37. H. Kojima, in *Ferromagnetic Materials*, edited by E. P. Wohlfarth (North-Holland, Amsterdam, 1982), Vol. 3, Chap. 5, p. 305.
38. A. Labarta, X. Batlle, B.Martínez, and X. Obradors, Phys. Rev. B **46**, 8994 (1992).
39. X. Batlle, X. Obradors, J. Rodríguez-Carvajal, M. Pernet, M. V. Calanas, and M. Vallet, J. Appl. Phys. **70**, 1614 (1991).
40. P. Görnert, E. Sinn, and M. Rössler, Key Eng. Mater. **58**, 129 (1991).
41. P. Görnert, E. Sinn, H. Pfeiffer, W. Scüppel, and M. Rössler, J. Magn. Soc. Jpn. **15**, 699 (1991).
42. X. Batlle, M. García del Muro, A. Labarta, and P. Görnert, J. Magn. Magn. Mat. **157-158**, 191 (1996).
43. J. L. Tholence, Solid State Commun. **35**, 113 (1980).
44. A. P. Malozemoff and E. Pytte, Phys. Rev. B **34**, 6579 (1986).
45. D. P. Fisher, J. Appl. Phys. **61**, 3672 (1987).
46. P. C. Hohenberg and B. I. Halperin, Rev. Mod. Phys. **49**, 435 (1977).
47. L. Lundgreen, P. Svendlindh, P. Nordblad, and O. Beckman, Phys. Rev. Lett. **51**, 911 (1983).
48. A. Labarta, R. Rodríguez, L. Balcells, J. Tejada, X. Obradors, and F. J. Berry, Phys. Rev. B **44**, 691 (1991).
49. E. Tronc, A. Ezzir, R. Cherkaoui, C. C. eac, M. Nogués, H. Kachkachi, D. Fiorani, A. M. Testa, J. M. Grenèche, and J. Jolivet, J. Magn. Magn. Mat. **221**, 63 (2000).
50. A. E. Berkowitz, W. J. Shuele, and P. J. Flanders, J. Appl. Phys. **39**, 1261 (1968).
51. K. Haneda, H. Kojima, A. H. Morrish, P. J. Picone, and K. Wakai, J. Appl. Phys. **53**, 2686 (1982).
52. D. Lin, A. C. Nunes, C. F. Majkrzak, and A. E. Berkowitz, J. Magn. Magn. Mat. **145**, 343 (1995).
53. F. Gazeau, E. Dubois, M. Hennion, R. Perzynski, and Y. L. Raikher, Europhys. Lett. **40**, 575 (1997).
54. F. Gazeau, J. C. Bacri, F. Gendron, R. Perzynski, Y. L. Raikher, V. I. Stepanov, and E. Dubois, J. Magn. Magn. Mat. **186**, 175 (1998).
55. M. P. Morales, C. J. Serna, F. Bødker, and S. Mørup, J. Phys.: Condens. Matter **9**, 5461 (1997).
56. S. Linderoth, P. V. Hendriksen, F. Bödker, S. Wells, K. Davis, S. W. Charles, and S. Mörup, J. Appl. Phys. **75**, 6583 (1994).
57. J. Nogués and I. K. Schuller, J. Magn. Magn. Mat. **192**, 203 (1999).
58. R. L. Stamps, J. Phys. D: Appl. Phys. **33**, R247 (2000).
59. R. H. Kodama, A. E. Berkowitz, E. J. McNiff, and S. Foner, J. Appl. Phys. **81**, 5552 (1997).
60. Y. A. Koksharov, S. P. Gubin, I. D. Kosobudsky, G. Y. Yurkov, D. A. Pankratov, L. A. Ponomarenko, M. G. Mikheev, M. Beltran, Y. Khodorkovsky, and A. M. Tishin, Phys. Rev. B **63**, 012407 (2000).
61. D. A. Dimitrov and G. M. Wysin, Phys. Rev. B **50**, 3077 (1994).
62. D. A. Dimitrov and G. M. Wysin, Phys. Rev. B **51**, 11947 (1995).

63. H. Kachkachi and M. Dimian, J. Appl. Phys. **91**, 7625 (2002).
64. H. Kachkachi and M. Dimian, Phys. Rev. B **66**, 174419 (2002).
65. Y. Labay, O. Crisan, L. Berger, J. M. Greneche, and J. M. D. Coey, J. Appl. Phys. **91**, 8715 (2002).
66. K. N. Trohidou and J. A. Blackman, Phys. Rev. B **41**, 9345 (1990).
67. D. Kechrakos and K. N. Trohidou, J. Magn. Magn. Mat. **177-181**, 943 (1998).
68. X. Zianni, K. N. Trohidou, and J. A. Blackman, J. Appl. Phys. **81**, 4739 (1997).
69. K. N. Trohidou, X. Zianni, and J. A. Blackman, J. Appl. Phys. **84**, 2795 (1998).
70. X. Zianni and K. N. Trohidou, J. Appl. Phys. **85**, 1050 (1999).
71. L. Neél, J. Phys. Radium **15**, 225 (1954).
72. D. Altbir, J. d'Albuquerque e Castro, and P. Vargas, Phys. Rev. B **54**, 6823 (1996).
73. D. V. Berkov, J. Magn. Magn. Mat. **117**, 431 (1992).
74. D. V. Berkov, Phys. Rev. B **53**, 731 (1996).
75. O. Iglesias, Ph.D. thesis, Universitat de Barcelona, 2002.
76. R. Ribas and A. Labarta, J. Appl. Phys. **80**, 5192 (1996).
77. S. Mørup and E. Tronc, Phys. Rev. Lett. **72**, 3278 (1994).
78. J. L. Dormann, L. Bessais, and D. Fiorani, J. Phys. C **21**, 2015 (1988).
79. A. Labarta, O. Iglesias, L. Balcells, and F. Badia, Phys. Rev. B **48**, 10240 (1993).
80. J. J. Préjean and J. Souletie, J. Phys. (Paris) **41**, 1335 (1980).
81. R. Omari, J. J. Préjean, and J. Souletie, J. Phys. (Paris) **45**, 11809 (1984).
82. O. Iglesias, F. Badia, A. Labarta, and L. Balcells, J. Magn. Magn. Mat. **140-144**, 399 (1995).
83. O. Iglesias, F. Badia, A. Labarta, and L. Balcells, Z. Phys. B **100**, 173 (1996).
84. L. Balcells, O. Iglesias, and A. Labarta, Phys. Rev. B **55**, 8940 (1997).
85. X. Batlle, M. García del Muro, and A. Labarta, Phys. Rev. B **55**, 6440 (1997).
86. J. García-Otero, M. Porto, J. Rivas, and A. Bunde, Phys. Rev. Lett. **84**, 167 (2000).
87. M. Porto, Eur. Phys. J. B **26**, 229 (2002).
88. M. Ulrich, J. García-Otero, J. Rivas, and A. Bunde, Phys. Rev. B **67**, 024416 (2003).
89. O. Iglesias and A. Labarta, J. Appl. Phys. **91**, 4409 (2002).
90. O. Iglesias and A. Labarta, Comput. Mater. Sci. **25**, 577 (2002).
91. P. Görnert, H. Pfeiffer, E. Sinn, R. Müller, W. Scüppel, M. Rössler, X. Batlle, M. García del Muro, J. Tejada, and S. Galí, IEEE Trans. Magn. **30**, 714 (1994).
92. G. G. Kenning, D. Chu, and R. Orbach, Phys. Rev. Lett. **66**, 2923 (1991).
93. P. Jönsson, S. Felton, P. Svedlindh, P. Nordblad, and M. F. Hansen, Phys. Rev. B **64**, 212402 (2001).
94. J. O. Andersson, C. Djurberg, T. Jonsson, P. Svedlindh, and P. Nordblad, Phys. Rev. B **56**, 13983 (1997).
95. H. Kachkachi, M. Nogués, E. Tronc, and D. A. Garanin, J. Magn. Magn. Mat. **221**, 158 (2000).
96. H. Kachkachi and D. A. Garanin, Physica A **300**, 487 (2001).
97. J. P. Bucher and L. A. Bloomfield, Phys. Rev. B **45**, 2537 (1992).
98. D. Altbir, P. Vargas, and J. d'Albuquerque e Castro, Phys. Rev. B **64**, 012410 (2001).

99. O. Iglesias, A. Labarta, and F. Ritort, J. Appl. Phys. **89**, 7597 (2001).
100. O. Iglesias, F. Ritort, and A. Labarta, in *Magnetic Storage Systems Beyond 2000*, Vol. 41 of *NATO ASI Series II*, edited by G. C. Hadjipanajyis (Kluwer Academic Press, Dordrecht, The Netherlands, 2001), p. 363.
101. K. Binder and P. C. Hohenberg, Phys. Rev. B. **9**, 2194 (1974).
102. D. P. Landau, Phys. Rev. B. **76**, 255 (1976).
103. M. N. Barber, in *Phase Transitions and Critical Phenomena*, edited by C. Domb and J. L. Lebowitz (Academic Press, New York, 1983), Vol. 8, p. 145.
104. K. Binder and P. C. Hohenberg, Phys. Rev. B. **6**, 3461 (1972).
105. H. E. Stanley, *Introduction to Phase Transitions and Critical Phenomena* (Oxford University Press, New York, 1987).
106. P. V. Hendriksen, S. Linderoth, and P. A. L. rd, Phys. Rev. B **48**, 7259 (1993).
107. K. Binder, H. Rauch, and V. Wildpaner, J. Phys. Chem. Solids **31**, 391 (1970).
108. Z. X. Tang, C. M. Sorensen, K. J. Klabunde, and G. C. Hadjipanayis, Phys. Rev. Lett. **67**, 3602 (1991).
109. B. Martínez and R. E. Camley, J. Phys.: Condens. Matter **4**, 5001 (1992).
110. J.-H. Park, E. Vescovo, H.-J. Kim, C. Kwon, R. Ramesh, and T. Venkatesan, Phys. Rev. Lett. **81**, 1953 (1998).
111. D. Zhao, F. Liu, D. L. Huber, and M. G. Lagall, Phys. Rev. B **62**, 11316 (2000).
112. D. H. Han, J. P. Wang, and H. L. Luo, J. Magn. Magn. Mat. **136**, 176 (1994).
113. A. H. Morrish and K. Haneda, J. Appl. Phys. **52**, 2496 (1981).
114. J. Z. Jiang, G. F. Goya, and H. R. Rechenberg, J. Phys.: Condens. Matter **11**, 4063 (1999).
115. , .
116. K. Binder and A. P. Young, Rev. Mod. Phys. **58**, 801 (1986).
117. H. Maletta, in *Excitations in Disordered Solids*, edited by M. Thorpe (Plenum, New York, 1981).
118. E. C. Stoner and E. P. Wohlfarth, Philos. Trans. R. Soc. London A **240**, 599 (1948), reprinted in *IEEE Trans. Magn.* **27**, 3475 (1991).
119. P. Gaunt, J. Appl. Phys. **59**, 4129 (1986).
120. P. Gaunt, Philos. Mag. B **48**, 261 (1983).
121. M. P. Morales, C. de Julián, J. M. G. alez, and C. J. Serna, J. Mater. Res. **9**, 135 (1994).
122. C. J. Serna, F. Bødker, S. Mørup, M. P. Morales, F. Sandiumenge, and S. Veintesmillas-Verdaguer, Solid State Commun. **118**, 437 (2001).
123. C. Salling, R. O'Barr, S. Schultz, I. McFadyen, and M. Ozaki, J. Appl. Phys. **75**, 7989 (1994).
124. M. Jamet, W. Wernsdorfer, C. Thirion, D. Mailly, V. Depuis, P. Mélion, and A. Pérez, Phys. Rev. Lett. **86**, 4676 (2001).

EFFECT OF SURFACE ANISOTROPY ON THE MAGNETIC RESONANCE PROPERTIES OF NANOSIZE FERROPARTICLES

Régine Perzynski

Laboratoire des Milieux Désordonnés et Hétérogènes, Université Pierre et Marie Curie, Case 78, Site de Boucicaut, 140 rue de Lourmel, 75015, Paris, France

Yuriy L. Raikher

Institute of Continuous Media Mechanics, Ural Division of the Russian Academy of Sciences, 1 Korolyov St., 614013, Perm, Russia

1. Introduction

Fine particles are unique physical objects with remarkable magnetic properties which differ greatly from their parent massive ferromagnets or ferrites. The source of this difference, which has numerous and diverse manifestations, is the fact that in fine particles surface effects are not at all negligible in comparison with the bulk ones. Notably, just this condition suffices to outline the range of spatial scales where the notion of *a fine magnetic particle* turns up.

All throughout this paper we discuss, estimate and analyze the equilibrium and dynamic properties of low-dimensional objects, thin films and fine particles made of a ferromagnet or ferrite. That is why from the very beginning we introduce the most important notations for "global" usage. In particular, we adopt the macroscopic (continuum) approach to magnetodynamics assuming simultaneously that the temperature is well below the Curie point. This framework had been proposed by Brown [1] and since then is known as *micromagnetism*. In this framework one may treat the magnetization of the ferromagnetic material as a vector of a constant length and denote it as

$$M = Me, \qquad (1.1)$$

thus introducing e, the unit vector of magnetization. Another unit vector, n, we associate with the uniaxial magnetic anisotropy whose direction is imposed either by the crystallographic structure of the particle or by the geometric anisometricity of its body. However, we suppose that this anisometricity is never too strong. That is why for estimates we take that each of the three dimensions

141

of the particle is close to d, the mean particle diameter; thence for the particle volume one gets $v \simeq \pi d^3/6$. Although we formally exclude the cases of acicular or disc-like particles, the estimation given below, if necessary, may be easily modified for these specific cases.

In the absence of an external field the volume density of the free energy of a magnetic material with a uniaxial internal anisotropy is written as

$$F = \frac{1}{2}\alpha M^2 (\nabla e)^2 - K_v v(e \cdot n)^2 - \frac{1}{2}M(e \cdot H_d). \qquad (1.2)$$

Here α is the so-called *exchange rigidity constant* which by the order of magnitude equals 10^{-12} cm^2; K_v is the volume density of the magnetic anisotropy energy, and is normally $K_v \sim 10^4 - 10^6$ erg/cm^3. The first term in Eq. (1.2) yields the macroscopic form of the non-uniform exchange interaction while the second renders the contribution of the spin-orbit interaction (the crystal field) that determines the energetically favorable directions of the magnetization inside the ferromagnetic (ferrite) sample. The last term yields the contribution of the so-called demagnetizing field H_d that is due to the "magnetic poles" emerging at the surface of any finite body either magnetized by an external field or possessing a spontaneous magnetization.

Therefore by choosing the bulk energy density in the form of Eq. (1.2) we imply that a sample of finite size is considered. Indeed, the presence of the demagnetizing, also called *the magnetostatic*, term is solely due to the assumed existence of a surface or interface where the magnetic poles appear. In the review paper [2] its authors justly propose to divide all the specific effects in fine particles in two classes. One class is formed by just the finite-size effects which are due exclusively to imposition of some cutoff length. Another class unites the effects which are due to the specific properties of the surface that differ from those of the bulk due to broken atomic bonds, deficiency of neighboring atoms, etc. In this context the very appearance of the magnetostatic term in Eq. (1.2) is a finite-size effect, see Ref. [1] for example.

By its definition, the demagnetizing term in Eq. (1.2) is non-zero for the cases where the particle is magnetized in such a way that it creates in the outer space a magnetic (*stray*) field. We remark that unlike the first two terms in Eq. (1.2), the magnetostatic term does not require that magnetization would be uniform over the sample or would be globally parallel to some internal axis. It is the interplay of the non-uniform exchange and the anisotropy energy on the one hand and the magnetostatic energy on the other hand which entails the most well-known magnetic surface effect: the existence of single-domain particles. Indeed, estimating the demagnetizing field as $H_d \sim -M$ (we recall that H_d is the field from the surface magnetic poles, so that near the surface its amplitude equals M), we find that due to the stray fields the particle energy is augmented by the decrement $\delta \mathcal{F}_{ms} \sim -H_d M v \sim \mu M \sim M^2 d^3$, where $\mu = Mv$ is the full magnetic moment of the particle.

Meanwhile, a ferromagnetic sample with a closed magnetic flux has a zero full magnetic moment and thus $\delta\mathcal{F}_{ms} = 0$. However, the spatial distribution of magnetization in such a particle is not uniform so that the energy increment originates from the non-uniform exchange term as

$$\delta\mathcal{F}_{ex} \sim \alpha(M^2/d^2)\, v \sim \alpha M^2 d; \qquad (1.3)$$

note that here we assume that $M^2 \gg K_v$ so that the sample is effectively isotropic. Apparently, the particle will be magnetized non-uniformly (split into domains) as long as $\delta\mathcal{F}_{ex} < \delta\mathcal{F}_{ms}$. Under the reversed condition, the state of the particle with minimal energy corresponds to a uniform magnetization, i.e., the particle becomes a single domain. Estimation for the critical size of single domain for a magnetically isotropic particle yields

$$d_* = \sqrt{\alpha} \sim 10^{-6}\,\mathrm{cm} = 10\,\mathrm{nm}. \qquad (1.4)$$

Estimation for the case of magnetically hard particles ($M^2 \ll K_v$) differs from the preceding one by the fact that the main term opposing the magnetostatic one is the uniaxial anisotropy. In this case, as the first step one needs to estimate the surface tension of the domain wall. On doing that, see Ref. [3], for example, instead of Eq. (1.4) one gets

$$d_* = \sqrt{\alpha K}/M. \qquad (1.5)$$

At $K \sim 10^6\,\mathrm{erg/cm}^3$ and $M \sim 10^3\,\mathrm{G}$ one gets for the critical size d_* practically the same numerical value as in Eq. (1.4). Therefore we arrive at the conclusion that most often the term *fine magnetic particles* may be considered as a verbal equivalent of the condition $d \lesssim 10\,\mathrm{nm}$, which means dealing with nanosize particles.[1]

Denoting the interatomic distance by a_0, we get a well-known estimate for the fraction of the atoms residing at the particle surface. For $d \simeq 10\,\mathrm{nm}$ and a_0 about 0.2 nm, one has

$$c \sim 6a_0/(d - 6a_0) \simeq 14\%. \qquad (1.6)$$

It is clear that in a particle made of a ferromagnet or ferrite c equals also the fraction of spins residing at the particle surface and thus deprived of a full set of neighbors.

Incomplete and distorted atomic surrounding creates for a surface spin a situation where the crystal field contributions, which self-average for a bulk spin,

---

[1] We remark that the given example of estimating the reference size of a single-domain particle is rather heuristic. The rigorous approach to the problem was outlined by Brown in 1969 [4]. Concerning the modern developments of the problem, see Refs. [5, 6] and references therein.

remain non-zero. Due to this enhanced crystal field experienced by a surface spin, its single-ion magnetic anisotropy is expected to be higher than that for a bulk one. This mechanism does not reduce to just geometrical restrictions and, as implied by the classification of Ref. [2], represents the situation where the surface possesses some specific properties different from those of the bulk. The existence of surface magnetic anisotropy has numerous experimental verifications in both films and fine particles but its origin is still not completely clear. In fact, the modern phenomenology did not proceed much beyond the fundamental concept worked out by Néel in 1954 [7]. In this approach the surface anisotropy is described at the macroscopic level by a specific energy parameter K_s and the direction of the local easy (or hard) magnetization axis so that the surface energy density is written as

$$F_s = -K_s \left(e \cdot \nu \right)^2 , \qquad (1.7)$$

where ν is the unit vector of the surface easy axis. From the symmetry considerations Néel [7] and, after him, Brown [1] associated ν with the direction of the normal to the particle surface. According to the same work by Néel, the surface energy density is estimated as $K_s \sim \pi \ell M^2$, where ℓ is some characteristic length of the order of 10 nm. For a well-known ferrite γ-Fe_2O_3 that has $M \simeq 400$ G this gives $K_s \sim 5 \times 10^{-2}$ erg/cm^2.

Taking into account the surface anisotropy of fine particles entails important consequences. In particular, with regard to it a widely exploited idea of a perfect single-domain particle looks as an over-simplified model. Indeed, the uniform internal magnetization cannot match without discontinuities the assumed radial distribution of easy axes at the particle surface. The search for equilibrium magnetization in this case will lead to a non-uniform distribution $e(r)$ so that the simple relation $\mu = Mv$ would be no longer valid. Another apparent consequence of the surface anisotropy is its effect on the magnetization phase behavior. Affecting the orientation of the surface spins, which in a particle with $d \lesssim 10$ nm is quite a considerable number, the surface anisotropy may become comparable with the exchange energy thus shifting considerably or even smearing out the Curie point of a nanosize ferromagnet. The same holds for the case where the normal to the surface is not the easy but the hard magnetization axis, i.e., the favorable direction of magnetization is tangent to the particle surface.

Comparing fine particles and thin films, one notes an obvious "topological" feature. For a film there is no real difference between the local and global anisotropy directions: all the normals to a flat surface are mutually parallel. But for a particle the situation is different. There the existence of a local axis means the absence of a global one and thus requires use of a local coordinate framework. It is mainly due to this complication (rather geometric that physical) that the possibilities of obtaining analytical solutions for the equilibrium magnetization distribution in a particle are minimal. This shifts emphasis towards numerical simulations. The existing amount of such works is substantial, quite a number of them are named in the review papers [2, 8]. In general, the studies mostly follow two lines. One is the numerical simulation of a magnetic phase transition in a finite spin assembly. For this purpose a Heisenberg model on some lattice in a finite sample is taken and non-equal values of the exchange integrals and local anisotropies are ascribed to the spins residing in the bulk and at the surface. A statistical thermodynamic problem is solved, see Refs. [9–11], for example, in order to understand what happens to the Curie point, what are the details of the magnetic ordering establishing, how does the sample magnetization depend on temperature below the Curie point, etc. The other line of studies addresses magnetization and re-magnetization under an external field; usually, it is done in the adiabatic limit $\omega \to 0$. The aim is to explain the actual shapes of hysteresis loops obtained for fine particle composites and magnetic superlattices. Of the number of results obtained we would like to point out an important qualitative conclusion [12] that for a perfectly symmetrical particle with the radial surface anisotropy the effect of the latter completely self-averages. This means that the particle outer layer, where the surface anisotropy is established, behaves as a "dead" substance with respect to the magnetic processes which concern the magnetic moment of the particle as a whole.

A special case of surface anisotropy differing from that of Néel was introduced at the end of the 1980's by Aharoni [13, 14]. According to this hypothesis the local surrounding of the surface spins is distorted in such a way that only the value of the anisotropy constant (single-ion anisotropy) changes in comparison with a bulk spin whereas the direction of the axis does not. Therefore the Aharoni case is described formally by the expression

$$F_s = -K_s \left(e \cdot n \right)^2, \tag{1.8}$$

that looks like the Néel one (1.7) except that the vector n is a "global" parameter and does not depend on the position of a point at the particle surface. Virtually the same model was proposed independently as a result of numerical modeling in Ref. [12]. There the authors, having concluded on the "death" of the surface layer under condition of a perfect radial anisotropy, pointed out a possibility of a global uniaxial surface anisotropy due to the presence of an alien substance (e.g. oxide) at the particle.

It is clear that with respect to magnetic films the Néel and Aharoni models coincide as soon as one takes K_s the same. However, in the case of a particle the Aharoni model is much more convenient for theoretical treatment. First of all, the anisotropy function described by Eq. (1.8) returns to the particle a possibility to be true single-domain. Indeed, in the absence of an external field the particle remains uniformly magnetized as soon as its magnetization aligns with the easy axis n. This brings the case of the particle closer to the case of a film. Below, we use this closeness in order to analyze the spin-wave resonance under surface anisotropy without too many of formal difficulties.

This brief introduction is by no means an exhaustive review on the problem as here we have addressed only the macroscopic aspects of the surface anisotropy problem. Besides, we mention only the works which are either of general interest or concern directly the work on the surface magnetic properties of fine γ-Fe_2O_3 particles which are the particular subject of this paper. As a concluding issue of this section we note a qualitative consideration given in Ref. [15] and quoted since then many times. There the authors propose to estimate the equivalent internal magnetic field H_{as} that transmits the action of surface anisotropy on the net magnetic moment of a particle in the following way. The magnetic anisotropy energy of a particle is written as a sum of the bulk and surface contributions:

$$E_a = E_a^{\text{bulk}} + E_a^{\text{surf}}, \tag{1.9}$$

and each of the terms is estimated assuming that the particle is a sphere of diameter d. This leads to

$$E_a^{\text{bulk}} = \pi K_v d^3/6, \qquad E_a^{\text{surf}} = \pi K_s d^2, \tag{1.10}$$

where the definition of the bulk magnetic anisotropy is taken according to Eq. (1.2). Substituting Eq. (1.10) in (1.9) and introducing an effective bulk anisotropy constant as $K_{\text{eff}} = E_a/v$, one gets

$$K_{\text{eff}} = K_v + \frac{6K_s}{d}, \qquad H_{as}^{\text{eff}} = \frac{2K_v}{M} + \frac{12K_s}{Md}. \tag{1.11}$$

For the case of a magnetic film with a unilateral surface anisotropy the same estimation gives

$$K_{\text{eff}} = K_v + \frac{K_s}{d}, \qquad H_{as}^{\text{eff}} = \frac{2K_v}{M} + \frac{2K_s}{Md}, \tag{1.12}$$

where now d stands for the film thickness. Adopting in principle the above presented scheme, the authors of Ref. [16] justly remark that the actual value of the numerical coefficients alongside the ratio K_s/d may vary considerably depending on the number of defects located at the surface.

Specifying the continuum limit of the tendency established by Eq. (1.6), Eqs. (1.11) and (1.12) show that the fraction of surface spins increases with the diminution of the sample size and due to that enhances the effect of the surface spins on the magnetic behavior of the bulk ones. Using these formulas, one can estimate the size of the sample below which the surface anisotropy will certainly dominate the bulk contribution. This gives: $d < 6K_s/K_v$ for particles and $d < K_s/K_v$ for films thus indicating that the relative amount of surface spins in particles is greater. One may also conclude from Eqs. (1.11) and

(1.12) that the linear dependence of the effective internal field on the inverse reference size is a "fingerprint" of the surface anisotropy domination on magnetic properties of a particle or a film. In the next section, performing a more thorough analysis, we show that in reality the situation with the dimensional dependence of the effective anisotropy field is somewhat more complicated so that the $1/d$ law is just the so-called *weak pinning* limit of a more general case.

2. Spin perturbations in fine particles. Interplay of the exchange and surface energies

Here we consider the magnetization distribution in a nanosize magnetic sample with a pronounced surface anisotropy, introduce the main parameters which govern the situation and provide some estimations by the order of magnitude. We begin with a remark that unlike the bulk anisotropy, the surface one acts (at least in the continuum approximation) in a layer of zero thickness and due to that does not yield straightforwardly any additional term in the internal bulk field. Thence it cannot affect directly the perturbations of the spin order, e.g. spin-waves. Instead the surface anisotropy influences the bulk magnetization by creating some "inconvenient" boundary conditions. The latter work in such a way that if in a particle or in a film once emerges a magnetic perturbation which is not collinear to the ground state, then by necessity the solution of the static and/or dynamic equations is a spatially non-uniform function $m(r)$. As a result, the non-uniform exchange term in the energy density increases and becomes a source of certain additional internal field that can be measured. In particular, in ferromagnetic resonance experiments on fine particle systems such a contribution is registered as an additional bias field.

In the exchange approximation, i.e., neglecting magnetostatic terms that is possible when dealing with nanoparticles, and omitting for simplicity the internal bulk anisotropy (crystallographic and other) the volume density (1.2) of the free energy reduces to the single-term expression

$$F_{ex} = \frac{1}{2}\alpha M^2 \left(\nabla e\right)^2, \tag{2.1}$$

and for the surface energy density we use the representation (1.7):

$$F_s = -K_s \left(e\nu\right)^2. \tag{2.2}$$

To get a qualitative notion of the situation where magnetization experiences an interplay of the bulk and surface effects it suffices to consider a one-dimensional problem. Let a sufficiently long (in the direction $z > 0$) magnetic sample, see Fig. 2.1a, be bounded at $z = 0$ by a flat surface infinite in the Ox and Oy directions. At this interface the easy magnetization direction is ν. Denoting by $\vartheta(0)$ the angular deviation of magnetization from the surface easy

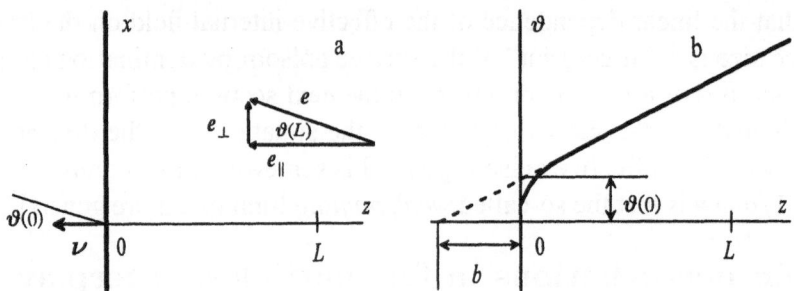

Figure 2.1.   Scheme explaining the definition of the extrapolation length.

axis and counting this angle from the negative direction of the Oz axis, see
Fig. 2.1a, one transforms Eq. (2.2) in

$$F_s = -K_s \cos^2 \vartheta(0) \;\Rightarrow\; K_s \left[\vartheta(0)\right]^2, \tag{2.3}$$

where the symbol $\Rightarrow$ means that we have omitted the orientation-independent
terms and expanded F_s with respect to small $\vartheta(0)$. The total energy of the
sample (of length L) calculated per unit area of its end-wall boundary takes
the form

$$\frac{1}{S}\mathcal{F} = K_s \left[\vartheta(0)\right]^2 + \frac{1}{2}\alpha M^2 \int_0^L \left(\frac{de}{dz}\right)^2 dz, \tag{2.4}$$

which shows that the equilibrium state of the sample is a uniform magnetization parallel to Oz. Accordingly, we set the vector e in equilibrium to be
directed along the negative direction of Oz.

   Now we perturb the spin order by deviating at a distance L from the surface
the unit vector of magnetization e from its direction $\vartheta = 0$ by the angle $\vartheta(L)$
and fix e in this position. Without loss of generality, we direct the Ox axis
of the coordinate framework along the thus created static perturbation $e_\perp$, see
Fig. 2.1a. For not so large deviation angles one may set $e_\perp(z) \approx \vartheta(z)$. The
equation of equilibrium in the bulk of the sample is derived by the conventional
variation of $\mathcal{F}_{\mathrm{ex}}$, see Eq. (2.1), and this case takes an easily solvable form

$$\frac{d^2}{dz^2}\vartheta = 0, \qquad \vartheta(z) = C_1 z + C_2, \tag{2.5}$$

with $C_{1,2}$ being integration constants. The obtained distribution of the angular
perturbation is shown schematically in Fig. 2.1b, where the solid line marks
the true function $\vartheta(z)$. It is clear, however, that in the close vicinity of the
boundary the exact magnetization distribution is determined by subtle details
of interaction between the surface and bulk spins and anyway cannot be described in the framework of a simple continuum approach. Accordingly, we

simplify the description and assume that the linear solution (2.5), which is shown in Fig. 2.1b by the dashed line, holds until the very boundary. Then for the tangent coefficient of the line (2.5) one has

$$C_1 = \frac{d\vartheta}{dz} = \frac{\vartheta(L) - \vartheta(0)}{L}. \tag{2.6}$$

Using approximation (2.6), we transform expression (2.4) for the energy of the sample to the form

$$\frac{1}{S}\mathcal{F} = K_s\,[\vartheta(0)]^2 + \frac{1}{2}\alpha M^2 L \left[\frac{\vartheta(L) - \vartheta(0)}{L}\right]^2, \tag{2.7}$$

and find its minimum with respect to $\vartheta(0)$. This yields

$$\vartheta(0) = C_2 = \frac{\alpha M^2}{2K_s}\left(\frac{d\vartheta}{dz}\right), \tag{2.8}$$

thus evaluating the second integration constant in the solution of Eq. (2.5) and leading to the spatial distribution of the angle in the form

$$\vartheta(z) = \left[\frac{d\vartheta}{dz}\right]\left(z + \frac{\alpha M^2}{2K_s}\right). \tag{2.9}$$

Determining from the last equation the point where the linear function $\vartheta(z)$ crosses the ordinate axis, we introduce a parameter with the dimensionality of length:

$$b = \frac{\alpha M^2}{2K_s}. \tag{2.10}$$

Figure 2.1b clarifies the role played by the spatial scale b, whose analogue in the physics of liquid crystals is known as *the extrapolation length*. It equals the distance at which a fictitious boundary may by set behind the true one so that at this imaginary wall the boundary condition is purely rigid. Indeed, at $z = -b$ the deviation of magnetization from the easy axis is zero.

Since the extrapolation length is inversely proportional to the surface anisotropy constant, it is apparent that large b means soft boundary conditions and vice versa. To decide on that for a sample of a finite size L, one should consider the dimensionless ratio

$$p = \frac{L}{b} = \frac{2K_sL}{\alpha M^2}, \tag{2.11}$$

which is called *the pinning parameter*, so that $p \ll 1$ means "magnetically soft" and $p \gg 1$ means "magnetically rigid" border. Let us substitute expression (2.8) for the boundary angle in equation (2.4) rendering the sample energy. Using the above-introduced notations we find

$$\frac{1}{S}\mathcal{F} = \frac{1}{2}\alpha M^2(b + L)\left(\frac{d\vartheta}{dz}\right)^2 = \frac{1}{2}\alpha M^2 b\,(1 + p)\left(\frac{d\vartheta}{dz}\right)^2. \tag{2.12}$$

Thus, under strong pinning ($p \gg 1$) the main share of the magnetic energy is associated with the volume distortions whilst at the boundary the magnetization vector practically does not deviate from the easy axis. Under weak pinning ($p \ll 1$) one encounters an opposing situation: the main energy is stored in the distortions which are "ousted" to the surface thus preserving the bulk magnetization practically uniform.

As already mentioned, a very close analogy to the static perturbation problem under study can be found in the theory of liquid crystals. Due to their remarkable molecular structure, those substances have anisotropic surface tension that results in the surface phenomenon known as *anchoring* [17]. This means that for the director of a liquid crystal at the surface that it abuts on, there exist easy and hard directions of alignment. If there is only one wall (e.g. a liquid crystal fills a half-space), then, due to its orientational elasticity (whose energy term is functionally very much like the non-uniform exchange interaction) the director of the liquid crystal will align uniformly with the direction of the surface easy axis. If then some field (e.g. magnetic) is imposed on the liquid crystalline sample non-collinearly to its alignment or a second wall is inserted with the easy axis that is not parallel to the first one, then the system seeks to optimize its distortion energy exactly as it is in the magnetic problem in question. This was in fact the liquid crystal theory where the extrapolation length b was first introduced and defined [17]. It acts as follows [18]. If $L/b \ll 1$ (weak anchoring) all the distortions concentrate at the surface leaving the inner region of the sample orientationally uniform. In the opposite limit of strong anchoring ($L/b \gg 1$) the surface is in the state of almost minimum energy whilst orientation distortions are smeared over the bulk of the sample.

Under a distortion of magnetic order the effective internal field induced by surface pinning and experienced by the bulk spins may be estimated as $H_{as} \sim \mathcal{F}/SLM$ with $d\vartheta/dz \sim k$, where k is the wave vector of the distortion. Then from Eq. (2.12) one finds

$$H_{as} \simeq \frac{\alpha M b}{L}(1+p)k^2 = \alpha M \frac{1+p}{p} k^2. \tag{2.13}$$

For rigid boundary conditions the minimal wave vector in a sample of size L is $k \sim \pi/L$. Thus for the minimal value of the effective field of the magnetic surface anisotropy we get

$$H_{as} \simeq \alpha M \left(\frac{\pi}{L}\right)^2 \qquad (p > 1). \tag{2.14}$$

In the case of soft boundary conditions where the sample size L is much smaller than the extrapolation length b, it is the latter that determines the minimal wave vector yielding $k \sim \pi/b$ so that Eq. (2.13) yields

$$H_{as} \simeq \frac{K_s}{M}\left(\frac{\pi}{L}\right) \qquad (p < 1). \tag{2.15}$$

Comparing expression (2.15) with Eqs. (1.11) and (1.12) we see that the range of their validity is determined by the condition of soft pinning. We also note that if we describe the spatial non-uniformity by the relation $k \sim \pi/(L+b)$, then from Eq. (2.13) we get an interpolation formula

$$H_{as} \simeq \frac{\pi^2 \alpha M}{L(L+b)},$$ (2.16)

which is capable of rendering both limiting cases of the considered dependence.

Let us estimate the values of the contributions imparted by surface pinning to the internal field. We specify this for the maghemite (γ-Fe$_2$O$_3$) nanoparticles studied in our papers [19–23] and described in some detail in Section 4. For the exchange rigidity parameter αM^2, following Refs. [24, 25], we take 5×10^{-7} erg/cm.^{2} The characteristic range for the surface anisotropy density is that pointed out by Néel [7]. The later studies confirmed [27, 28] that K_s may hardly exceed 1 erg/cm^2 and often is just a small fraction of this value [15, 29]. In particular, for nanoparticles of γ-Fe$_2$O$_3$ the authors of Ref. [29] in static measurements have found 4.2×10^{-2} erg/cm^2. Therefore, setting $K_s \sim 10^{-2} - 1$ erg/cm^2, we find from Eq. (2.10)

$$b \sim 2 \times (10^{-7} - 10^{-5})\, \text{cm} = 2 - 200\, \text{nm}.$$ (2.17)

Choosing the particle size range as $L \sim 4 - 10$ nm, we find that depending on the actual value of the surface anisotropy, the obtained pinning parameter interval by the order of magnitude is 10^{-2}–10. This means that both limiting cases—weak and strong pinning—are conceivable. Then with the aid of Eqs. (2.14) and (2.15) one finds

$$H_{as}(p > 1) \sim 10\,\text{kOe}, \qquad H_{as}(p < 1) \sim 0.1 - 10\,\text{kOe}.$$

for strong and weak pinning, respectively. According to the last equation, for the chosen particle size range the quantitative difference between the absolute values of the bias fields is not very large. However, if, for example, we change the particle size just two times either side, the differences would be much more pronounced. This reveals the essential distinction: in the weak and strong pinning cases the internal field contributions H_{as} obey different power laws with respect to the sample dimension. As Eqs. (2.14) and (2.15) establish, for strong pinning the effective field scales with the sample size as $H_{as} \propto L^{-2}$ whilst in the weak pinning case the scaling law is $H_{as} \propto L^{-1}$.

---

[2] We mention also the earlier works [14, 26], which place the exchange rigidity parameter of maghemite at $\simeq 10^{-7}$ erg/cm that is five times lower than the one we use. This is however irrelevant for the results on maghemite which we present below in Sections 5 and 6 since for the found value $K_s \sim 10^{-2}$ erg/cm^2 it leaves unchanged the estimate justifying the weak pinning for the particles used.

Summarizing the afore-presented considerations we remark that a convenient criterion which distinguishes the cases of weak and strong pinning in fine particles or in thin films is the extrapolation length b. As Eq. (2.10) shows, it is a combination of the material parameters of a given magnetic sample and thus is a universal characteristics of a given ferromagnet or ferrite. In the nano-size range depending on the actual particle size d and/or the method of particle manufacturing one may encounter magnetic grains with weak ($d < b$) or strong ($d > b$) pinning as well as the border case $d \sim b$. Apparently in the case of closeness $d \sim b$ one should expect to find for the internal field H_{as} an intermediate (crossover) size dependence: between d^{-1} and d^{-2}.

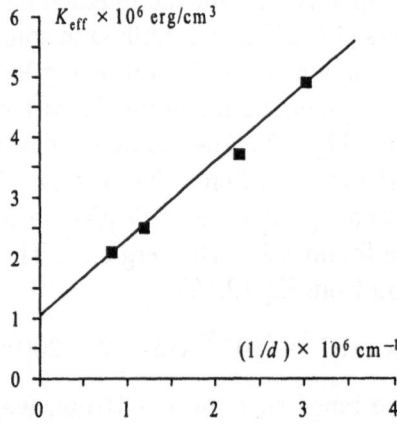

Figure 2.2.   Size dependence of the effective bulk anisotropy constant found for iron nanoparticles in Ref. [30]; points are the experimental data, line shows fitting with the $1/d$ law.

Good examples for testing the concept (2.13) provide two papers, Refs. [30, 31] published at the beginning of the 1990's. Ref. [30] is a detailed experimental study of fine iron particles obtained by a vapor deposition technique. Table IV of this paper contains numerical data obtained by static magnetization method yielding the dependence of the effective bulk anisotropy on the particle size for the diameter range 3.3 to 12.1 nm. Even a preliminary analysis shows immediately that the data scales very close to the $1/d$ law. Using this hypothesis for fitting, we arrive at the results shown in Fig. 2.2. One sees that the achieved agreement is fairly good. The bulk anisotropy of the iron particles was found in Ref. [30] to be uniaxial with the K_v values in the interval (2–5)$\times 10^6$ erg/cm^3. So we are interested only in the surface term not accounted for. Having determined the slope of the $1/d$ line from fitting, we then use the estimate (1.11) and obtain $K_s \simeq 0.2$ erg/cm^2. Estimating the pinning parameter by Eq. (2.11), one finds that even for the largest particles ($d \simeq 12$ nm) the value of p does not exceed unity. This once more justifies the use of Eq. (1.11).

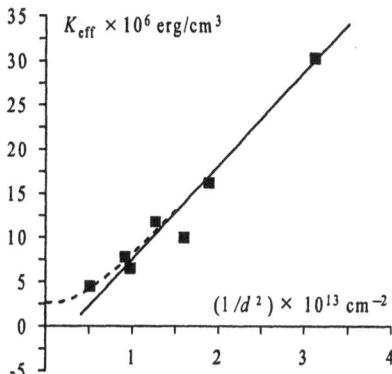

Figure 2.3.   Size dependence of the effective bulk anisotropy constant in cobalt nanoparticles after Ref. [31]; points are the experimental data, thin straight line shows approximate fitting with the d^{-2} law, dashed line indicates the presence of non-zero K_v.

In Ref. [31] nanosize cobalt particles prepared by a microemulsion method were studied. The anisotropy constant was evaluated by measuring the Néel blocking temperature of the samples. A size-dependent effect in the bulk anisotropy constant K_{eff} had been found but, as in Ref. [30], was not analyzed. In this case the preliminary scaling analysis definitely refutes the possibility of the $1/d$ law. As an alternative, we made an attempt to interpret the data assuming another limiting case, i.e., $\propto d^{-2}$. The result is presented in Fig. 2.3. There the dashed curve is an eyeguide reminding that at larger diameters the importance of the surface effect diminishes and the effective anisotropy must tend to its bulk value; the latter, as given in Ref. [31], is about 2.7×10^6 erg/cm^3. Adopting the slope (solid line) as an estimate by the order of magnitude for the coefficient in expression (2.14), one gets the exchange rigidity parameter $\alpha M^2 \simeq 2 \times 10^{-7}$ erg/cm that is quite a correct value. In Ref. [31] rather high values of K_{eff} were found. In this connection we note that very strong anisotropy imparted to the surface by chemisorbed atoms was reported more than once, see Refs. [13, 32].

3. Spin-wave resonance in the presence of a uniaxial surface anisotropy

In this section we analyze the magnetic behavior of nanoparticles with surface anisotropy in the radio-frequency range where the magnetic resonance is excited. Along with passing from the quasi-static to dynamic description we make the formal framework for the problem more rigorous. As the essential starting point we recall the rigorous boundary condition for magnetization in a particle with uniaxial surface anisotropy.

3.1 BOUNDARY CONDITIONS FOR THE MAGNETIZATION

Following Brown [1], we write the volume density of the magnetic energy as

$$F_v = \tfrac{1}{2}\alpha M^2(\nabla e)^2 + w_A - \tfrac{1}{2}Me\cdot H_d - Me\cdot H_0, \qquad (3.1)$$

cf. Eq. (1.2). Here the volume density of the magnetic anisotropy energy is written in a general form as w_A and the presence of an external field H_0 is admitted.

For a finite sample of volume v the energy functional comprises the volume contribution based on Eq. (3.1):

$$\mathcal{F}_v = \int_v F_v\, dv = \int_V dV\left[\tfrac{1}{2}\alpha M^2(\nabla e)^2 + w_A - \tfrac{1}{2}Me\cdot H_d - Me\cdot H_0\right],$$
$$(3.2)$$

and the surface one

$$\mathcal{F}_s = \int_S F_s\, dS, \qquad (3.3)$$

that is based on the representation

$$F_s = -K_s\,(e\cdot\nu)^2, \qquad (3.4)$$

introduced by Eq. (2.2). Note that the unit vector ν of the surface easy axis in a general case is completely independent of the volume one, n.

Variation of the total energy with respect to e gives

$$\delta\left(\mathcal{F}_v + \mathcal{F}_s\right) = \int_v dv\left[\alpha M^2\nabla e\cdot\delta\nabla e + \frac{\partial w_A}{\partial e}\cdot\delta e - \tfrac{1}{2}Me\cdot\delta H_d\right.$$
$$\left. - \tfrac{1}{2}MH_d\cdot\delta e - MH_0\cdot\delta e\right] - 2\int_S K_s\left[(\nu\cdot\nu)\nu\delta e\right]dS; \quad (3.5)$$

here keeping K_s under the integral allows for variations of the type and strength of surface anisotropy along the particle body. (For example, the problem of such a type was considered in Ref. [33], where a particle was assumed to be magnetically anisotropic only at a part of its surface.) To make δe a common factor, one rearranges the exchange contribution with the aid of the formula

$$\nabla e\cdot\delta\nabla e = \nabla e\cdot\nabla\delta e = \nabla[\nabla e\cdot\delta e] - \nabla^2 e\cdot\delta e. \qquad (3.6)$$

On substitution in Eq. (3.5), the divergence term after integration transforms in a surface one whose integrand is the vector $\nabla e\cdot\delta e$ multiplied by the outer normal N to the surface. The magnetostatic term in Eq. (3.5) is transformed by the theorem of the same name given in Ref. [1], see Eq. (3.47) there:

$$\int_v H_d\cdot\delta M\, dv = \int_v M\cdot\delta H_d\, dv. \qquad (3.7)$$

With allowance for Eqs. (3.6) and (3.7) the energy variation (3.5) takes the form

$$\delta(\mathcal{F}_v + \mathcal{F}_s) = -\int_v dv \, \delta e \cdot \left[\alpha M^2 \nabla^2 e - \frac{\partial w_A}{\partial e} + M H_d + M H_0\right]$$
$$+ \int_S dS \, \delta e \cdot [(N \cdot \nabla)e - 2K_s \, (e \cdot \nu)\nu]. \quad (3.8)$$

Note that now the surface integral depends on two units vectors: N that is the outer normal to the sample surface and ν that is the surface magnetic easy axis.

As the vector e has a unit length, its variation may be presented identically as $\delta e = \delta\varphi \times e$, where $\delta\varphi$ is the angle of rotation of e. Thence Eq. (3.8) transforms into

$$\delta(\mathcal{F}_v + \mathcal{F}_s) = -\int_v dv \, \delta\varphi \cdot \left(e \times \left[\alpha M^2 \nabla^2 e - \frac{\partial w_A}{\partial e} + M H_d + M H_0\right]\right)$$
$$+ \int_S dS \delta\varphi \cdot \left(e \times [(N \cdot \nabla) e - 2K_s \, (e \cdot \nu)\nu]\right). \quad (3.9)$$

According to the usual rules, setting the volume and surface variations (3.9) independently to zero, we get two equations of equilibrium. One is inside the particle:

$$e \times \left(\alpha M^2 \nabla^2 e - \frac{\partial w_A}{\partial e} + M H_d + M H_0\right) = 0, \quad (3.10)$$

and another is to hold at the particle surface:

$$e \times [\alpha M^2 (N \cdot \nabla)e + 2K_s \, (e \cdot \nu)\nu] = 0. \quad (3.11)$$

On presenting the volume anisotropy energy in the customary uniaxial form

$$w_A = -K_v(e \cdot n)^2, \quad (3.12)$$

as in Eq. (1.2), the volume equilibrium equation transforms to

$$e \times [\alpha M^2 \nabla^2 e + 2K_v(e \cdot n) \, n + M H_d + M H_0] = 0. \quad (3.13)$$

If the surface anisotropy is ignored, i.e., one assumes that $K_s = 0$ identically, then Eq. (3.11) transforms to the so-called "natural" boundary condition for magnetization:

$$e \times (N \cdot \nabla) e = 0. \quad (3.14)$$

Being a by-product of a formal variational procedure, Eq. (3.14) accounts rather for geometrical restrictions—the finiteness of the particle body—and does not recognize any special physical properties of its surface. It does not contain any material parameters and its physical meaning is rather simple: (1)

it is satisfied automatically (is trivial) for all the situations where the magnetization is uniform, (2) for non-uniform configurations the required zero of the derivative means that magnetization perturbations attain maximum (spin waves have loops) at the surface. To reflect the special properties of the surface one has to keep terms $\propto K_s$ or alike in the energy expression.

To obtain a closed set of equilibrium equations, to Eqs. (3.11) and (3.13) one should add the relationship connecting the demagnetizing field H_d with the external one H_0 and magnetization $M = Me$. In a general case the dependence $H_d(H_0, M)$ is determined by adding the set of magnetostatic equations for a nonconductive medium. However, as in what follows we would not consider geometrical objects of other than spherical or ellipsoidal shape, it suffices to use the demagnetizing tensor representation. Denoting it as $\mathbb{N}$, one has

$$H_d = -\mathbb{N} \cdot M = -M\mathbb{N} \cdot e. \qquad (3.15)$$

Writing together the final form of the equilibrium equations

$$e \times \left[\alpha M^2 \nabla^2 e + 2K_v(e \cdot n)\, n - M^2 \mathbb{N} \cdot e + M H_0\right] = 0, \quad (3.16)$$
$$e \times \left[\alpha M^2 (N \cdot \nabla)e + 2K_s\, (e \cdot \nu)\nu\right] = 0, \qquad (3.17)$$

we emphasize that this set is equally valid for the case when the surface easy axis vector ν is a local variable like $\nu = N$, as is assumed in the Néel model, or it is a "global" variable $\nu = n$, as is taken in the Aharoni hypothesis [13]. Note the transformation of the surface boundary condition (3.17) under the change of the anisotropy constant. At $K_s \to 0$ it reduces to the "natural" boundary condition (3.14) that means magnetically isotropic surface, whereas at $K_s \to \infty$ it yields $e = \pm \nu$ that means an infinitely strong (rigid) pinning of magnetization at the surface.

3.2   THE MAGNETODYNAMIC EQUATION

The phenomenological equation that governs the motion of the particle magnetic moment in a non-dissipative case follows from classical electrodynamics as

$$\frac{d}{dt} M = -\gamma\,(M \times H_{\text{eff}}), \qquad H_{\text{eff}} = -\frac{\partial F_v}{\partial M} = -\frac{1}{M}\frac{\partial F_v}{\partial e}, \qquad (3.18)$$

where γ is the gyromagnetic ratio and the volume energy density F_v is given by Eq. (3.1) with allowance for Eqs. (3.12) and (3.15). As it should be, in the stationary (equilibrium) state Eq. (3.18) turns into the static condition (3.16). Accordingly, one may interpret Eq. (3.18) as a requirement that under equilibrium the local magnetization is parallel to the local effective field H_{eff}.

As seen from Eq. (3.18), a ferromagnet is a gyrotropic medium where any torque, e.g. exerted by some external field H_1, causes not the motion of the system to the new equilibrium position but precession around the direction of the new effective field $H_{\text{eff}} + H_1$. Formally, in Eq. (3.18) this (Larmor) precession exists *ad infinitum* since it does not cost any energy. To allow for dissipation (spin-lattice relaxation), which inevitably affects all the real processes, Eq. (3.18) is modified by adding a damping term. Conventionally the dissipation term is introduced either in the Landau-Lifshitz [34] or in the Gilbert [35] form. Each one has its methodical merits and drawbacks but basically the modified equations coincide under an appropriate change of notations, see Ref. [36]. In this paper, to be specific, we use the Landau-Lifshitz form, so that the magnetodynamic equation transforms to

$$\frac{d}{dt} e = -\gamma (e \times H_{\text{eff}}) - \lambda\gamma (e \times (e \times H_{\text{eff}})), \qquad (3.19)$$

where λ is the phenomenological damping parameter.

Let us re-write Eq. (3.19) with allowance for the fact that the effective field contains algebraic (local) as well as differential (short-range non-local) terms. In order not to make the equation too long, we pack all the local terms[3] into the uniform field

$$H_u \equiv H_0 + H_d + (2K_v/M)(e \cdot n)\, n. \qquad (3.20)$$

Thus we obtain

$$\frac{d}{dt} e = -\alpha\gamma M \left(e \times \nabla^2 e \right) - \gamma (e \times H_u)$$
$$- \lambda\alpha\gamma M \left(e \times (e \times \nabla^2 e) \right) - \lambda\alpha\gamma (e \times (e \times H_u)). \qquad (3.21)$$

To reveal the spatiotemporal behavior of magnetization we expand the latter in the Fourier series as

$$e(r,t) = \sum_m \sum_\ell e_{m\ell} \exp[i\omega_m t + i(k_\ell \cdot r)]. \qquad (3.22)$$

and substitute this in Eq. (3.21). Then one sees that oscillatory motions (precession) of a non-uniform as well as of a uniform ($k_\ell = 0$) kind are possible. If the ground state of the system is uniform, then a uniform rf field, which is usually used for weak-amplitude ferromagnetic resonance probing, can excite there only a uniform precession at the Larmor frequency $\omega \simeq \gamma H_u$. This follows from the fact that the non-uniform perturbations, see Eq. (3.1), invariably have higher energies than the uniform one. Thus, one concludes that to

---

[3] Generally speaking, the demagnetizing field depending on the distribution of magnetization in the sample as a whole implies integration, that is a long-range non-locality. However, in the demagnetizing tensor approximation this term effectively appears to be a local one.

excite non-uniform modes in a linear probing regime it requires either a non-uniform ground state (then the probing field may be uniform) or the exciting field should have an appropriately spatial modulation of its amplitude.

So was the account of the situation before 1958 when Kittel [37] in his pioneering paper had shown that as soon as some spins of the sample experience the internal field that is not identical to that acting on the others, then a uniform rf probing is well capable of exciting *non-uniform* modes on the background of a *uniform* equilibrium state of the system. As a physically clear example of such non-equivalence in the spin environment, Kittel pointed to the surface magnetic anisotropy.

3.3   SPIN-WAVE RESONANCE IN A MAGNETIC FILM WITH A UNIAXIAL SURFACE ANISOTROPY: STRONG PINNING LIMIT

The importance of Kittel's work [37] is that it is the first example where a linear but non-uniform response of a system with surface magnetic anisotropy to a uniform probing field is considered. Getting acquainted with this case seems to be the most comprehensive way to approach the magnetodynamics of fine particles with surface anisotropy, and so in this subsection we rederive in brief the central result of Kittel.[4]

Let a planar film of a ferromagnetic material be stretched infinitely in Ox and Oy directions. Along the Oz axis the film is bounded by the planes $z = 0$ and $z = L$ and is magnetized in this direction by a constant field H_0. For simplicity in this calculation the internal anisotropy and demagnetizing field are neglected. As well we omit the dissipation terms in the magnetodynamic equation (3.21). At each face the film possesses identical surface anisotropy of a uniaxial type, see Eq. (3.4), with the easy axis ν directed parallel / antiparallel to Oz, that is along the magnetizing field. Due to that, in the ground state the film magnetization is uniform:

$$e_0 \parallel H_0 \parallel Oz.$$

The non-equilibrium situation is introduced with the aid of e_x and e_y, small deviations of magnetization in x and y directions. In the linear approximation those terms do not change the length of the main vector: $e_{0z} = 1$. In the absence of magnetostatic effects there are no perturbations of the field. Then the linearized equation (3.21) reduces to the set

$$\frac{\partial e_x}{\partial t} = -\gamma e_y H_0 + \alpha \gamma M \nabla^2 e_y, \qquad \frac{\partial e_y}{\partial t} = \gamma e_x H_0 - \alpha \gamma M \nabla^2 e_x. \quad (3.23)$$

---

[4]We remark that the spin-wave resonance in thin magnetic films is rather a developed branch of applied physics, see, for example, Refs. [38,39] and references therein. Here we recall just some basic things from it.

With allowance for the uniaxial symmetry of the system the oscillating perturbations are taken to depend only on the z coordinate:

$$e_x = e_x(z) \exp(i\omega t), \quad e_y = e_y(z) \exp(i\omega t),$$

which reduces Eqs. (3.23) to

$$\alpha\gamma M \frac{d^2 e_x}{dz^2} - \gamma H_0 e_x = -i\omega e_y, \qquad \alpha\gamma M \frac{d^2 e_y}{dz^2} - \gamma H_0 e_y = i\omega e_x. \quad (3.24)$$

Substituting there the coordinate dependence in the form e^{ikz} and using the
compatibility condition for the resulting homogenous set of equations we arrive at

$$\omega = \gamma H_0 + \alpha\gamma M k^2. \quad (3.25)$$

Equation (3.25) is a well known dispersion relationship for spin waves in
an infinite medium showing that the spectrum begins at the Larmor frequency
γH_0, i.e., that of the uniform ferromagnetic resonance (FMR). In our case,
however, on the faces of the film the boundary condition (3.17) must be imposed, which selects some discrete values of k. With allowance for the considered one-dimensional geometry, from Eqs. (3.24) one finds

$$\left(\frac{d}{dz} \mp \frac{2K_S}{\alpha M^2}\right) e_x \bigg|_{z=0,L} = 0, \qquad \left(\frac{d}{dz} \mp \frac{2K_S}{\alpha M^2}\right) e_y \bigg|_{z=0,L} = 0. \quad (3.26)$$

Assuming that

$$e_x, \; e_y = \exp(i\omega t) \left[A \sin(kz) + B \cos(kz)\right], \quad (3.27)$$

and substituting this in Eq. (3.26) at $z = 0$ we get

$$\frac{B}{A} = \frac{\alpha M^2 k}{2K_S} = kb, \quad (3.28)$$

where b is the extrapolation length introduced by Eq. (2.10). This relationship couples the partial amplitudes of the free (B) and pinned (A) modes of
the equivalent "string", see Eq. (3.27), and presents the essential result of the
above-cited work by Kittel. Using Eq. (2.17) we estimate the ratio (3.28) as

$$B/A \sim k \times (10^{-7} - 10^{-5}). \quad (3.29)$$

The consideration of Ref. [37] was aimed at magnetic films of small $(L \gtrsim 10^{-4}$ cm) but not at all nanoscopic scale. Due to that Kittel justly inferred
that the ratio B/A estimated by Eq. (3.29) remains small up to a sufficiently
large number of allowed wave numbers $k \sim \pi n/L$, where n is the integer
enumerating the modes. From that he concluded that the magnetization at

the surface, being pinned up by the local anisotropy, responds to the external excitation to a much smaller extent than that in the inner regions of the sample. Therefore the magnetization along Oz behaves similarly to a string with fixed ends ($B \rightarrow 0$) so that the spatial profiles of the magnetic oscillations $e_x(z)$ and $e_y(z)$ take the form of standing waves comprising any number of half-periods with the nodes positioned at the faces of the film. Recalling definition (2.11) for the pinning parameter and estimating it with the same accuracy as in Eq. (3.29), we get

$$p \sim 10 - 10^3 \quad \text{for} \quad L \gtrsim 10^{-4}\,\text{cm.} \tag{3.30}$$

Due to that, Kittel's description may be called the rf magnetodynamics of a single-domain sample, in *the strong pinning limit*.

Let us estimate the obtained shift of the resonance field (frequency). If across the film a spin wave stands which comprises n half-waves, then the wave vector equals $k = \pi n/L$. Substituting this in Eq. (3.25) one finds

$$\omega = \gamma \left[H_0^{(n)} + \alpha M \left(\frac{\pi n}{L} \right)^2 \right]. \tag{3.31}$$

Therefore, upon sweeping the magnetizing field H_0 down from the value ω_0/γ, where ω_0 is the spectrometer frequency, a sequence of resonances will turn up at the positions $H_0^{(n)}$ with $n = 1, 2, \ldots$ given by condition (3.31). Taking $L \sim 3 \times 10^{-4}$ cm, as used by Kittel, with $M \simeq 400\,\text{G}$ (estimation for γ-Fe_2O_3) and $\omega_0/\gamma \simeq 3 \times 10^3$ Oe (a standard X-band spectrometer), the relative field shift between the $(n+1)$-st and n-th resonances is

$$\frac{\gamma}{\omega_0} \left[H_0^{(n)} - H_0^{(n+1)} \right] = (2n+1) \frac{\pi^2 \alpha \gamma M}{\omega_0 L^2} \sim n \times 10^{-4}. \tag{3.32}$$

Hence, in the strong pinning limit, the FMR spectrum of a thin film emerges as a comb-like sequence of narrow (infinitesimally narrow, as long as relaxation is neglected) lines with interdistances $\Delta H_0^{(n)}$ varying proportionally to n. With the reference magnetizing field about 3 kOe, Eq. (3.32) predicts the differences in the resonance fields to be less than 1 Oe.

3.4 SPIN-WAVE RESONANCE IN A MAGNETIC FILM WITH A UNIAXIAL SURFACE ANISOTROPY: WEAK PINNING LIMIT

The problem of high-frequency magnetic oscillations in a fine particle that we are going to address is formulated exactly as Kittel's with just two differences. First, working on the nanoscopic spatial scale (not in its quantum part but close to that) and considering particles of the size of a few nanometers, we cannot any longer treat the lengths of order of 10 nm as infinitesimal. On the contrary, it is the reference distance at which the changes we are interested in take place. At the same time, not entering the quantum domain allows us to use the continuum description. Reducing the spatial scale two or three orders of magnitude, we have to deal with the sample sizes $L \sim 10^{-7} - 10^{-6}$ cm. Repeating estimation (2.11) for this scale, one sees that with respect to pinning, the nanoparticles are typically in the situation where the parameter p does not exceed unity. This means that in nanoscopic objects the problem of excitation of spin waves by a uniform rf field corresponds to *the weak pinning limit*. Due to that, it should be considered under the main assumption opposite to that used in the Kittel theory.

Another difference stems from the geometric distribution of the easy axes. It is a rare occasion that nanoparticles are formed as platelets or flakes. Although it is possible, see Ref. [25], for example, much more typical are the particles of rounded shapes close to a spherical one. As mentioned in Section 1, if the surface anisotropy is of the Néel kind, then the easy axis is a local characteristic, and the case of a particle is topologically different from that of a film. However, realistic enough is the Aharoni model [13] that ascribes to the particle a single global easy axis, which may or may not be correlated with the directions of the internal bulk anisotropy. Although a particle with the Aharoni surface anisotropy is not geometrically isomorphic to a planar film, those two situations are akin. In both cases a uniform ground state of a system is possible and with respect to the FMR properties the results of a planar film model and a spherical particle are qualitatively the same. Meanwhile, the mathematical considerations for a planar object are much easier. Therefore, in order not to obscure the analysis of the phenomena by details of calculation, we shall model a particle with the Aharoni surface anisotropy by a thin film that is uniaxially anisotropic at one face ($z = L$) and isotropic at the other ($z = 0$). Regarding a spherical particle, the boundary $z = 0$ imitates the interior (center) and the one at $z = L$ refers to the outer surface. The results of more rigorous treatment of the spherical particle problem are described in Section 5.

To understand the main difference between the new situation and the strong pinning limit, let us first consider the case of the film with both faces magnetically isotropic, i.e. the zero pinning limit. Once again we present the magnetization distribution in the spin-wave form

$$e_x = \exp(i\omega t)\,[A\sin(kz) + B\cos(kz)]. \tag{3.33}$$

Imposing the boundary conditions

$$(de_x/dz)_{z=0,L} = 0, \tag{3.34}$$

one finds

$$kA = 0, \qquad k\,[A\cos(kL) - B\sin(kL)] = 0. \tag{3.35}$$

From the set of the solutions of this equation we single out the one which corresponds to $k = 0$ and thus describes the uniform precession. Accordingly, see Eq. (3.25), the frequency of this mode does not depend on the size of the sample. All the other solutions of Eqs. (3.35) emerge on setting $A = 0$. Then the second of them yields

$$\sin(kL) = 0, \qquad k = \pi n/L, \quad n = 1, 2, \ldots. \tag{3.36}$$

Substitution of these wave vectors in Eq. (3.25) yields an equation that is identical with Eq. (3.31) obtained in the strong pinning limit. The found coincidence of the spectra for strong and weak pinning is not surprising. It is easy to understand that the energy of a standing spin wave comprising n half-wavelengths does not depend on whether it has a node or a loop at the surface.

Quite expectedly, when passing from the case of strong surface anisotropy to that of weak anisotropy the solutions of higher order ($n \geq 1$) hardly change in shape, except for a shift by a quarter of the wavelength. This means, in particular, that they are equal in energy. However, the situation with the effect of pinning on the uniform solution is different. Indeed, under pinning, however weak, the uniform precession is forbidden, and thus the lowest level of the spectrum must transform into a spin-wave, i.e., become non-uniform. This problem is solved by subjecting the general solution (3.33) to the set of boundary conditions

$$\left.\frac{de_x}{dz}\right|_{z=0} = 0, \qquad \left(\frac{d}{dz} + \frac{1}{b}\right) e_x \bigg|_{z=L} = 0, \tag{3.37}$$

where the first one corresponds to the magnetically isotropic surface, which imitates the inner region of a particle. Substitution of Eq. (3.33) in the first of Eqs. (3.37) once again yields $A = 0$ whilst from the second one we get the transcendental equation for the wave vector:

$$k = \frac{1}{b} \cot(kL). \tag{3.38}$$

Multiplying it by L and denoting $kL \equiv y$ we transform Eq. (3.33) in the dimensionless form

$$y = p \cdot \cot(y), \tag{3.39}$$

where p is the pinning parameter introduced by Eq. (2.11).

In the weak pinning limit—contrary to the Kittel case—the spatial modulation of the spin wave of the lowest order is weak so that the excited perturbations are rather long-wave: $y = kL \ll 1$. This means, as mentioned, that instead of a distinctive standing wave we deal with *an almost uniform precession*. With allowance for p, $kL \ll 1$ Eq. (3.39) expanded to the lowest order gives

$$k^2 = \frac{2K_s}{\alpha M^2 L} = \frac{p}{L^2} \quad \text{or} \quad kb = \frac{1}{\sqrt{p}}. \tag{3.40}$$

Substituting this in the dispersion equation (3.25) which determines the frequency of the spin wave, we obtain the resonance frequency in the form

$$\omega = \gamma H_0 + \alpha \gamma M k^2 = \gamma \left(H_0 + \frac{2K_s}{ML} \right) \tag{3.41}$$

—compare with Eq. (3.31).

Equation (3.41) shows that FMR probing of the samples with a weak surface pinning should give the results which may be readily interpreted in terms of some additional internal field acting inside the sample along with the external one. This field looks very much like the internal (bulk) anisotropy as its direction is determined by the orientation of the sample (uniaxial symmetry). A specific property of this field is that its magnitude varies in inverse proportion to the reference size of the sample. For a film it is its thickness, for a particle it is its diameter. Note that the surface contribution in Eq. (3.40) in the case of a film is exactly the same as predicted by estimation (1.12).

Let us assert the corresponding resonance field shift. Under the same values of M and ω_0/γ which we used for the strong pinning case, from Eq. (3.41) one gets

$$\frac{\gamma}{\omega_0} \Delta H \approx \frac{2\gamma K_s}{M\omega_0 L} \sim \frac{8 \cdot 10^{-7}}{L} \simeq 0.8, \tag{3.42}$$

where we set $L \sim 10 \, \text{nm}$. Thus, under weak pinning the resonance field shift of the quasi-uniform precession mode in nanosize particles turns out to be orders of magnitude greater than the "step" in the resonance field in the strong pinning limit. At first sight it seems correct to apply the estimate (3.42) for any L. For example, assuming, as in Section 3.3, that $L \gtrsim 10^{-4} \, \text{cm}$ one finds a reasonable value $\gamma \Delta H/\omega_0 \sim 10^{-2}$. However, this is not so. Indeed, changing L while keeping the extrapolation length b constant (the same material) means changing the pinning parameter $p = L/b$. Then even for the greatest b from Eq. (2.17), at $L \sim 10^{-4}$ the pinning parameter becomes large, and the condition that justifies usage of Eqs. (3.40)–(3.42) fails. This means that a correct description for the case is given by Eqs. (3.31) and (3.32), where the field shift is not 10^{-2} but 10^{-4}.

As for the other roots of Eq. (3.39), it is easy to see [for example, schematizing Eq. (3.39) graphically] that they may be approximately described by the sequence $k \simeq n\pi/L$ for $n = 1, 2, \dots$ that is of course the same as Eq. (3.36) in the case of zero pinning. The fact of its coincidence with the case of strong pinning was already commented on above. Here we just mention the following. As estimation (3.32) is valid for the higher modes in the weak pinning limit as well, then setting $L \sim d \sim 10\,\mathrm{nm}$, and $n = 1$ we get for the second possible resonance:

$$\frac{\gamma}{\omega_0} H_0^{(2)} = \frac{\pi^2 \alpha \gamma M}{\omega_0 d^2} \sim 10. \tag{3.43}$$

As the reference magnetizing field of the X-band spectrometer is $\omega_0/\gamma \sim 3\,\mathrm{kOe}$, then Eq. (3.43) means that in a fine particle to observe the spin-wave resonance peak which is the "closest neighbor" to the quasi-uniform precession one will need the magnetizing field of several teslas that is practically impossible to achieve with the ordinary technique.

In short, the above-presented theoretical results are as follows. Along with the exchange rigidity parameter αM^2, the surface anisotropy constant K_s is a fundamental characteristic of ferromagnetic materials. For a given magnetic substance by combining those material parameters the reference extrapolation length b can be evaluated; usually it ranges tens of nanometers. Then, taking a magnetic sample (e.g., a particle) whose reference size d is below b, one ensures for the system a weak pinning situation $p = d/b < 1$. Perturbing slightly the particle magnetization, one finds that under given conditions the Larmor precession is almost uniform so that the rf magnetic spectrum of the particle (or of an assembly of the latter) by its shape resembles that of a massive sample of the same material. The main distinction is that in a fine particle there exists a strong size effect. Namely, the resonance frequency of a free precession contains an additional contribution inversely proportional to the particle size d. Moreover, unlike usual spin-wave resonances, a rather wide frequency band around this quasi-uniform magnetic resonance is free of any satellite lines. All this evidences that the weak pinning limit realized with an assembly of fine particles opens a unique way to measure the surface magnetic anisotropy with the aid of the standard FMR technique. The only fundamental restriction on the proposed study is that FMR measurements should be performed at low temperatures; otherwise the neat "dynamic" picture will be substantially distorted by superparamagnetic, i.e., thermofluctuational, effects, see [23]. In particular, as the theory predicts [40], the damping rate of precession may increase greatly virtually destroying the periodical motion.

In practice, in the majority of cases it is K_s which one needs to evaluate. The first step towards this is to determine the scaling law of the effective bulk anisotropy constant K_{eff} with respect to the particle size. Methods of measuring K_{eff} may vary: approaching saturation in high static fields [30], monitoring low-temperature hysteresis loops [16], analyzing the dependence of the Néel relaxation time [31], but this does not matter much. As soon as the scaling law $K_{\mathrm{eff}}(d)$ is found, it points out the appropriate limit of pinning that, in turn, leads to the estimate for K_s. As noted, the best conditions for measuring it are provided in the weak pinning limit, i.e., on using sufficiently small particles.

4. Experimental

4.1 MAGNETIC FLUIDS AS SAMPLE OBJECTS FOR FMR EXPERIMENTS

Magnetic fluids or ferrofluids, being assemblies of nanosize magnetic particles suspended in a liquid matrix, are well known as the objects of unique qualities for micromagnetic studies. They both provide experimental objects for determining the magnetic properties of fine particles and serve as plausible test systems to verify the developed theoretical approaches. The number of papers on this subject is quite numerous. We just refer to the proceedings of the last three International Conferences on Magnetic Fluids [41] noting that each issue contains a cumulative bibliographic list for several years.

Below follows a brief description of the ionic magnetic fluids containing γ-Fe$_2$O$_3$ (maghemite) particles, which were used in the magnetic resonance investigations [20–23]. The system in question is a colloidal suspension containing a great number of particles of a nanometer size coexisting in a liquid matrix. The method to synthesize ionic magnetic fluids was developed by Massart [42] at the beginning of the 1980's. First, magnetite (Fe$_3$O$_4$) particles are synthesized by coprecipitation in an alkaline medium of an aqueous mixture of ferric and ferrous salts. Further oxidation yields stable maghemite particles. The oxide γ-Fe$_2$O$_3$ is known to have an inverse spinel structure with Fe^{3+} ion vacancies in the octahedral metal sublattice. The mean size of the magnetic particles is controlled by the synthesis parameters such as temperature, molar ratio Fe^{3+}/Fe^{2+} and pH. The colloidal stability is ensured [43–45] by screened electrostatic repulsions between the particles due to a preliminary sodium citrate adsorption. The stability is confirmed in a wide temperature range and in external fields up to 10 kOe by SANS [46–48] and optical diffraction [43].

The spinel structure of the maghemite particle cores is verified by X-ray and neutron diffraction. The Curie point of the bulk maghemite is about 590 °C and thus far above the room temperature [49]. Given that, one may be sure that at helium temperatures M remains of the order of the bulk value. Static magnetization measurements confirm this conclusion indicating that the ferrimagnetic order, at least for the particle core, is preserved even for rather small (~ 4 nm) grains [49]. The bulk anisotropy of massive γ-Fe$_2$O$_3$ deduced from microwave measurements [50] is known to be cubic with the crystallographic anisotropy constant $K_v = -4.7 \times 10^4$ erg/cm^3, i.e., the easy directions correspond to [110] axes. Due to the low value of the magnetocrystalline anisotropy K_v, other anisotropies such as shape-induced or surface anisotropy are expected to completely dominate it. We remark that a significant though not drastic (less than 20%) dependence of the magnetization on the temperature and the particle volume was observed at 50 kOe [49, 51]. This fact should be considered along with the evidence first noted about thirty years ago [52] of the spin canting effect in the outer layer of maghemite nanoparticles. This yields the well known absence of saturation in γ-Fe$_2$O$_3$ fine particle assemblies, see Refs. [25, 51, 53, 54]. The spin-glass aspect of this problem we consider in Section 6.

Electron micrographs of evaporated ferrofluid samples show rock-like particles. Their shape is usually approximated by spheres. The size histograms are satisfactorily described by a log-normal distribution of the particle diameters d:

$$P(d) = \frac{1}{\sqrt{2\pi}sd} \exp\left[-\frac{1}{2s^2}\left(\ln\frac{d}{d_0}\right)^2\right], \qquad (4.1)$$

where d_0 is the most probable particle diameter and s is the standard deviation of $\ln(d)$. The size distribution in ferrofluids was tested by SANS experiments in the dilute regime by measuring the weight average diameter and gyration radius [44]. As an independent complementary test, the magnetic moment distribution may be deduced from magnetization measurements in the liquid state at room temperature. In the independent-grain regime (particle volume fraction < 2%), the shape of the magnetization curve in a liquid suspension is well described as a direct superposition of single-particle contributions. Because of the rotational degrees of freedom of the particles in the liquid carrier, the response of each isolated grain is given by the Langevin function $L(\xi)$ of the argument $\xi = MvH/kT$. Within the experimental accuracy ($\sim 10\%$), nuclear neutron size determinations and magnetization ones are equivalent [46, 47]. Standard laboratory-synthesized ferrofluids have d_0 about 7–8 nm and s about 0.35–0.4. On the one hand, such a value of the polydispersity parameter s is much lower than that encountered under condensation technique, e.g. in Ref. [30] for iron particles the width s was greater than 1.2. On the other hand, even $s \sim 0.4$ implies that in a suspension in comparable quantities coexist grains of diameters 3 nm and 14 nm, which are characterized by an essentially different dynamic behavior.

To avoid this mixing of properties and to be able to trace the effect of the particle size, in Ref. [20] a size-selection process was performed based on the following. If one applies to a magnetic fluid the conventional term "smart material", then we may say that an ionic ferrofluid is a smart material not only with respect to the magnetic field but as well it is a system which has a strong response to small changes of the ionic strength of its matrix. Taking advantage of this effect, a phase separation in the initial sample was initiated by an increase of the ionic strength, i.e., screening out the electrostatic repulsions. As the phase diagram of the colloidal suspension is dependent on the size of the particles, a succession of controlled phase separations allowed "slicing" of the initial size distribution of the colloid [55, 56]. As a result, for each fraction, the distribution width s was reduced to 0.1–0.2. Under such conditions, the histograms of the extracted samples became more narrow than the difference between the mean particle sizes of the samples in the series. This ensured a unique opportunity to investigate the size dependencies of the magnetic effects at one and the same temperature. Table 4.1 gives the size distributions of the "sliced" samples as determined from the magnetization curve at room temperature; for comparison the characteristics of the initial preparation are also given. We remark that the size-sorting process warrants that aggregates are practically absent. The majority of FMR experiments of Refs. [20, 23] were carried out on glycerol-carrier ferrofluids (freezing temperature $T_g = 210\,\mathrm{K}$) in the low concentration range ($\lesssim 1\%$).

Table 4.1. Histogram characteristics of ferrofluid samples 2–6 from Ref. [20].

Sample	d_0, nm	s	$v = (\pi/6)\langle d^3\rangle$ $= (\pi/6)\,d_0^3\exp(4.5s^2)$, nm^3
Initial preparation	7	0.35	312
2	10	0.2	627
3	9.3	0.1	441
4	7.7	0.1	250
5	6.5	0.15	159
6	4.8	0.2	69

FMR measurements were performed with Varian E102 spectrometer at the frequency $\omega_0/2\pi = 9.26\,\mathrm{GHz}$ (X-band). The modulation field had the frequency 100 kHz and amplitude 10 Oe. The first derivative dW/dH of the power absorption W was recorded as a function of the applied field H in the range 0–20 kOe. Very small quantities (microliters) of highly dilute ferrofluids were used in order to prevent distortion of the radio-frequency field. Sample cooling was performed with the aid of an Oxford Instruments cryostat in the temperature range 3.5 to 300 K. The cooling process was essential because of the orientational mobility of the particles in the carrier liquid; freezing fields H_{fr} up to 15 kOe were used.

For a fixed frequency of the spectrometer, the dynamic susceptibility of the sample was recorded as a function of the strength and the direction of the applied field. With the spectrometer used, the reference resonance field is $H_0 = \omega_0/\gamma = 3.3\,\mathrm{kOe}$. For a given sample the field H_{res} was registered as the field for which $dW/dH = 0$. The distance between the bounding peaks of the first derivative of the absorption line was taken as the experimental width of the resonance line. The recorded signal was found to be proportional to the microwave power up to 100 mW with no sign of saturation which ensured that contribution from nonlinear effects is negligible.

168   R. Perzynski and Yu. L. Raikher

4.2   MAGNETIC SPECTRA OF FROZEN MAGNETIC FLUIDS: RANDOM ASSEMBLIES AND ORIENTATIONAL TEXTURES

The liquid matrix of magnetic fluids provides a unique possibility to prepare samples with controlled orientational structure. Namely, imposing an external constant field H_{fr} when cooling a magnetic fluid below the freezing temperature of its liquid matrix, one can produce a sample where a preferred direction of internal magnetic anisotropy axes (bulk or surface) is fixed. After that, performing FMR measurements with the magnetizing field directed along, across or under some angle to the axis of the prepared structure, valuable information about the anisotropic properties of the particles may be obtained. Another reference configuration—an orientationally random structure—is created by freezing a sample in a zero field.

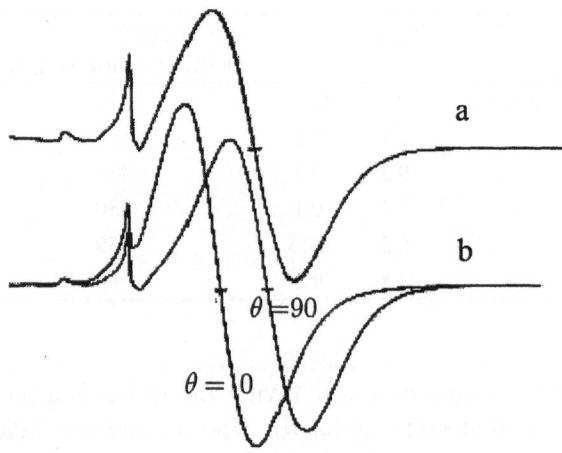

Figure 4.1.   X-band FMR spectra taken at 3.5 K with magnetic fluid sample 2 of Ref. [20], see Table 4.1; (a) the system is frozen without applied field; (b) the system is frozen under $H_{fr} = 10\,\text{kOe}$, the orientations of the magnetizing field while measuring with respect to $H_{fr}(0)$ are marked on the curves. The first peak is due to residual impurities in the glass micropipe.

Examples of the experimental results of the above-mentioned kind are presented in Figs. 4.1 and 4.2. There θ is the angle between the directions of the freezing field H_{fr} and the actual magnetizing field at which FMR is measured. Figure 4.3 illustrates variations of H_{res} as a function of θ in the field-frozen sample at 3.5 K.

The absorption line of field-frozen samples shifts monotonically to higher resonance fields as the magnetizing field is turned away from the direction of the freezing field. This is shown in Figs. 4.1 and 4.3 for sample 2 cooled under $H_{fr} = 10\,\text{kOe}$. The minimum and maximum of H_{res} are found, respectively, at $\theta = 0$ and 90 degrees. For randomly oriented samples, the resonance field falls in between these extreme values. Above the glycerol freezing temperature T_g,

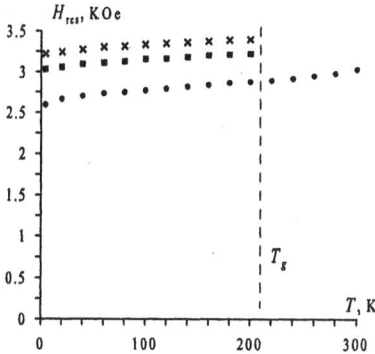

Figure 4.2.   Temperature dependence of the resonance field for sample 2 frozen under $H_{\text{fr}} = 10\,\text{kOe}$ and without field. Dashed vertical indicates the melting point of the liquid matrix; $\theta = 0$ (circles), $\theta = 90°$ (crosses), frozen with $H_{\text{fr}} = 0$ (squares). After Ref. [20].

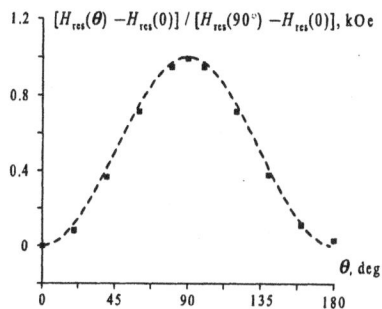

Figure 4.3.   Angular variation of the resonance field for the sample 2 at $T = 3.5\,\text{K}$. Squares are experiment, the dashed line renders the dependence $\sin^2 \theta$; after Ref. [20].

the line loses its anisotropy. Indeed, the particles in the liquid matrix are free to rotate towards the direction of the magnetizing field, and thus the resonance field values collapse to the curve for $\theta = 0$. The angular dependence of H_{res} shown in Fig. 4.3 is fitted with a good accuracy by the function

$$H_{\text{res}}(\theta) = H_{\text{res}}(0) + [H_{\text{res}}(90°) - H_{\text{res}}(0)] \cdot \sin^2 \theta. \qquad (4.2)$$

From the FMR theory [36] the resonance condition for a single-domain particle depends upon the orientation of its anisotropy axes with respect to the magnetizing field. The spectrum of the textured suspension, being a superposition of the signals from individual grains, should reflect this tendency. However, the obtained behavior of H_{res} turns out to be incompatible with the hypothesis that the particle anisotropy is that of the bulk ferrite. Indeed, bulk maghemite is known to be cubic. Meanwhile, the dependency given in Fig. 4.3 being well fit by Eq. (4.2), evidences a uniaxial magnetic anisotropy. This follows from the fact that the empirical expression (4.2) coincides with the theoretical condition [57, 58]:

$$H_{\text{res}} = \omega_0/\gamma - H_a P_2(\cos\theta), \tag{4.3}$$

where P_2 is the second Legendre polynomial. Note that when comparing Eqs. (4.2) and (4.3) we identified the direction of the particle principal axis with the direction of texturation of the system so that the angle θ retains its meaning.

According to Eq. (4.3), the particular values of H_{res} for $\theta = 0$ and $90°$, assuming perfectly aligned particles, are $H_{\text{res}}(0) = \omega_0/\gamma - H_a$ and $H_{\text{res}}(90°) = \omega_0/\gamma + H_a/2$. Thus the increment

$$\Delta H_{\text{res}} = H_{\text{res}}(90°) - H_{\text{res}}(0) = 3H_a/2$$

should yield the sign and the amplitude of the anisotropy field. Experimentally, for sample 2 of Ref. [20], $\Delta H_{\text{res}} = 600\,\text{Oe}$ at $3.5\,\text{K}$ thus meaning that the sample has a uniaxial anisotropy with $H_a = 400\,\text{Oe}$. The equality, however, would take place only for the perfect alignment of grain axes. In order to determine H_a, it is necessary to account for the orientational distribution of the particles which depends on the amplitude of the freezing field and on the ability of the grains to align with its direction.

Under an applied field, the orientational distribution of the particle anisotropy axes n in a ferrofluid sample results from a competition between the magnetic energy tending to orient the magnetic moment $\mu = Mve$, the anisotropy energy which binds μ to the anisotropy axis and the thermal energy which tends to disorient both μ and n. In thermal equilibrium, the distribution function of the anisotropy axes in the freezing point of the matrix is rendered by the expression [58]:

$$f(n, H_{\text{fr}}, T_g) = \frac{1}{Z} \int \exp\left[\frac{Mv}{kT_g}(e \cdot H_{\text{fr}}) + \frac{E_a}{kT_g}(e \cdot n)^2\right] de. \tag{4.4}$$

Here E_a is the particle magnetic anisotropy energy which is assumed to be uniaxial with the easy axis n. Introducing the dimensionless parameters

$$\xi = MvH/kT, \qquad \sigma = E_a/kT, \tag{4.5}$$

with the use of them we rewrite the partition function Z in Eq. (4.4) as

$$Z = 16\pi^2 \frac{\sinh\xi_g}{\xi_g} \int_0^1 \exp(\sigma_g y^2)\,dy, \tag{4.6}$$

where index g means that the values of these parameters are taken at $T = T_g$.

Figure 4.4 illustrates the evolution of the orientational distribution function (4.4) for a particle of diameter $10\,\text{nm}$ at several freezing fields and for $E_a =$

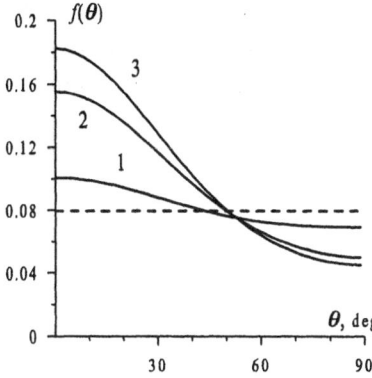

Figure 4.4.   Orientation distribution function of the particle anisotropy axes for various freezing fields: $H_{\mathrm{fr}} = 0.4\,\mathrm{kOe}$ (1), $2\,\mathrm{kOe}$ (2), and $10\,\mathrm{kOe}$ (3). Evaluation is done for a maghemite particle of diameter $10\,\mathrm{nm}$ at $H_a = 0.5\,\mathrm{kOe}$ and $T_g = 200\,\mathrm{K}$; the dashed line shows the case of isotropic particles $f = 1/4\pi$.

MvH_a assuming that $H_a = 0.5\,\mathrm{kOe}$. For this model system major changes take place in the range $H_{\mathrm{fr}} < 2000\,\mathrm{Oe}$, further increase of the freezing field changes the distribution in a much weaker way. This conclusion agrees with the experimental variations of H_{res} shown in Fig. 4.5. Indeed, for sample 2 ($d_0 = 10\,\mathrm{nm}$) saturation occurs already above $H_{\mathrm{fr}} = 3\,\mathrm{kOe}$.

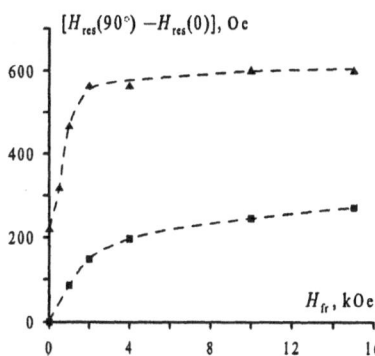

Figure 4.5.   Experimentally found differences in the resonance fields of orientationally textured maghemite ferrofluids: sample 2, $d_0 = 10\,\mathrm{nm}$ (▲); sample 6, $d_0 = 4.8\,\mathrm{nm}$ (■); measurements made at T=3.5 K; dashed lines are for eyeguide only.

On diminution of the particle size, as it can be seen from Eqs. (4.4) and (4.6) and Fig. 4.4, the orienting effect of the field at a given temperature, e.g. T_g, decreases. Indeed, however perfectly would the external field orient the particle magnetic moments, the orientation of the particle axes depends on the value of the anisotropy energy, which in dimensionless form is rendered by the parameter σ. Note that at $\sigma = 0$ the distribution function (4.4) of the particle axes becomes isotropic regardless of the field-dependent argument ξ.

To extract quantitative information from the particle size dependence of the angular variations of H_{res}, a simple model is used. Instead of averaging the complete expression for the susceptibility, one assumes that the angular dependence of H_{res} in a partially textured suspension is reduced by the orientational distribution function as

$$H_{\text{res}}(90°) - H_{\text{res}}(0) = \frac{3}{2} H_a \overline{P_2(\cos\theta)}$$

$$= \frac{3H_a}{2H_{\text{fr}}^2} \int \left[\frac{3}{2}(n \cdot H_{\text{fr}})^2 - \frac{1}{2} \right] f(n, H_{\text{fr}}, \sigma_g)\, dn. \quad (4.7)$$

The facts that experimental spectra are symmetrical and have angular variations $\propto \cos^2\theta$ justify this expression.

In Eq. (4.7) the parameter σ_g which enters the distribution function f is the only unknown quantity. The results of its evaluation from the measurement data are given in Table 4.2 under the assumption that $\sigma_g = K_v v/kT_g$.

Table 4.2. Values of the bulk anisotropy constant of maghemite nanoparticles as deduced from the FMR data at 3.5 K. The orientation distribution factor $\overline{P_2(\cos\theta)}$ is taken at $T_g = 210\,K$ for $H_{\text{fr}} = 10\,kOe$.

Sample	d_0, nm	$\sqrt[3]{\langle d^3\rangle}$	$H_{\text{res}}(90°) - H_{\text{res}}(0)$, Oe	$\overline{P_2(\cos\theta)}$	K_{eff}, $\frac{\text{erg}}{\text{cm}^3}$
2	10.0	10.7	641	0.46	1.5×10^5
3	9.3	9.4	512	0.34	1.7×10^5
4	7.7	7.8	382	0.19	2.0×10^5
5	6.5	6.7	307	0.13	2.3×10^5
6	4.8	5.1	240	0.0073	3.3×10^5

5. FMR in a spherical particle with the Aharoni surface anisotropy

In attempting to interpret the size dependence of the FMR resonance field (Table 4.2) in terms of bulk magnetic anisotropy, one comes up with a conclusion that K_v is particle size-dependent. This does not seem possible, however. The more so that all the samples (and, thus, all the particles) are produced simultaneously from one and the same mixture of iron salts and are size-separated only after that. Far more natural, with regard to the considerations presented in Sections 2 and 3, is to attribute the effect to the developed surface anisotropy of the particles. As we have seen, this very effect causes in fine particles a size-dependent contribution to the internal field whose dependence on the particle size varies from $\propto d^{-2}$ for strong pinning to $\propto d^{-1}$ for weak pinning.

The first result which may be obtained from the data in Table 4.2 is the determination of the slope of the dependence of K_{eff} on the particle diameter. Although in each sample we deal with a polydisperse assembly of particles, the small width of each lognormal distribution evidenced by closeness of d_0 and $\sqrt[3]{\langle d^3 \rangle}$ allows one to take any of those as the measure of the mean particle diameter in a given sample. A plot in double logarithmic coordinates done in Ref. [20] proved that with a good accuracy the dependence $K_{\text{eff}}(d)$ obeys the hyperbolic law $\propto d^{-1}$. With allowance for the above-given considerations this means the weak pinning limit and the possibility of macroscopic description in terms of the surface anisotropy energy density K_s.

In above when analyzing the spin-wave excitation problem we derived the scaling laws and other principal dependencies assuming a sample of a plane configuration (a film). This allowed us to easily obtain the relationships, which were used then to discuss the main issue of this work, i.e., ferrite nanoparticles whose shape is close to a spherical one. Being admissible in qualitative aspect, such an approach may, however, fail in the part which concerns exact numerical coefficients. In order to augment precision, a more rigorous treatment is necessary. As already noted, in the case of a radial anisotropy the problem does not have an analytical solution even for a perfect sphere. This perfectness, however, makes it very probable that the radial contributions to K_s will self-average. Thence the main role goes to a uniaxial contribution of the Aharoni type. Remarkably, Aharoni himself had developed a formal scheme which yields the analytic solution of the problem [59,60].

In Ref. [59] he studied spin-wave excitations in a spherical particle whose size is small enough to make the exchange increment of the energy dominate the magnetostatic one. Numerically this is expressed by the condition $d < \sqrt{\alpha}$ that for maghemite yields $d \lesssim 20$ nm. Consider a typical FMR situation, where a particle is subjected to a strong external magnetic field H_0, which in combination with the internal fields (that of the bulk anisotropy H_{av} and the demagnetizing field H_d) determines the equilibrium orientation of the particle magnetization $M = Me$. This state is perturbed by a weak probing rf field that is perpendicular to H_0. In the approximation adopted one neglects nonuniform demagnetizing fields created by the perturbations of M in the transverse (with respect to H_0) direction. As a result, the dynamic equation to be solved takes the form of Eq. (3.22). Directing Oz along the external magnetizing field H_0 and omitting at the first step the relaxation term, for the transverse components of the dimensionless magnetization e we get

$$\frac{\partial e_x}{\partial t} = -\gamma e_y H_u + \alpha \gamma M \nabla^2 e_y, \qquad \frac{\partial e_y}{\partial t} = \gamma e_x H_u - \alpha \gamma M \nabla^2 e_x, \quad (5.1)$$

where the uniform field along Oz is

$$H_u = H_0 - 4\pi M/3 + 2K_v/M. \qquad (5.2)$$

Equation (5.1) is solved in spherical coordinates (polar axis along H_0) with the aid of the expansions proposed in Refs. [59, 60]:

$$e_x = \sum_\ell A_\ell e^{i\omega t} P_\ell(\cos\vartheta) j_\ell\left(\frac{\varkappa r}{R}\right), \quad e_y = \sum_\ell B_\ell e^{i\omega t} P_\ell(\cos\vartheta) j_\ell\left(\frac{\varkappa r}{R}\right),$$

$$(5.3)$$

where A and B are determined by the initial conditions, P_ℓ are the Legendre polynomials, j_ℓ are spherical Bessel functions, $R = d/2$ is the particle radius and $\varkappa$ are the eigenvalues to be evaluated. As follows from Eqs. (5.3), the distribution of magnetization is axially symmetrical and the profile of the standing spin wave that occurs in the particle is characterized by a product of two functions each of which oscillates with respect to its argument: r and ϑ. Substituting Eqs. (5.3) in (5.1) and requiring that the set has a nonzero solution one arrives at the dispersion equation [59, 60]:

$$\omega = \gamma\left[H_0 + \frac{2K_v}{M} + \frac{\alpha M}{R^2}\,\varkappa_\ell^2(R)\right], \qquad (5.4)$$

that expresses the magnetic resonance condition in a nanosphere.

The last term on the right-hand side of Eq. (5.4) presents a general form of the internal field increment owing to the effect of the particle surface. Particular forms of this term become clear only after each eigenvalue $\varkappa_\ell$ is found with the aid of the boundary condition. The first case is the "natural" condition (3.15) which in the present notations is written as

$$\left[\frac{d}{dr} j_\ell\left(\frac{\varkappa_\ell r}{R}\right)\right]_{r=R} = 0, \qquad (5.5)$$

and, as mentioned, corresponds to a particle with a magnetically isotropic surface. Apart from the trivial $\varkappa = 0$ which pertains to uniform precession, the roots of this equation are positive numbers greater than 2 whatever ℓ [59, 60]. Thence the resonance condition (5.4) acquires a contribution that is proportional to R^{-2}. The obtained situation becomes completely clear as soon as one recalls that it is a direct "spherical" analog of the case of higher spin-wave modes in the zero pinning situation in a film, see Section 3.4. As is shown there, with respect to the energy of the higher spin-wave modes the cases of strong and weak pinning are practically identical. This means that in those regimes it is impossible to measure K_s but the situation is promising for evaluation of the exchange parameter.

The case of finite surface anisotropy with regard to the exchange spin-wave modes was studied in Ref. [60]. In formal aspect, the only change that is necessary is to replace the boundary condition (5.5) by the equation

$$\left[b \frac{d}{dr} j_\ell \left(\frac{\varkappa_\ell r}{R} \right) + j_\ell \left(\frac{\varkappa_\ell r}{R} \right) \right]_{r=R} = 0, \qquad (5.6)$$

where $b = \alpha M^2 / 2K_s$ is the extrapolation length defined by Eq. (2.10). In Ref. [60] such a study has been undertaken with the focus on the spectrum of the spin waves proper ($\ell \geq 1$). Equation (5.6) was solved numerically for various b and appearance of an effective dependency $\varkappa(R)$ was demonstrated. The latter means that the combination $\varkappa/R^2$ ceases to be a simple power law and becomes a more complicated function. In Ref. [60] some examples of this behavior are given and the theory is shown to fit qualitatively the size dependence of the spin-wave eigenfrequencies observed on submicron particles [61].

In Ref. [60] the quasi-uniform mode ($\ell = 0$), which is a successor of the Larmor precession in the weak pinning limit, was not touched at all. Meanwhile for nanoparticles this mode is: (1) the only one accessible with the conventional technique and (2) the only mode which readily allows one to extract from the measurement data the value of the particle surface anisotropy constant. The pertinent work was performed in Refs. [21, 62]. There a full Landau-Lifshitz equation with allowance for the relaxation term was solved in a spherical particle with a weak surface pinning and the profiles of the standing spin waves were obtained for the cases where the magnetizing field H_0 is either parallel or perpendicular to the surface and bulk magnetic anisotropy axes which were assumed to have coinciding directions.

Not repeating the details of calculations done in Refs. [21, 62], let us show how the resonance condition for the longitudinal situation where the magnetizing field H_0 is parallel to the common easy axis n of the bulk and surface turns up. Weak pinning means a long-wave limit, that is $\varkappa \ll 1$. Setting $\ell = 0$ in Eq. (5.6) and expanding the spherical Bessel function j_0 at small values of its argument we get $\varkappa^2 = 3R/b$. Substituting this in Eq. (5.4) yields

$$\omega = \gamma(H_0 + H_{av} + 6K_s/MR) = \gamma(H_0 + H_{av} + 12K_s/Md), \qquad (5.7)$$

where $H_{av} = 2K_v/M$ is the effective field of the bulk uniaxial anisotropy. Thus one sees that the surface contribution obtained rigorously, exactly equals the one predicted by the semi-quantitative estimation (1.11). The latter fact, however, is rather a coincidence than a fundamental circumstance.

The expressions for the resonance frequency and relaxation time of the quasi-uniform precession under a strong field H_0 in the longitudinal ($H_0 \parallel n$) and transverse ($H_0 \perp n$) cases obtained in Refs. [21, 62] are, respectively:

$$\omega_\parallel = \gamma \left[H_0 + H_{av} + 6K_s/MR \right], \quad \tau_\parallel = (\gamma\lambda)^{-1} \left[H_0 + H_{av} + 6K_s/MR \right]^{-1};$$
$$\omega_\perp = \gamma \left[H_0 - H_{av} - 3K_s/MR \right], \quad \tau_\perp = (\gamma\lambda)^{-1} \left[H_0 - H_{av} - 3K_s/MR \right]^{-1}.$$
$$(5.8)$$

Transforming these formulas into conditions which determine the resonance fields and then combining the latter to account for the resonance field difference measured on the textured (field-frozen) samples, one gets

$$(2/3)\,[H_{\text{res}}(90°) - H_{\text{res}}(0)] = 6K_s/MR + H_{av}. \qquad (5.9)$$

This dependence was used to interpret the data of Ref. [20] presented here in Table 4.2. The results are shown in Fig. 5.1 for two variants of fitting. The dashed line corresponds to Eq. (5.9) with $K_v = 0$, it yields the Aharoni anisotropy constant $K_s = 2.4 \times 10^{-2}$ erg/cm^2. The solid line corresponds to the same Eq. (5.9) but under the condition of the best linear fit, its parameters are $K_s = 2.8 \times 10^{-2}$ erg/cm^2 and $2K_v/M = -210$ Oe. With the accuracy of the experimental determination of H_{res}, the difference between the found values of the surface anisotropy is irrelevant, its reference value agrees well with the estimate $(2.7 \pm 0.1) \times 10^{-2}$ erg/cm^2 given for the same samples in Ref. [20]. We remark that the found value of K_s satisfies well the weak pinning approximation thus ensuring the self-consistence of the calculation carried out.

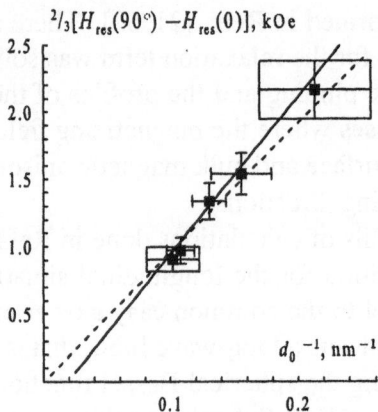

Figure 5.1. Difference of resonance fields in magnetic fluid samples orientationally textured under the field 1 T as a function of the most probable particle diameter. Points result from the FMR data taken at 3.5 K and corrected with respect to the imperfectness of texturation; dashed line corresponds to $K_v = 0$ and $K_s = 2.4 \times 10^{-2}$ erg/cm^2; solid line to $K_s = 2.8 \times 10^{-2}$ erg/cm^2 and $2K_v/M = -210$ Oe.

As it appears from the plot, with the existing accuracy one cannot get any trustworthy evaluation for the bulk anisotropy parameter $2K_v/M$. In Fig. 5.1 the possible "fan" of lines, that the set of experimental points admits—see errorboxes drawn around the end points—spans over quite a range of $2K_v/M$. However, the plots of Fig. 5.1 definitely support the conclusion that for the studied quasi-spherical maghemite nanoparticles the bulk anisotropy field is small in comparison with the surface term which plays the dominating part.

In Ref. [63] the formalism of the Aharoni exchange modes was applied for interpretation of the magnetic rf spectra of larger (25–250 nm) spherical particles of Co-Ni and Fe-Ni-Co alloys of varying content. On those systems the so-called intrinsic FMR was measured [61, 63, 64] which results from the Larmor precession that takes place in the absence of an external magnetizing field. In the experiments, several spin-wave modes were resolved with the resonance frequencies depending on the particle size. The theory involved [63] dealt with two limiting cases—zero and rigid (infinite) pinning. Due to that the uniform precession mode was not modified while the higher modes displayed a number of size dependencies of the type (5.4), where $H_0 = 0$ and $\varkappa$ is a number. From fitting the measurement results, the bulk anisotropy constants were deduced in satisfactory agreement with the reference data. In Refs. [65, 66] the problem of intrinsic FMR in a fine spherical particle with magnetically isotropic surface was solved numerically in full formulation, i.e., with allowance for the non-uniform demagnetizing fields. The obtained dependencies were successfully used to interpret the measurements of Refs. [61, 64], and to evaluate the exchange parameter for Ni and Co-Ni microspheres.

6. FMR in a spherical particle with the rotatable exchange anisotropy

High relative fraction of unsaturated or broken atomic bonds at the surface of fine particles together with the presence of impurities and defects leads to spin frustration. This, in turn, implies that the magnetic structure of the surface becomes irregular and its arrangement resembles that of a spin glass. The occurring magnetic disorder, however, seems not to go too far inside the bulk and is located mostly at the surface. This justifies the core-shell model where the particle is presented as a magnetically ordered bulk surrounded by the outer layer with irregular spin arrangement.

The fact of distorted magnetic structure at the surface of fine particles was noted [52] long ago. Known as *spin canting* this effect was many times since then confirmed by both Mössbauer spectroscopy [67, 68] and magnetization measurements in high constant fields [69]. On closer examination it turned out, however, that just the assumption of spin canting is not enough. Although it accounts for the fact that in fine particles magnetization does not saturate even in very strong (tens of teslas) applied fields, it is insufficient to explain the lack of reversibility that is observed in fine particle assemblies under cyclic re-magnetizing. To account for the latter, one has to suppose that the occurring spin canting is so irregular that the structure of the surface layer in a fine particle is much like that of a spin glass [25, 70, 71]. Nanoparticles where this spin-glass-like layer is formed on the surface of the magnetically ordered core display magnetic behavior of a specific type. While the main contribution to the magnitude of the particle magnetic moment μ originates from that of the core, the anisotropy experienced by μ and thus its orientational behavior is mainly controlled by its interaction with the surface layer and not with the anisotropy axis of the core. In other words, the magnetization of the surface layer does not matter much; what really matters is the exchange interaction across the core-shell border, i.e., the interface anisotropy.

A well-known example of such an interaction is the *exchange* (or unidirectional) anisotropy discovered by Mieklejohn and Bean at the end of the 1950's [72, 73] in the case of the ferromagnet–antiferromagnet border. A spin glass with its great number of spins looking opposing directions structurally resembles an antiferromagnet (AF). Thence one should expect the exchange anisotropy to be inherent to some or other extent to any interface between a ferromagnet (FM) and a spin glass (SG). This conclusion is well proven, and the exchange (unidirectional) anisotropy since long ago is a customary effect for spin-glass films and sandwich structures, especially for those superthin ones known as *superlattices*.

In the classical model, the exchange anisotropy does not change the direction of its axis inside the particle. As a result, if one makes a cyclic remagnetization of the ferromagnetic part of the AF-FM pair, the emerging hysteresis loop is shifted with respect to the axis $H = 0$. The exchange anisotropy occurring in an SG-FM sandwich behaves in a different way. Above the spin-freezing temperature the magnetically frustrated part is very labile and readily changes its orientational structure in response to motion of magnetization in the ferromagnetic counterpart. These adjustments lower the potential energy of the system and stabilize the orientation of the magnetization in FM. When describing the effect of SG reorientation in terms of effective internal fields, it should be attributed to an internal anisotropy field which is always parallel to the actual direction of the magnetization vector of the FM part of the sandwich. Hence, a hysteresis loop, if existing, would not display any asymmetry.

As the temperature goes down, the response time of the frustrated part grows and finally, on passing transition to the low-temperature state, the disorder freezes [19]. As a result of creation of a magnetically rigid SG a permanent direction of the exchange anisotropy axis is established. Under these conditions the SG-FM sandwich should display a shifted hysteresis loop as in the classical case. Moreover, it is possible to tune the direction of the exchange anisotropy by choosing appropriate freezing conditions under field. Apparently these considerations equally apply to the case where instead of a flat sandwich one has a spin-frustrated layer (shell) wrapped around a magnetically ordered core of a metallic or ferrite nanoparticle. For example, existence of a surface layer capable of transition from a ferromagnetic to SG state under freezing was demonstrated in Refs. [74, 75] where the state of the surface in nanosize iron particles was investigated with the aid of AC susceptibility, remanence, magnetic relaxation measurements, and Mösssbauer spectroscopy.

The above-described behavior of a spin-disordered surface layer in a fine particle has a striking resemblance to the effects which are inherent to so-called *reentrant* spin glasses, where the ferromagnetic ordering competes with the spin-glass one [76, 77]. Indeed, in reentrant spin glasses, contrary to a typical scenario, the spin-glass state lies below the ferromagnetic one with respect to temperature. In such systems two characteristic temperatures are observed. First, the point where on diminution of T the ferromagnetic is replaced by a labile spin-glass-like structure and another, lower temperature, after reaching which the spin-glass arrangement becomes "rigid". As shown experimentally including, among others, the FMR technique [77], in the labile spin-glass-like state the internal field contains a contribution which has the form of the anisotropy field but whose axis seems to be always aligned with the equilibrium direction of the particle magnetic moment. Further investigations [78, 79] revealed that at low temperatures in reentrant spin glasses one encounters both *dynamic* and *rigid* components of the exchange anisotropy. The first is the one whose axis readily follows the equilibrium magnetization whereas the second is tightly bonded to some direction inside the particle. In some cases, as in NiMn or NiFe films [79–81], the dynamic (rotatable) component turns out to entirely dominate the rigid one.

A macroscopic model that takes into account the presence of this dynamic (or rotatable) anisotropy in nanoparticles was constructed in Ref. [22]. It is based on the phenomenological expression for the surface density of the exchange anisotropy [72]:

$$F_s = -K_{ea}\,(e \cdot l), \tag{6.1}$$

where K_{ea} is the pertinent constant and e is the unit vector of the particle magnetic moment. The unit vector l in a classical model of exchange anisotropy where a ferromagnet abuts on an antiferromagnet, coincides with the *antiferromagnetic vector* at the interface. For an antiferromagnet with two sublattices, their magnetizations being m_1 and m_2, vector l is defined as $l = (m_1 - m_2)/|m_1 - m_2|$. The sign of the anisotropy constant K_{ea} in Eq. (6.1) is chosen in such a way that its positive value favors parallelism of e and l. In the case of a core-shell interface in a fine particle one does not have so rigorous a definition for l, and we replace it by implication that it is some vector imposed by the surface. Fortunately, nothing else is necessary since when building a model for the rotatable anisotropy the second essential assumption is that with respect to low-frequency magnetization processes vector l aligns with the particle magnetic moment. In a standard situation of FMR measurement a strong magnetizing field H_0 (about 3 kOe) is imposed on a particle to align its magnetic moment μ. As this is done, then, according to the hypothesis of rotatable anisotropy, vector l readily orients parallel to μ. At the same time, with respect to the high-frequency process (the Larmor precession) the exchange anisotropy should be considered as a constant (unmovable) vector.

As the exchange anisotropy in a nanoparticle is a surface effect, then similarly to the treatment in Section 2 one may define the extrapolation length b and the pinning parameter $p = d/b$. Assuming weak pinning one can study FMR in a particle with rotatable anisotropy by exactly the same formal scheme as the one developed in Ref. [21] for the case of usual (rigid) surface anisotropy, see Section 5. The obtained solutions for the resonance frequency and the relaxation time of the precession take the following form

$$\omega = \gamma \left[H_0 + H_{av} P_2(\cos \theta) + 6K_{ea}/Md \right], \quad \tau = 1/(\lambda \omega), \qquad (6.2)$$

where θ is the angle between the freezing field H_{fr} and the magnetizing field H_0. Note that the angular-dependent factor enters only the bulk anisotropy term: in the simplest approximation the rotatable anisotropy does not depend on the position of the particle axes. As for the case of the surface anisotropy considered in the preceding section, we see that with respect to the resonance frequency the presence of exchange anisotropy manifests itself as some additional field shifting the resonance value.

Estimating the validity of expression (6.2) we remark that it is most appropriate at low temperatures, first, because in this region formation of a spin-glass-like layer at ferrite grains [70, 71] and, in particular, at γ-Fe_2O_3 (maghemite) nanoparticles [19, 25] is favored; second, because at low temperatures the superparamagnetic effects, which Eq. (6.2) does not account for, are weak. With allowance for these limitations, in Ref. ([22]) Eq. (6.2) was used to interpret the FMR measurements [20] carried out at 3.5 K with size-sorted maghemite ferrofluids, see Table 4.1. The X-band spectra were recorded in a standard way: under constant frequency $\omega_0/2\pi = 9.3$ GHz by sweeping the magnetizing field. Its resonance value for a single particle according to Eq. (6.2) is

$$H_{res}(\theta) = \omega_0/\gamma - 6K_{ea}/Md - H_{av} P_2(\cos \theta). \qquad (6.3)$$

Solidification of a magnetic fluid sample in a zero field is equivalent to isotropic angular averaging of Eq. (6.3). This eliminates the uniaxial anisotropy and for a zero-field frozen (ZFF) case yields the angular-independent relation

$$H_{iso} = \omega_0/\gamma - H_{res}(ZFF) = 6K_{ea}/Md. \qquad (6.4)$$

Thence the values of H_{iso} plotted as a function of $1/d$ should give a line ascending from the coordinate origin, its slope rendering the mean value of $6K_{ea}/M$. In Fig. 6.1 reproduced from Ref. [22] this conclusion is tested with the data of Ref. [20] on five ZFF samples. Their particle size distributions were identified, see Table 4.1, as lognormal ones with the most probable diameters d_0 ranging from 4.8 to 10 nm and the width parameters between 0.1 and 0.2. The latter (in dimensional units) are shown as the errorbars along the abscissa. Due to sufficiently small widths of the size distributions, the resulting plot practically does not change whatever statistical size measure is associated with the theoretical diameter d when fitting.

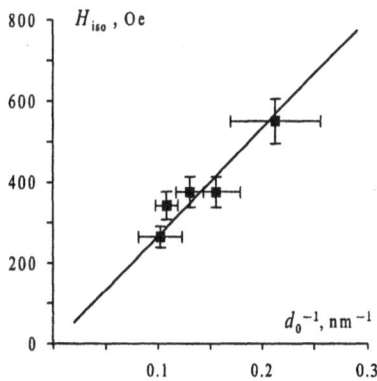

Figure 6.1. Isotropic contribution to the resonance field of ZFF magnetic fluid samples; points are FMR data after Ref. [20] taken at 3.5 K; the line corresponds to Eq. (6.4) with $K_{ea} = 1.4 \times 10^{-2}$ erg/cm^2.

The line fit shown in Fig. 6.1 yields for the exchange anisotropy density $K_{ea} = 1.1 \times 10^{-2}$ erg/cm^2 if $M \approx 400$ G, as in massive maghemite, or $K_{ea} = 1.4 \times 10^{-2}$ erg/cm^2 if the magnetization of the dispersed maghemite is set to $M = 300$ G as was estimated in Ref. [19]. Both these values do not differ greatly from the uniaxial surface anisotropy constant for the same particles, which, see Section 5, was evaluated as $K_s \approx 2.8 \times 10^{-2}$ erg/cm^2.

Direct measurements of K_{ea}, like the one reported above or done in Ref. [82] for MnFe$_2$O$_4$, are very important since until now implicit quantitative estimations for the exchange anisotropy are quite uncertain. For example, let us try the estimate K_{ea} based on the results of random field modeling of partially disordered ferromagnetic-antiferromagnetic interfaces on the mesoscopic level. This approach was described in Ref. [25] and applied there to oblate maghemite nanoparticles of nanosize range, the case close enough to ours. Rewriting the estimate in the units adopted here, one gets $K_{ea} \sim M\sqrt{\alpha K_{AF}}/6$. Setting, as in Ref. [25], the antiferromagnetic anisotropy constant $K_{AF} = 4 \times 10^5$ erg/cm^3 and $\alpha M^2 = 5 \times 10^{-7}$ erg/cm, one finds $K_{ea} \sim 0.1$ erg/cm^2, that is an order of magnitude greater than the value following from Fig. 6.1. Whatever explanation for this discrepancy could be given, it is clear that at present theoretical models for surface spin disorder in fine particles are yet too weak to really have predictive force. Meanwhile, simple phenomenological approaches like the above-described one are rather helpful for finding correlations between empirical facts and for systemizing the accumulating experimental data.

7. Concluding remarks

In the above in rather a schematic way spin we consider wave perturbations in thin films and fine particles. The macroscopic approach is used, so that magnetodynamics is described with the aid of the Landau-Lifshitz equation. The essential feature of the problem is that magnetization is pinned at the sample borders. The magnetic properties of the surface are characterized by just two phenomenological parameters: the energy density of surface anisotropy and the direction of the surface easy axis. Despite its obvious simplicity, the model is rather helpful and quite applicable to nanoscale objects. It allows for an arbitrary intensity of spin pinning and delivers a reasonable qualitative explanation for the dependence of the effective magnetic anisotropy constant on the particle size.

Application of the model to standard FMR measurements in nanoparticle assemblies opens a direct way to evaluate the surface anisotropy constants and to reveal the presence of spin-glass-like state in the surface layers. We demonstrate this possibility using, as an example, the theoretical interpretation of FMR data on size-sorted maghemite ferrofluids. Note that a unique feature of ferrofluids as test objects is that they may be orientationally texturized by first magnetization and then freezing under external field.

Notably, in the presented case of maghemite ferrofluids the surface anisotropy turns out to completely dominate the bulk contribution. This suggests a hypothesis able to explain a well known practically but rarely, if ever, discussed theoretically fact. Namely, a very typical situation is that a ferrofluid, whose particles are made of a magnetic compound with a cubic lattice, behaves under external field as if the particles had unambiguous uniaxial anisotropy. Formerly, two explanations were in use. First, it was assumed that the nanoparticles possess some nonsphericity so that the demagnetizing (shape) effect provides the necessary uniaxial anisotropy. Another explanation, admitting the initial sphericity of the particles, implied that, due to magnetic interactions, they unite in small chains. As such a chain is an anisometric object, it acquires uniaxial anisotropy owing to the same magnetostatic mechanism. Both hypotheses face logical difficulties when in reality the tests indicate rock-like shape of individual particles without any pronounced anisometricity and no signs of aggregation [48].

With the surface mechanism the model may be qualitatively formulated as follows. The cubic magnetic anisotropy of the particle core is intact but makes, however, just a small part of the orientation-dependent energy of the particle magnetic moment. A much larger contribution originates from the surface effect. In the latter, the components of the crystal field responsible for the local easy-axis or easy-plane anisotropy (the Néel type) self-average and fall out of consideration. This gives way to the uniaxial terms, and as a result the surface anisotropy of the Aharoni type is established. With allowance for the fact that the surface anisotropy per spin is much greater than the bulk one, in such a system, no visual anisometricity will be registered. Note that this hypothesis derives the commonly observed single-particle uniaxial magnetic anisotropy merely from the fact that a nanograin is a finite-size magnetic object.

In support of the above-proposed, we remark the study described in Ref. [83]. There the presence of uniaxial surface anisotropy is proven by direct measurements on isolated cobalt and iron clusters (1000 spins). The anisotropy effect is quantified in terms of effective bulk values, similar to that done in our Table 4.2. Experimental results for cobalt and iron are alike and range as high as $(1-3) \times 10^6$ erg/cm^3. In numerical simulations of cluster formation it is shown that this may be due to natural asymmetry arising while atomic layers settle on the cluster facets. Remarkably, the occurring anisometry remains weak: less than 10%. For cobalt clusters of about 1500 atoms the authors of Ref. [83] have achieved quantitative agreement between modeling and measurement.

In nanoparticle systems like ferrofluids the surface-induced magnetic anisotropy effect may of course combine with many-body (aggregation) mechanisms, and the effective anisotropy is caused by this joint action. To build a physically reasonable picture in these "hybrid" cases is a challenging but very difficult work.

In general, the main goal of this paper is to present evidence to the effect that magnetic surface science is a challenging field from both theoretical and experimental viewpoints. Moreover, regarding the latter, the FMR technique, being a routine with massive crystals, appears in the role of a powerful instrumental method. As a good example of a FMR-based detailed investigation of nanoparticles and the magnetic properties of their surface we mention the work reported recently in Refs. [84, 85]. There it is shown that studies of temperature dependence of FMR linewidth in maghemite nanoparticles in the low-temperature range bring independent proofs of a spin-glass-like layer at the particle surfaces.

Y.L.R. acknowledges partial financial support of the work from the International Association for the Promotion of Cooperation with Scientists from the New Independent States of the Former Soviet Union (INTAS) under Grant No. 01–2341 and by Award No. PE–009–0 of the U.S. Civilian Research & Development Foundation for the Independent States of the Former Soviet Union (CRDF).

References

[1] Brown, Jr., W.F. (1963) *Micromagnetics*, Wiley–Interscience, New York–London.

[2] Batlle, X. and Labarta, A. (2002) *J. Phys. D: Appl. Phys.* **35**, R15.

[3] Landau, L.D. and Lifshitz, E.M. (1984) *Electrodynamics of Continuous Media*, Pergamon Press, Oxford.

[4] Brown, Jr., W.F. (1969) *Ann. N.Y. Acad. Sci.* **147**, 463.

[5] Aharoni, A. (1996) *Introduction to the Theory of Ferromagnetism*, Clarendon Press, Oxford.

[6] Aharoni, A. (2001) *J. Appl. Phys.* **90**, 4645.

[7] Néel, L. (1954) *J. Phys. Radium* **15**, 225.

[8] Dormann, J.L., Fiorani, D., and Tronc, E. (1997) *Adv. Chem. Phys.* **98**, 283.

[9] Kaneyoshi, T. (1991) *J. Phys.: Condens. Matter* **3**, 4497.

[10] Dimitrov, D.A. and Wysin, G.M. (1994) *Phys. Rev. B* **50**, 3077.

[11] Dimitrov, D.A. and Wysin, G.M. (1995) *Phys. Rev. B* **51**, 11947.

[12] Zianni, X., Trohidou, K.N., and Blackman, J.A. (1997) *J. Appl. Phys.* **81**, 4739.

[13] Aharoni, A. (1987) *J. Appl. Phys.* **61**, 3302.

[14] Aharoni, A. (1988) *J.Appl. Phys.* **63**, 4605.

[15] Bødker, F., Mørup, S., and Linderoth, S. (1994) *Phys. Rev. Lett.* **72**, 282.

[16] Respaud, M., Broto, J.M., Rakoto, H., Fert, A.R., Thomas, L., Barbara, B., Verelst, M, Snoeck, E., Lecante, P., Mosset, A, Osuna, J., Ould Ely, T., Amiens, C., and Chaudret, B. (1998) *Phys. Rev. B* **57**, 2925.

[17] de Gennes, P.G. and Prost, J. (1993) *The Physics of Liquid Crystals*, Clarendon, Oxford.

[18] Lavrentovich, O.D. and Palffy-Muhoray, P. (1995) *Liquid Crystals Today* **5**, 5.

[19] Gazeau, F., Dubois, E., Hennion, M., Perzynski, R., and Raikher, Yu.L. (1997) *Europhys. Lett.* **40**, 575.

[20] Gazeau, F., Bacri, J.C., Gendron, F., Perzynski, R., Raikher, Yu.L., Stepanov, V.I., and Dubois, E. (1998) *J. Magn. Magn. Mater.* **186**, 175.

[21] Shilov, V.P., Bacri, J.C., Gazeau, F., Gendron, F., Perzynski, R., and Raikher, Yu.L. (1999) *J. Appl. Phys.* **85**, 6642.

[22] Shilov, V.P., Raikher, Yu.L., Bacri, J.C., Gazeau, F., and Perzynski, R. (1999) *Phys. Rev. B* **60**, 11902.

[23] Gazeau, F., Shilov, V.P., Bacri, J.C., Dubois, E., Gendron, F., Perzynski, R., Raikher, Yu.L., and Stepanov, V.I. (1999) *J. Magn. Magn. Mater.* **202**, 535.

[24] Zhang, K. and Fredkin, D.R. (1996) *J. Appl. Phys.* **79**, 5762.

[25] Martínez, B., Obradors, X., Balcells, Ll., Rouanet, A., and Monty, C. (1998) *Phys. Rev. Lett.* **80**, 181.

[26] Eagle, D.F. and Mallinson, J.C. (1967) *J. Appl. Phys.* **38**, 995.

[27] Quach, H.T., Freidmann, A., Wu, C.Y., and Yelon, A. (1978) *Phys. Rev. B* **17**, 312.

[28] Henrich, B., Celinski, Z., Cockran, J.F., Arrot, A.S., and Myrtle, K. (1991) *J. Appl. Phys.* **70**, 5769.

[29] Dormann, J.L., D'Orazio, F., Lucari, F., Tronc, E., Prené, P., Jolivet, J.P., Fiorani, D., Cherkaoui, R., and Nogués, M. (1996) *Phys. Rev. B* **53**, 14291.

[30] Gangopadhyay, S., Hadjipanayis, G.C., Dale, B., Sorensen, C.M., Klabunde, K.J., Papaefthymiou, V., and Kostikas, A. (1992) *Phys. Rev. B* **45**, 9778.

[31] Chen, J.P., Sorensen, C.M., Klabunde, K.J., and Hadjipanayis, G.C. (1994) *J. Appl. Phys.* **76**, 6316.

[32] Berkowitz, A.E., Lahut, J.A., Jacobs, I.S., Levinson, L.M., and Forester, D.W. (1975) *Phys. Rev. Lett.* **34**, 594.

[33] Zhang, K. and Fredkin, D.R. (1999) *J. Appl. Phys.* **85**, 6187.

[34] Landau, L.D. and Lifshitz, E.M. (1935) Reprinted in D. ter Haar (ed.), *Collected Works of L.D. Landau* (1967), Gordon and Breach, New York, p.101.

[35] To honor the essential contribution from the author with a correct dating we can but reproduce the reference from Chap.3 of the book [1], which is contemporary to the work: "Gilbert T.L. (1955) *Phys.Rev.* **100**, 1243 only an abstract; full text: *Armour Res. Found. Project AO59* — Suppl. Rep. 1956, May 1". We note also that the indicated *Physical Review* page is not available by means of PROLA.

[36] *Ferromagnetic Resonance*, S.V. Vonsovskii (ed.) (1966) Pergamon Press, Oxford.

[37] Kittel, C. (1958) *Phys. Rev.* **110**, 1295.

[38] Wigen, P.E. (1984) *Thin Solid Films* **114**, 135.

[39] Speriosu, V.S., Parkin, S.S.P., and Wilts, C.H. (1987) *IEEE Trans. Magn.* **23**, 2999.

[40] Raikher, Yu.L. and Shliomis, M.I. (1994) *Adv. Chem. Phys.* **87**, 595.

[41] Magnetic Fluids Bibliography. (1995, 1999, 2003) *J. Magn. Magn. Mater.* **149**, Nos.1–2; *ibid.* **201**, Nos.1–3; *ibid.* **252**.

[42] Massart, R. (1981) *IEEE Trans. Magn.* **17**, 1247.

[43] Bacri, J.C., Perzynski, R., Salin, D., Cabuil, V., and Massart, R. (1990) *J. Magn. Magn. Mater.* **85**, 27.

[44] Dubois, E., Cabuil, V., Boué, F., and Perzynski, R. (1999) *J. Chem. Phys.* **111**, 7147.

[45] Dubois, E., Perzynski, R., Boué, F., and Cabuil, V. (2000) *Langmuir* **16**, 5617.

[46] Bacri, J.C., Boué, F., Cabuil, V., and Perzynski, R. (1993) *Colloids and Surfaces A* **80**, 11.

[47] Dubois, E., Cabuil, V., Boué, F., Bacri, J.C., and Perzynski, R. (1997) *Prog. Colloid Polym. Sci.* **104**, 173.

[48] Gazeau, F., Dubois, E., Bacri, J.C., Boué, F., Cebers, A., and Perzynski, R. (2002) *Phys. Rev. E* **65**, Art.No. 031403.

[49] Aquino, R., Dubois, E., Depeyrot, J., Tourinho, F.A., and Perzynski, R. (in press).

[50] Birks, J.B. (1950) *Proc. Phys. Soc. B* **63**, 65.

[51] Gazeau, F. (1997) Thèse, Université Denis Diderot, Paris.

[52] Coey, J.M.D. (1971) *Phys. Rev. Lett.* **27**, 1140.

[53] Prené, P., Tronc, E., Jolivet, J.P., Livage, J., Cherkaoui, R., Nogués, M., and Dormann, J.L. (1994) *Hyperfine Interactions* **93**, 1409.

[54] Tronc, E., Prené, P., Jolivet, J.P., D'Orazio, F., Lucari, F., Fiorani, D., Godinho, M., Cherkaoui, R., Nogués, M., and Dormann, J.L. (1995) *Hyperfine Interactions* **95**, 129.

[55] Dubois, E. (1997) Thèse, Univesité Pierre et Marie Curie, Paris.

[56] Massart, R., Dubois, E., Cabuil, V., and Hasmonay, E. (1995) *J. Magn. Magn. Mater.* **149**, 1.

[57] Raikher, Yu.L. and Stepanov, V.I. (1992) *J. Experim. and Theor. Phys.* **75**, 764.

[58] Raikher, Yu.L. and Stepanov, V.I. (1994) *Phys. Rev. B* **50**, 6250.

[59] Aharoni, A. (1991) *J. Appl. Phys.* **69**, 7762.

[60] Aharoni, A. (1997) *J. Appl. Phys.* **81**, 830.

[61] Viau, G., Fiévet-Vincent, F., Fiévet, F., Toneguzzo, Ph., Ravel, F., and Acher, O. (1997) *J. Appl. Phys.* **81**, 2749.

[62] Shilov, V.P. (1999) Thèse, Université Denis Diderot, Paris.

[63] Mercier, D., Lévy, J.C.S., Viau, G., Fiévet-Vincent, F., Fiévet, F., Toneguzzo, P., and Acher, O. (2000) *Phys. Rev. B* **62**, 532.

[64] Toneguzzo, Ph., Acher, O., Viau, G., Fiévet-Vincent, F., and Fiévet, F. (1997) *J. Appl. Phys.* **81**, 5546.

[65] Voltairas, P.A., Fotiadis, D.I., and Massalas, C.V. (2000) *J. Magn. Magn. Mater.* **217**, L1.

[66] Voltairas, P.A., Fotiadis, D.I., and Massalas, C.V. (2000) *J. Appl. Phys.* **88**, 374.

[67] Haneda, K. and Morrish, A.H. (1988) *J. Appl. Phys.* **63**, 4258.

[68] Linderoth, S., Hendriksen, P.V., Bødker, F., Wells, S., Davies, K., Charles, S.W., and Morup, S. (1994) *J. Appl. Phys.* **75**, 6583.

[69] Martínez, B., Roig, A., Obradors, X., Molins, E., Rouanet, A., and Monty, C. (1996) *J. Appl. Phys.* **79**, 2580.

[70] Kodama, R.H., Berkowitz, A.E., McNiff, E.J., and Foner, S. (1996) *Phys. Rev. Lett.* **77**, 394.

[71] Kodama, R.H., Berkowitz, A.E., McNiff, E.J., and Foner, S. (1997) *J. Appl. Phys.* **81**, 5552.

[72] Mieklejohn, W.H. and Bean, C.P. (1957) *Phys. Rev.* **105**, 904.

[73] Mieklejohn, W.H. (1958) *J. Appl. Phys.* **29**, 454.

[74] Bonetti, E., Del Bianco, L., Fiorani, D., Rinaldi, D., Caciuffo, R., and Hernando, A. (1999) *Phys. Rev. Lett.* **83**, 2829.

[75] Del Bianco, L., Hernando, A., and Fiorani, D. (2002) *Phys. Status Solidi A* **189**, 533.

[76] Campbell, I.A., Senoussi, S., Varret, F., and Hamzić, A. (1983) *Phys. Rev. Lett.* **50**, 1615.

[77] Campbell, I.A., Hurdequint, H., and Hippert, F. (1986) *Phys. Rev. B* **33**, 3450.

[78] Webb, D.J., and Bhagat, S.M. (1984) *J. Magn. Magn. Mater.* **42**, 109.

[79] Öner, Y. and Sari, H. (1994) *Phys. Rev. B* **49**, 5999.

[80] Ozdemir, M. Aktaş, B., Oner, Y., Sato, T., and Ando, T. (1996) *J. Magn. Magn. Mater.* **164**, 53.

[81] McMichael, R.D., Stiles, M.D., Chen, P.J., and Egelhoff, Jr., W.F. (1998) *Phys. Rev. B* **58**, 8605.

[82] Bakuzis, A.F., Morais, P.C., and Pelegrini, F. (1999) *J. Appl. Phys.* **85**, 7480.

[83] Jamet, M., Wernsdorfer, W., Thirion, C., Dupuis, V., Mélignon, P., Pérez, A., and Mailly, P. (2004) *Phys. Rev. B* **69**, Art.No. 024401.

[84] Koksharov, Yu.A., Gubin, S.P., Kosobudsky, I.D., Beltran, M., Khodorkovsky, Y., and Tishin, A.M. (2000) *J. Appl. Phys.* **88**, 1587.

[85] Koksharov, Yu.A., Gubin, S.P., Kosobudsky, I.D., Yurkov, G.Yu., Pankratov, D.A., Pono-
      marenko, L.A., Mikheev, M.G., Beltran, M., Khodorkovsky, Y., and Tishin, A.M. (2000)
      Phys. Rev. B **63**, Art.No. 012407.

SURFACE-DRIVEN EFFECTS ON THE MAGNETIC BEHAVIOR OF OXIDE NANOPARTICLES

R. H. Kodama
Department of Physics, University of Illinois at Chicago
Chicago, IL 60607, USA

A. E. Berkowitz
Physics Department & Center for Magnetic Recording Research,
University of California, San Diego, La Jolla, CA 92093, USA

1. Introduction

The special role of surfaces in the magnetic behavior of fine particles has been the subject of experimental and theoretical investigation for approximately 50 years. Much of the magnetic behavior of fine particles can be understood in terms of a "giant-spin model", which assumes that all spins participate in a single domain, and that the magnetic energy is determined by the volume and shape of the particle. The phenomenology of "surface anisotropy", "dead layers", "spin canting", "spin pinning", etc. has been used to explain anomalous magnetic behavior, aside from what can be understood in terms of the simple, giant-spin model. Recently, we have shown that, by constructing a specific, realistic model for the atomic-level spin interactions, much of the anomalous behavior can be calculated in detail, at least for ionic materials such as oxides. In addition to accounting for the unusual magnetic behavior of ferrite nanoparticles such as the lack of saturation and the high-field irreversibility, the calculations have shed light on long-standing theoretical dilemmas such as the influence of surface anisotropy on a spherical particle. It is also possible that the surface-driven spin frustration and disorder present in these nanoparticles are closely related to the poorly understood nucleation centers believed to cause the low nucleation field for magnetization reversal of particles or wires (Brown's paradox). In this chapter, we will discuss applications of the modeling technique to ferrimagnetic nanoparticles, $NiFe_2O_4$ and γ-Fe_2O_3, and antiferromagnetic nanoparticles, NiO and CoO, as well as experimental studies of the same.

Finite size effects dominate the magnetic properties of nano-sized particles, and become more important as the particle size decreases. In many cases, they arise because of the competition between surface magnetic properties and core magnetic properties.

189

These effects are of intense technological interest because of their relevance to the stability of information stored in the form of magnetized particles or crystallites, which compose rigid disk, floppy disk, and tape recording media. The data storage industry is driving towards higher densities of stored "bits", which necessitates the use of smaller particles or grains in the media, making a basic understanding of finite size effects critical. More generally, surface and interface effects such as the spin disorder which we find in nanoparticles have relevance to thin film devices in the new field of magneto-electronics, e.g. spin valves, spin transistors, spin dependent tunneling devices. Since spin transport through magnetic/non-magnetic interfaces plays an important role in these devices, the state of interfacial magnetic moments will impact device performance.

The basis for much of our understanding of the magnetic behavior of nanoparticles is the so-called "giant-spin model". This is appropriate for particles in the nanometer size regime because they are well below the critical size for single-domain behavior for most materials. Thus, the atomic moments are expected to rotate in unison as a large, single spin. The zero-temperature limit is generally described by a Stoner-Wohlfarth-type model[1], in which the magnetization reversal is determined by the average magnetic anisotropy of the particle. The theory of thermally activated magnetic fluctuations of ensembles of such particles is called "superparamagnetism"[2]. These theories have been tested extensively and are largely successful in describing observed behavior for weakly interacting samples of nanoparticles.

Our recent work [3-6] has been an attempt to go beyond the giant-spin model, in order to describe anomalous behavior that cannot be explained within this framework. Consideration of the special role of particle surfaces is a logical extension of the basic model. Most previous studies had considered the surface as an ideal termination of a bulk continuum, or used non-specific crystal lattices such as simple cubic. While those approaches have the virtue of generality, it is our belief that they do not capture the essential physics required to explain experiments. We have advocated the use of bulk crystal structures for particular materials, where the surface is treated simply as an atomic-level termination of the bulk crystal. While in principle, a "perfect" description of single-particle magnetic behavior should be possible by specifying all atomic positions, exchange, and anisotropy parameters, this information is generally not available. Even if all atomic positions are known, the theoretical capability to predict exchange and anisotropy parameters (e.g. via first principles electronic structure calculation) is still limited. Furthermore, for nanoparticle systems we expect inhomogeneity in actual samples, because even for a specific particle size there can be variation in the surface atomic structure. Nevertheless, useful information about magnetic behavior can be obtained, although our ability to quantitatively predict behavior is still hampered by limited input data.

The experimental evidence suggests a degree of spin disorder in ferrite nanoparticles. Our model of $NiFe_2O_4$ nanoparticles showed that for a simple surface termination, significant spin disorder was found for very small particles, <200 atoms. This was driven by the inherent frustration in the bulk magnetic structure (i.e. antiferromagnetic exchange paths within a magnetic sublattice). In order to make a better connection to larger particles investigated in experiments, we considered two phenomenological parameters for surface structure: Surface Vacancy Density and Broken Bond Density. These further reduce the average coordination number for surface atoms, increasing the spin disorder, and extending the effect to larger particles.   The surface model is somewhat arbitrary, but reasonable given the limited structural data. In this way, we reproduced phenomena observed experimentally for similar sized nanoparticles.

To set the context for our discussion of thermally activated or quantum tunneling (MQT) magnetization reversal, we will briefly review the relevant theory. Using the formalism of thermal activation, we can express the frequency of thermally activated reversals as

$$f = f_0 e^{-E/k_B T}$$

where f_0 is the "attempt frequency" which has been estimated for various materials to be in the range from 10^9 to 10^{13} s^{-1}, and E is the energy barrier to reversal. Following the analysis by Street and Woolley [7] and more recently by Barbara [8], the magnetization as a function of time after changing the applied field is the following:

$$M(t) = M_0 \left(1 - \int_0^{E_C} e^{-tf} n(E) dE \right) \tag{1}$$

where $n(E)dE$ is the fraction of particles having an energy barrier between E and $E+dE$. The exponential factor in Eq. (1) has an abrupt step as a function of E, near $k_B T \ln(tf_0)$. The narrow energy range spanned by this step can be described as the "experimental window" of the measurement. The experimental window sweeps over different parts of $n(E)$ as the temperature is changed. Specifically, if the barrier distribution has the form $n(E) = 1/E$ at low energies, then the relaxation rate (or "viscosity parameter") will become temperature independent at low temperatures. This type of barrier distribution is *not* consistent with $E=KV$ (where $K \equiv$ anisotropy constant, $V \equiv$ particle volume), as it would be for magnetocrystalline anisotropy and/or shape anisotropy within the giant-spin model. However, as we will discuss in Section 3, such a distribution could arise from spin-glass-like surface states of a particle.

Although the present work is in the context of transition metal oxide nanoparticles, we note some broader implications of this work. As shown by Brown [9], theoretical estimates of nucleation fields for magnetization reversal in ellipsoidal, ferromagnetic particles were far larger than was found experimentally. The discrepancy (known as "Brown's paradox") is generally thought to be due to surface imperfections or defects not included in the "micromagnetic" framework [10]. The most serious experimental effort to address Brown's paradox was DeBlois' study of Fe whiskers [11,12], whose surfaces are perfect enough to identify defects connected to nucleation. Even after thorough analysis [13,14], it was not possible to satisfactorily account for all the observed behavior. Although possibilities for improvements in micromagnetic modeling have not been exhausted [10], we suggest that our present considerations of disordered surface spins may play a role in nucleation of magnetization reversal, even in larger "microparticles". Certainly, in the low temperature regime we have shown the dramatic *increase* of the coercivity of a nanoparticle of "soft" magnetic material due to these disordered spins (see Fig. 3). The effects at high temperatures or with a weaker surface anisotropy are less clear, although it is likely that a weaker surface anisotropy (relative to the bulk anisotropy) will turn a "pinning center" into a "nucleation center".

Metal particle systems may exhibit similar effects as we find for oxides. Clearly, many nominally metallic materials studied have some surface oxide, including, for example, the Fe whiskers mentioned above. The predominance of antiferromagnetic, superexchange interactions in such oxides provides the potential for the type of surface spin disorder we will discuss below. Further, for clean metal systems, there is a possibility for qualitatively similar effects to occur. It is suggestive that non-collinear moments were predicted [15,16] for Ni-Fe alloys and thin films based on first-principles electronic structure calculation. However, little is known about the potential for non-collinear magnetism at metal surfaces and particularly not for particles. More theoretical and experimental work in this area is necessary to address this issue.

2. Atomic-Scale Magnetic Modeling

2.1 BACKGROUND

Finite element modeling of magnetic materials, commonly known as "micromagnetic modeling" [17] has been used extensively in recent years. It has been applied to study problems such as transition noise in magnetic recording media [18], reversal modes of magnetic particles [19,20], and domain structure in soft magnetic films [21,22]. Typically, the magnetic body is subdivided into several hundred or thousand volume elements which are considered to have uniform magnetization. Exchange, anisotropy, magnetostatic, and Zeeman energies are calculated based on the orientation of the magnetization of each element. The total energy is then minimized by some algorithm to obtain the magnetization distribution. Such techniques are particularly suited to the study of macroscopic systems, with dimensions on the order of microns, since the surface-to-volume ratio is relatively small, making a detailed consideration of surface microstructure and surface magnetism less important in determining the overall behavior. Practically, it is beyond current computing power to treat each atomic moment individually for systems of this size, since a cubic micron contains approximately 10^{11} atoms.

In the present work, we consider magnetic behavior of particles having diameters from 1 to 7 nm. On this scale, surface atoms make up at least 25% of the total number of atoms in a particle. The high surface-to-volume ratio makes a detailed consideration of surface microstructure and the behavior of individual atomic moments critical in understanding the overall behavior. Since the total number of magnetic atoms is less than 10^4 for these sizes, it also becomes practical to treat atoms individually in calculations of magnetization distributions. This section will present the basic approach of these calculations, with more detail on specific systems described in later sections. Many other examples of such atomic scale magnetic modeling exist in the literature. Some examples are work by Koon and Saslow [23,24] on interlayer exchange coupling and random anisotropy systems, work by Pappas et al. [25] on spin configurations in Gd clusters, as well as by other authors in the present volume.

2.2 BULK AND SURFACE PARAMETERS

The first step in each calculation is a survey of the experimental and theoretical literature on the bulk magnetic properties of the material in question. Exchange constants are typically found by fitting moment vs. temperature data to a mean field theory or fitting inelastic neutron scattering data to a spin Hamiltonian. Anisotropy constants are found by torque magnetometry or inelastic neutron scattering. We assume for the calculations that the pairwise exchange interactions have the same magnitude for bulk and surface atoms, but that the total exchange interaction is less for surface atoms because of their lower coordination number (i.e. fewer neighbors). As discussed above, we postulate the existence of "broken exchange bonds" due to oxygen vacancies or bonding with ligands other than oxygen at the surface. In short, we set the exchange constant for pairs of atoms equal to the bulk value, or equal to zero for some fraction of pairs of spins at the surface. This fraction of broken exchange bonds between surface atoms we term the "broken bond density".

The magnetocrystalline anisotropy reflects the symmetry of the neighbors of each atom, so it is reasonable to use bulk anisotropy for atoms in the core of the particle. We simply take the bulk anisotropy values in ergs/cm^3 and divide by the number of atoms per cm^3 and apply it as a single-ion anisotropy to all the atoms in the core. However, the large perturbations to the crystal symmetry at surfaces, should lead to magnetocrystalline anisotropy of different magnitude and symmetry for surface sites. Néel first proposed this phenomenon of surface anisotropy in 1954 [26]. For the purpose of our calculations, we assume that the symmetry of a surface site is uniaxial to lowest order, hence

$$E_A = -k_S \cos^2 \theta \qquad (2)$$

(lowercase "k" is used because it is in units of energy per cation, rather than energy per unit volume denoted by uppercase "K") and we assume that we have an easy *axis* anisotropy, i.e. $k_S > 0$, rather than an easy *plane* anisotropy. We define the easy axis $\hat{u}$ as the dipole moment of the nearest neighbor (oxygen ion) positions relative to a surface atom as follows:

$$\hat{u}_i \propto \sum_{j}^{n.n.} (\vec{P}_j - \vec{P}_i)$$

where $\mathbf{P}_i$ is the position of the i-th atom and the sum is over the nearest neighbors of the i-th atom. Since some of the neighbors are missing for a surface atom, $\hat{u}_i$ will be non-zero and directed approximately normal to the surface. The magnitude of the surface anisotropy has not been determined experimentally in magnetic oxides. Nevertheless, some indication of the magnitude of surface anisotropies can be obtained by examining the literature on EPR measurements of dilute magnetic ions in bulk crystals of non-magnetic oxides [27,28]. It is found that when the magnetic ions are substituted into sites having low symmetry, rather large anisotropies are obtained, even for ions such as Ni^{2+} and Fe^{3+} which have singlet ground states (i.e. low anisotropy) in cubic sites [29]. Values of k_S between 1 and 4 k$_B$/cation were used in the calculations, and are representative of the magnitudes found by EPR. The total spin Hamiltonian is then

$$\mathcal{H} = \sum_{i}^{\{all\ spins\}} -g_i \mu_B S_i \hat{S}_i \cdot \left[\vec{H} + \tfrac{1}{2} \vec{H}_{int,i} \right] + E_{A,i},$$

$$E_{A,i} = \begin{cases} E_{A,bulk}(\hat{S}_i), & \text{core cations} \\ -k_S(\hat{S}_i \cdot \hat{u}_i)^2, & \text{surface cations} \end{cases}$$

$$\vec{\mathbf{H}}_{\text{int}} = \sum_{j}^{\{n.n.\}} \frac{2J_{ij}S_j}{g_i\mu_B}\hat{\mathbf{S}}_j + \sum_{j\neq i}^{\{all\ spins\}} g_j\mu_B S_j \frac{3\hat{\mathbf{r}}_{ji}(\hat{\mathbf{r}}_{ji}\cdot\hat{\mathbf{S}}_j)-\hat{\mathbf{S}}_j}{\left|\vec{\mathbf{r}}_{ji}\right|^3} \tag{3}$$

where $g_i\mu_B S_i$ is the magnitude of the ionic moment and the unit vector $\hat{\mathbf{S}}_i$ gives its direction. The summation over {n.n.} denotes first and second nearest neighbors. The dipolar terms are included here for completeness, but they are neglected for the present nanoparticle calculations since the intra-particle dipolar interactions are included in experimental determinations of bulk magnetocrystalline anisotropy, and shape anisotropy will be small since we only consider nearly spherical particles or antiferromagnets which have very small net magnetization.

2.3 CONSTRUCTION OF A NANOPARTICLE

The model nanoparticle is generated by putting magnetic ions on lattice sites corresponding to the bulk crystal structure. The particle is initially defined as either an ellipsoid or platelet of fixed elliptical cross section, and lattice sites within this volume are occupied with the appropriate ions. A variety of distinct particles of the same shape can be obtained by varying the center position of the ellipsoid or platelet within the crystal unit cell. Exchange bonds are set up between neighbors using the bulk exchange constants obtained from the literature. Typically, this includes first and second nearest neighbors, and different exchange constants between different ion species in the case of the ferrites. We refer to those with lower than bulk coordination as "surface cations". Surface roughness is created by removing surface cations at random. The fraction of surface cations removed in this way we refer to as the surface vacancy density (SVD). After this procedure we remove any asperities, which we define as cations with fewer than 2 nearest cation neighbors. As indicated above, a fraction of exchange interactions between surface cations are removed from the first summation in Eq. (3), effectively breaking the exchange bond between them.

2.4 ENERGY MINIMIZATION

Energy minimization algorithms in common use for micromagnetic modeling arise from the Landau-Lifshitz-Gilbert equation [17]

$$\frac{d\mathbf{m}}{dt} = \gamma_0 \mathbf{m}\times\mathbf{H}_{eff} - \lambda\mathbf{m}\times(\mathbf{m}\times\mathbf{H}_{eff}) \tag{4}$$

where γ_0 is the gyromagnetic ratio and λ is the damping constant. The first term on the right hand side corresponds to precession about $\mathbf{H}_{eff}$ and the second term corresponds to a damping that brings the moment vector closer to $\mathbf{H}_{eff}$. For low frequency problems, the precession term is typically discarded. W. F. Brown noted [30] in 1962 that the concept of a "local field" such as $\mathbf{H}_{eff}$ is unreliable when the reversal of a magnetic system nucleates by a collective mode. Hence, we adopted an algorithm that goes beyond the "local field" and accounts for collective modes. The algorithm is a 3-dimensional generalization of one developed by G. F. Hughes [31] in 1983. Key points in this generalization were the choice of an appropriate coordinate system and the evaluation of energy derivatives. In our coordinate system, the spin unit vector was defined as a function of (α_i, β_i) which correspond to rotations in two orthogonal directions:

$$\hat{S}_i(\alpha_i, \beta_i) = \frac{\hat{S}_{0i} + \alpha_i \hat{e}_{\alpha i} + \beta_i \hat{e}_{\beta i}}{\sqrt{1 + \alpha_i^2 + \beta_i^2}} \tag{5}$$

where $\hat{S}_{0i}$ is the initial spin direction and $\hat{e}_{\alpha i}$ and $\hat{e}_{\beta i}$ are chosen to make $(\hat{e}_{\alpha i}, \hat{e}_{\beta i}, \hat{S}_{0i})$ a mutually orthogonal set. It is easy to show that the rotation angle corresponding to a nonzero α or β is $\arctan\alpha$ or $\arctan\beta$, respectively, so for small rotation angles we can think of α and β being the rotation angles in radians. This choice of coordinates gives the following expressions for the derivatives of E (evaluated at $\alpha_i = \beta_i = 0$):

$$\frac{\partial E}{\partial \alpha_i} = -g_i \mu_B S_i \hat{e}_{\alpha i} \cdot \vec{H}_{eff},$$

$$\vec{H}_{eff} = \vec{H} + \vec{H}_{A,i} + \sum_j^{\{n.n.\}} \frac{2 J_{ij} S_j}{g_i \mu_B} \hat{S}_j + \sum_{j \neq i}^{\{all\,spins\}} g_j \mu_B S_j \frac{3\hat{r}_{ji}(\hat{r}_{ji} \cdot \hat{S}_j) - \hat{S}_j}{|\vec{r}_{ji}|^3} \tag{6}$$

For uniaxial anisotropy (e.g. surface cations),

$$E_A = -k_u (\hat{S}_i \cdot \hat{u}_i)^2$$

$$\vec{H}_{A,i} = \frac{2 k_u (\hat{S}_i \cdot \hat{u}_i)}{g_i \mu_B S_i} \hat{u}_i$$

$$\frac{\partial^2 E}{\partial \alpha_j \partial \alpha_i} = \begin{cases} +g_i \mu_B S_i (\hat{S}_{0i} \cdot \vec{H}_{eff}) - 2k_u(\hat{e}_{\alpha i} \cdot \hat{u}_i)^2, \quad i = j \\ \\ -2J_{ij} S_i S_j \hat{e}_{\alpha i} \cdot \hat{e}_{\alpha j} - g_i g_i \mu_B^2 S_i S_j \left[\frac{3\hat{r}_{ji}(\hat{r}_{ji} \cdot \hat{e}_{\alpha j}) - \hat{e}_{\alpha j}}{|\vec{r}_{ji}|^3} \right] \cdot \hat{e}_{\alpha i}, \quad i \neq j \end{cases} \tag{7}$$

Expressions for other forms of anisotropy were given in Ref. 6. The equilibrium state is not affected by small errors in the second derivatives, although for accurate calculation of hysteresis behavior it is important to treat the largest second derivative terms exactly. For the nanoparticles, the exchange terms are the largest by 2-4 orders of magnitude, so it is reasonable to approximate the contributions of anisotropy to the second derivatives. As indicated above, the dipolar terms are also neglected in the nanoparticle calculations. If, however, this method is applied to a problem where the dipolar terms are dominant it is necessary to include all the off-diagonal second derivative terms in order to obtain the correct reversal fields.

Application of the conjugate directions algorithm (following Hughes [31]) was carried out as follows. The variation of the energy with changes in spin orientations is

$$\delta E = \vec{T} \cdot d\vec{\theta} + \frac{1}{2} d\vec{\theta} \cdot \vec{Q} \cdot d\vec{\theta}$$

$$d\vec{\theta} = ((\alpha_1, \beta_1), (\alpha_2, \beta_2), ...(\alpha_N, \beta_N))$$

$$\vec{T} = ((\frac{\partial E}{\partial \alpha_1}, \frac{\partial E}{\partial \beta_1}), (\frac{\partial E}{\partial \alpha_2}, \frac{\partial E}{\partial \beta_2}), ...(\frac{\partial E}{\partial \alpha_N}, \frac{\partial E}{\partial \beta_N}))$$

$$\ddot{Q} = \begin{pmatrix} \dfrac{\partial^2 E}{\partial \alpha_1^{\,2}} & 0 & \dfrac{\partial^2 E}{\partial \alpha_1 \partial \alpha_2} & \dfrac{\partial^2 E}{\partial \alpha_1 \partial \beta_2} & \cdots \\[2mm] 0 & \dfrac{\partial^2 E}{\partial \beta_1^{\,2}} & \dfrac{\partial^2 E}{\partial \beta_1 \partial \alpha_2} & \dfrac{\partial^2 E}{\partial \beta_1 \partial \beta_2} & \cdots \\[2mm] \dfrac{\partial^2 E}{\partial \alpha_1 \partial \alpha_2} & \dfrac{\partial^2 E}{\partial \beta_1 \partial \alpha_2} & \dfrac{\partial^2 E}{\partial \alpha_2^{\,2}} & 0 & \cdots \\[2mm] \dfrac{\partial^2 E}{\partial \alpha_1 \partial \beta_2} & \dfrac{\partial^2 E}{\partial \beta_1 \partial \beta_2} & 0 & \dfrac{\partial^2 E}{\partial \beta_2^{\,2}} & \cdots \\[2mm] \vdots & \vdots & \vdots & \vdots & \ddots \\ & & & & & \dfrac{\partial^2 E}{\partial \alpha_N^{\,2}} & 0 \\ & & & & & 0 & \dfrac{\partial^2 E}{\partial \beta_N^{\,2}} \end{pmatrix} \tag{8}$$

where $d\vec{\theta}$ is a vector specifying the rotations of all the spins. Its components (α_i, β_i) specify the rotation of the i-th spin. Similarly, $\vec{T}$ is the "torque" vector specifying the first derivatives of the energy with respect to (α_i, β_i), and $\ddot{Q}$ is the "Hessian" matrix specifying the second derivatives of the energy with respect to (α_i, β_i). The algorithm is an iterative one, by which the derivatives about the initial set of spin orientations are used to compute an optimal set of spin rotations (α_i, β_i) to reduce the energy. We use two auxiliary vectors $\mathbf{R}_m, \mathbf{P}_m$ and two auxiliary constants λ_m, ε_m, where m specifies the iteration number. Hughes chose to define $Q_\mu = Q + 1 \cdot \mu$, where $\ddot{1}$ is the unit matrix and μ is a Lagrange multiplier selected to make Q_μ positive definite (i.e. $\ddot{Q} \cdot d\theta > 0$ for any vector $d\theta$). This is because the iteration becomes unstable when λ_m changes sign. We recognized that the significance of $\ddot{Q} \cdot d\theta < 0$ is that the spin system is unstable with respect to rotation in the direction $d\vec{\theta}$, meaning that both the torque term and the second derivative term are negative. Instead of using the Lagrange multiplier to stabilize the iteration, we found it more satisfactory, when $\ddot{Q} \cdot d\theta < 0$, to allow the step in the unstable direction, since it represents a particularly good direction for reducing the energy, and simply terminate the iteration at that stage. In case the iteration produces a $d\vec{\theta}$ with very large rotation angles, we limit the maximum rotation to 0.1 radian in a final renormalization step. Our implementation of the conjugate directions algorithm is detailed in Ref. [6]. When the system is far from equilibrium, the algorithm converges in 1-2 iterations and the resultant step is roughly in the direction of the local torques, whereas when the system is closer to equilibrium it may take 5-15 iterations, but the resultant step typically gives a factor of 10 larger energy reduction than a step in the local torque direction. For zero temperature calculations, it can be useful to treat situations where the torque vector is identically zero, such as when the system is at a saddle point in energy space. In this case, we initialize the auxiliary vectors $\mathbf{R}_m, \mathbf{P}_m$ with a random vector in place of $\vec{T}$, and proceed as usual.

2.5 FINITE TEMPERATURE

Finite temperature is simulated by applying rotations to the spins in random directions between energy minimizations. The magnitude of rotations is adjusted to give a total energy change $\Delta E = N k_B T$, where N is the total number of spins. This is in contrast to the theory of superparamagnetism where each magnetic particle is given $k_B T$ of energy.

In that case, the intra-particle degrees of freedom are assumed to be "frozen-out", so that the particle behaves like a giant spin having only 2 rotational degrees of freedom. Since our model includes the intra-particle degrees of freedom, we must give $\frac{1}{2}k_BT$ of energy to each of the $2N$ rotational degrees of freedom. As the success of the superparamagnetic theory suggests, the "uniform mode" corresponding to uniform rotation of the particle moment is particularly important in describing the behavior. If we apply random rotations to each spin individually, the overlap of these rotations with the uniform mode approaches zero as N becomes large. Therefore, a better approach is to apply rotations to each spin resulting in $\Delta E = (N-1)k_BT$ plus a uniform rotation resulting in $\Delta E = k_BT$.

The details of the perturbation procedure are the following. First, we apply the uniform mode rotation as follows:

$$\hat{\mathbf{S}}_i^{\text{rot}} = \ddot{\mathbf{R}}(\hat{\mathbf{n}}_R,\theta)\cdot\hat{\mathbf{S}}_{0i} \tag{9}$$

where $\hat{\mathbf{n}}_R$ is a random unit vector, and $\ddot{\mathbf{R}}(\hat{\mathbf{n}}_R,\theta)$ is the rotation matrix that rotates a 3-vector about the axis $\hat{\mathbf{n}}_R$ by the angle θ [32]. The total energy is then calculated for the new set of spin orientations, and the angle θ is optimized with a few steps of a linear search routine to give $\Delta E \approx k_BT$. Next, a similar procedure is applied to each spin. Here, a different method is used to rotate the spin in order to avoid calculating many trigonometric functions:

$$\hat{\mathbf{S}}_i^{\text{rot}}(\alpha) = \frac{(1-|1-\alpha|)\hat{\mathbf{e}}_{\alpha i}+(1-\alpha)\hat{\mathbf{S}}_{0i}}{\sqrt{(1-|1-\alpha|)^2+(1-\alpha)^2}}, \qquad \hat{\mathbf{e}}_{\alpha i} = \frac{\hat{\mathbf{n}}_R\times\hat{\mathbf{S}}_{0i}}{\left\|\hat{\mathbf{n}}_R\times\hat{\mathbf{S}}_{0i}\right\|} \tag{10}$$

We define an approximate expression for the energy change with respect to rotation of the i-th spin:

$$\Delta E_{\text{local},i}(\alpha) \equiv g_i\mu_B S_i\left[\hat{\mathbf{S}}_{0i}-\hat{\mathbf{S}}_i^{\text{rot}}(\alpha)\right]\cdot\ddot{\mathbf{H}}_{\textit{eff}} \tag{11}$$

As we did for the uniform mode rotation, we optimize α with a linear search routine to give $\Delta E \approx k_BT$. The expression for $\Delta E_{\text{local},i}$ is exact if only the i-th spin is rotated, but correlation effects will make it inexact when all of the spins are simultaneously rotated. In practice, the correlation effects will give positive and negative contributions that roughly average to zero, making the total energy change approximately Nk_BT after simultaneously applying $\Delta E_{\text{local}} = k_BT$ to every spin.

Finite temperature spin perturbations equivalent to $T = 5$ K were applied between energy minimization steps during the hysteresis loop calculations discussed below. The algorithm, as detailed in Ref. 6, selects the lowest energy state after each perturbation/relaxation step. This algorithm is most appropriate for low temperature hysteresis loop calculations. The perturbation procedure can also be used to calculate activation energy distributions corresponding to transitions between metastable spin configurations as discussed below for $NiFe_2O_4$ nanoparticles.

Monte Carlo Modeling. The preceding discussion relates to very low temperature properties, particularly focusing on energy barriers to relaxation. To get an overall picture of the finite temperature properties, we developed an application of the Monte Carlo method [33]. We applied average rotations of between 0 and 135 degrees, dynamically adjusting the average rotation to obtain an acceptance probability of about 5%. We averaged over approx. 100 particle orientations and approx. 1000 Monte Carlo steps per orientation and temperature. The starting point was the minimum energy configuration. These techniques were applied to CoO nanoparticles, as discussed below.

3. Ferrimagnetic Nanoparticles

3.1 BACKGROUND

Spin canting in ball milled $NiFe_2O_4$ [34,35], chemically precipitated γ-Fe_2O_3 [36], and acicular γ-Fe_2O_3 recording media particles [37,38] has been demonstrated via Mössbauer spectroscopy, as a mechanism for moment reduction. Polarized neutron scattering experiments on ball milled $CoFe_2O_4$ particles were consistent with a core of aligned spins surrounded by a disordered shell [39]. Similar conclusions were made from a Mössbauer study of chemically precipitated $NiFe_2O_4$ particles [40]. We proposed that the canted spins are in a surface layer and that they freeze into a spin-glass-like phase at temperatures below 50 K [3,4]. Thus, the surface spins have multiple configurations for any orientation of the core magnetization. This model accounts for previously reported anomalous behavior, as well as the remarkable irreversibility and time dependent moment in high fields that we have reported [3,4]. The model also provides an alternative to macroscopic quantum tunneling (MQT) for interpretation of our magnetization relaxation measurements at low temperatures [41]. More recently, evidence for spin-glass-like behavior of surface spins of γ-Fe_2O_3 nanoparticles via quasi-elastic neutron scattering and FMR measurements was reported [42].

3.2 $NiFe_2O_4$ NANOPARTICLES

Fine particle samples were produced by grinding coarse powders (1-2 μm) of high purity $NiFe_2O_4$ in kerosene and oleic acid (organic surfactant). The milling was carried out for 1000 hours, and the fine particle component was extracted by centrifugation. The samples were washed of excess surfactant and dried. It was found that approximately one monolayer of oleic acid remained strongly bonded to the surface, and could not be removed by chemical means [35]. It was determined by TEM studies and X-ray diffraction line breadths that the average particle size is 65 Å, with a dispersion of about 50%. High resolution TEM studies on an identically prepared $CoFe_2O_4$ sample showed that the cubic spinel structure was preserved, and that the particles are for the most part equiaxed single crystals [43].

The nanoparticle samples exhibit remarkable high field irreversibility of the magnetization at temperatures below 50 K. Measurements [3], reproduced in Fig. 1, using a water-cooled Bitter magnet at 4.2 K showed open hysteresis loops, with positive and negative field sweeps separated, up to approximately 160 kOe. This separation implies that some of the magnetic spins have a "switching field" of 160 kOe. The temperature dependence of the high field magnetization behavior was investigated by a series of measurements on a SQUID magnetometer. We found that the high field differential susceptibility is roughly independent of temperature, which is consistent with the surface spin canting model, and not consistent with paramagnetism or

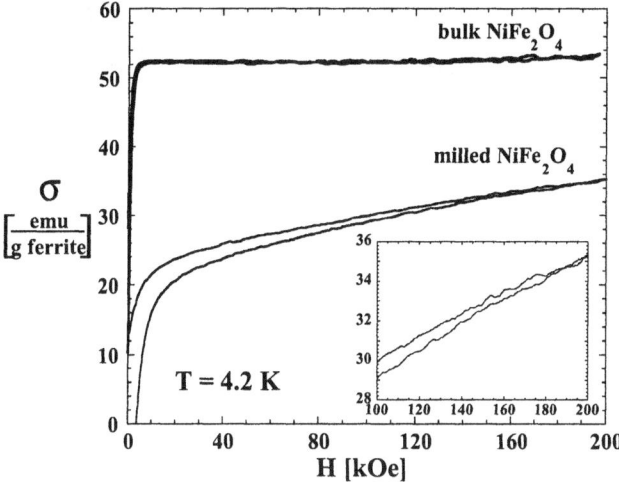

Figure 1. High field hysteresis loop of $NiFe_2O_4$ nanoparticles measured at 4.2 K (zero field cooled). Only the first quadrant of a full ±200 kOe cycle is shown. Reproduced with permission from Ref. [3].

superparamagnetism, which would show a stronger temperature dependence. We also found that the high field hysteresis decreases at higher temperatures, and is negligible at 50 K [6].

Low temperature hysteresis measurements [4] showed that the coercivity and loop shift decrease rapidly with increasing temperature, with the loop shift vanishing near 50 K. We associate the onset of the loop shift and high field irreversibility at about 50 K with a "freezing" of disordered surface spins. Since the coercivity has a similar temperature dependence as the loop shift, it suggests that the coupling with the frozen disordered surface spins makes core spin reversal more difficult.

Magnetization vs. time was measured at temperatures down to 0.4 K after application and removal of a 60 kOe field [41]. The time dependence of the remanent magnetization was fitted to a logarithmic function, where the viscosity parameter $(1/M_0)$ $dM/d(\ln t)$ is the prefactor of the logarithmic term. The magnetic viscosity vs. temperature extrapolated to a non-zero value at zero temperature, with roughly constant viscosity below 2 K.

3.3 γ-Fe_2O_3 NANOPARTICLES

Aqueous colloids of γ-Fe_2O_3 were prepared by coprecipitating a $Fe^{2+} + 2Fe^{3+}$ mixture with NH_3, as described in Ref. [44]. A solution of the polymer polyvinyl alcohol, PVA, was added, producing homogeneous and rigid films upon drying at room temperature. Characterization was done by chemical analysis, X-ray diffraction, TEM, and Mössbauer spectroscopy to determine the composition, structure and particle size [45]. We measured magnetization vs. field for a sample of 46 Å particles in ZFC and FC conditions using a SQUID magnetometer. Figure 2 shows the first quadrant of a full ±70 kOe hysteresis loop. The loop closes at a relatively low field (~4 kOe), indicating

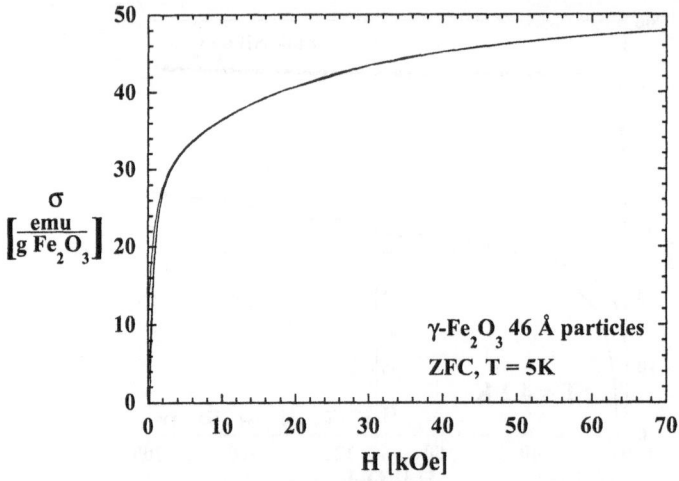

Figure 2. Hysteresis loop of γ-Fe_2O_3 nanoparticles measured at 5 K (zero field cooled). Only the first quadrant of a full ±70 kOe cycle is shown. Reproduced with permission from Ref. [6].

that there is no high field irreversibility such as we found for the $NiFe_2O_4$ nanoparticles. However, the moment is unsaturated up to the maximum field of 70 kOe, similar to what was found for the $NiFe_2O_4$, consistent with surface spin canting. Further, we find a hysteresis loop shift of 33 Oe, upon field cooling to 5 K in a +70 kOe field, whereas the average coercive field is 192 Oe.

3.4 DISCUSSION OF EXPERIMENT

Sufficiently small magnetic particles are usually regarded as single domains, with atomic spins completely aligned by exchange interactions. The rotational barriers due to magnetocrystalline, magnetoelastic, and shape anisotropy can trap such particles in two or more metastable orientations, giving rise to hysteresis. The persistence of hysteresis up to 160 kOe in the $NiFe_2O_4$ nanoparticles could be interpreted as resulting from anisotropy fields of 160 kOe. However, this is 400 times larger than the bulk magnetocrystalline anisotropy field. Our observation of shifted hysteresis loops suggests that the surface spins are spin-glass-like, having multiple configurations that become frozen below 50 K. Due to the exchange coupling between the surface and core spins, field cooling can select a surface spin configuration which favors the particle being magnetized in the field cooling direction, hence resulting in a shifted hysteresis loop below 50 K. The field required to force transitions between surface spin configurations can be very large since the exchange fields are approximately 5×10^6 Oe. Therefore, our interpretation is that the open hysteresis loop at high field is the result of irreversible changes between these surface spin configurations rather than reversals of particle magnetization as a whole.

Time-dependent magnetization of a fine particle system is usually modeled in terms of thermal activation of particles with two stable magnetization states. Within our surface spin disorder model, time-dependent magnetization may not *only* be due to particles reversing their orientation of magnetization, but may also result from thermally activated transitions between surface spin configurations. This has significance in

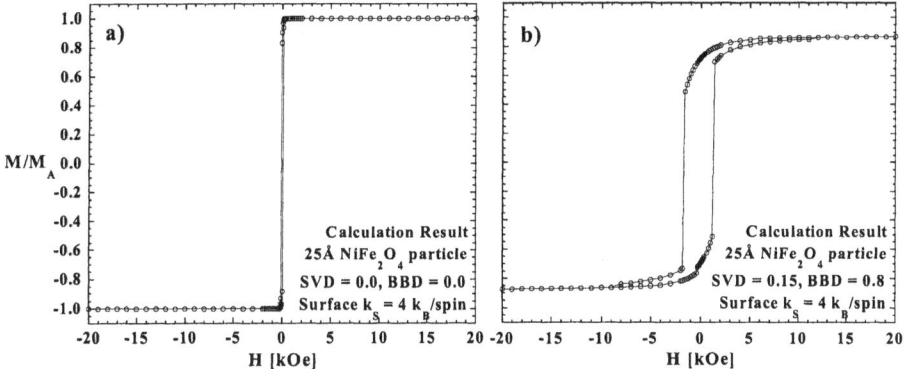

Figure 3. Calculated hysteresis loops for a 25 Å $NiFe_2O_4$ particle, with a surface anisotropy of 4 k_B/spin. a) Particle with no broken bonds or additional roughness, hence no surface spin disorder. b) Particle with surface vacancy density SVD = 0.15, and broken bond density BBD = 0.8, hence significant surface spin disorder. The combination of surface anisotropy and surface spin disorder leads to high field irreversibility. Reproduced with permission from Ref. 6.

interpreting our measurements of the time decay of remanent magnetization for the $NiFe_2O_4$ nanoparticles, for which we observed a temperature-independent viscosity parameter from 2.0 K down to 0.4 K. Such a crossover into a temperature-independent regime is predicted for MQT of single domain particles [46]; however it also has been shown that a distribution of energy barriers $n(E) \sim 1/E$ gives crossover behavior for thermal activation [47]. This type of barrier distribution is not consistent with single domain particles, but it *is* consistent with a spin-glass-like collection of surface spins, as demonstrated below.

Measurements on chemically precipitated γ-Fe_2O_3 nanoparticles show some but not all of the same features as the $NiFe_2O_4$ nanoparticles. The moment is unsaturated and has a large differential susceptibility at high field. This was noted by J.M.D. Coey in his seminal work on similarly prepared γ-Fe_2O_3 nanoparticles [36]. He additionally noted the lack of paramagnetism in the Mössbauer spectrum at low temperature, indicating spin canting rather than a non-magnetic surface layer. We also find a shifted hysteresis loop upon field cooling this sample, which can be explained on similar grounds as for the $NiFe_2O_4$ nanoparticles. What is missing is the high field irreversibility, which we will discuss in Section 3.5.

3.5 MODELING: $NiFe_2O_4$

The exchange constants [48] are the following (in units of K):

$$J_{AA} = -21.0, \quad J_{AB} = -36.0, \quad J_{AB'} = -28.1$$
$$J_{BB} = -22.0, \quad J_{BB} = +2.0, \quad J_{B'B'} = -8.6 \tag{12}$$

where A = (Fe^{3+}, tetrahedral site), B = (Ni^{2+}, octahedral site), B' = (Fe^{3+}, octahedral site). The small size of the exchange constants relative to the ordering temperature (838 K) of $NiFe_2O_4$ is due to the large coordination; A sites having 16 neighbors and B sites having 12 neighbors.

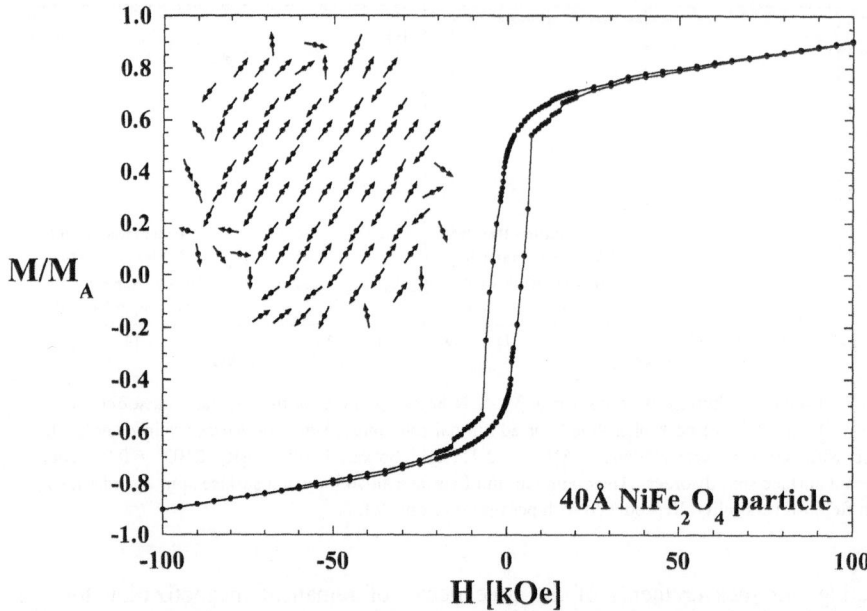

Figure 4. Calculated hysteresis loop for a 40 Å $NiFe_2O_4$ particle, with surface vacancy density SVD = 0.1/0.1/0.1, and broken bond density BBD = 0.8. The three SVD parameters are applied iteratively as described in the text. A surface anisotropy of 4 k_B/spin is included. The inset shows the spin configuration at H = +50 kOe, for a cross section of the particle.

Approximately spherical particles were generated by including all sites on a spinel lattice within a spherical volume. In the following discussion, "surface cations" are those with lower than bulk coordination. Surface roughness was created by removing surface cations at random. The fraction of surface cations removed in this way we refer to as the surface vacancy density (SVD). Following this procedure we removed any asperities, defined as cations with fewer than 2 nearest neighbors. As indicated in Section 2, a fraction of the exchange energy terms between surface cations was removed, breaking the exchange bond between them. The fraction of broken exchange bonds relative to the total number of neighboring pairs of surface cations we refer to as the broken bond density (BBD).

Surface Anisotropy. As discussed in Section 2, we expect large perturbations to the crystal field at surface sites resulting in surface anisotropy. Values of k_S between 1 and 4 k_B/spin were chosen for the calculations, and are representative of the magnitudes determined by EPR measurements of the anisotropy of dilute Ni^{2+} and Fe^{3+} in various oxide host crystals [27,28]. We treat this surface anisotropy as uniaxial, with the axis defined by the dipole moment of the neighboring ions. Hence, the easy axis for these ions is approximately radial. It is intuitively clear that if the spins were perfectly aligned (i.e. no surface spin disorder), the effect of a radially symmetric surface anisotropy would average to zero. This is demonstrated in Fig. 3a, which shows a calculated hysteresis loop for a 25 Å $NiFe_2O_4$ particle having no broken exchange bonds (BBD = 0) or additional roughness (SVD = 0), but with a surface anisotropy of 4 k_B/spin included. For this case, there is no surface spin disorder and we find that the effect of

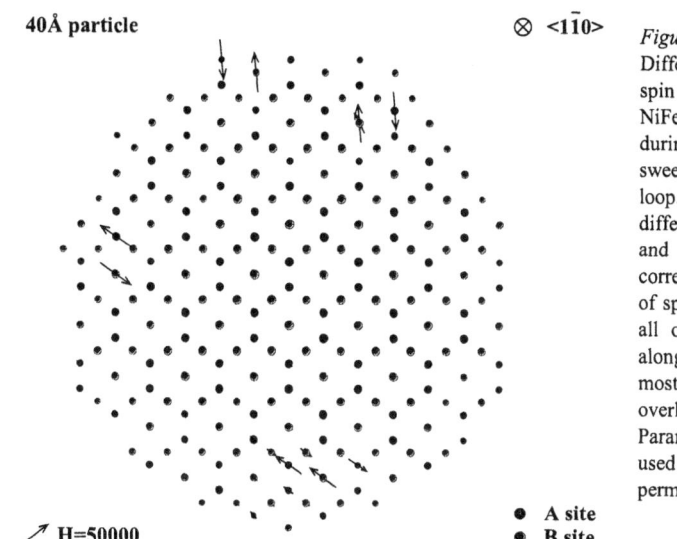

40Å particle   ⊗ <1$\bar{1}$0>

H=50000

● A site
◉ B site

Figure 5.
Difference between calculated spin configurations for a 40 Å $NiFe_2O_4$ particle at +50 kOe during up-sweep and down-sweep of the field in a hysteresis loop. Sites at which there is no difference are indicated by a dot, and the longest arrows correspond to complete reversals of spins. This is a projection of all of the sites in the particle along the <1$\bar{1}$0> axis, so in most cases there are several sites overlaid upon each other. Parameters are identical to those used in Fig. 4. Reproduced with permission from Ref. 6.

the surface anisotropy indeed averages to zero, hence the coercivity is vanishingly small. In contrast, Fig. 3b shows a calculated hysteresis loop for the same particle size with SVD = 0.15 and BBD = 0.8, plus a surface anisotropy of 4 k_B/spin. We find that when there is surface spin disorder the surface anisotropy results in an enhanced coercivity (1800 Oe in this example), as well as irreversibility up to about 10 kOe.

The effect of the surface anisotropy becomes more pronounced when more roughness is added. For example, Fig. 4 shows the hysteresis loop for a 40 Å $NiFe_2O_4$ particle with SVD = 0.1/0.1/0.1 and BBD = 0.8. This three-fold surface vacancy density indicates that first 10% of the surface cations are removed as described previously, then the cations are reclassified to determine which cations are on the surface and the procedure is repeated twice. Finally, any asperities are removed as described previously. This iterative procedure promotes a more irregular surface, since the roughness is no longer limited to the outermost monolayer of the initial sphere. The high field irreversibility is quite pronounced, the loop being open up to approximately 60 kOe. The spin configuration at +50 kOe, during the downward field sweep, is shown in the inset of Fig. 4. The nature of the high field irreversibility can be seen by comparing the spin configurations in the upward and downward field sweeps. Figure 5 shows the difference $S(H, \text{up sweep})_i - S(H, \text{down sweep})_i$ between the two configurations at +50 kOe. This figure is a projection of *all* of the spins in the particle along the <1$\bar{1}$0> axis, so in most cases there are several sites overlaid upon each other. The difference vectors are denoted by arrows whose lengths indicate the magnitude of the difference. Sites at which there is no difference are indicated by a small dot, and the longest arrows correspond to complete reversals of spins. A noteworthy feature of this figure is that there are often neighboring pairs of spins which both flip 180 degrees. This results in a slightly different net moment because the different, antiferromagnetically coupled, cations have different moments ($\mu(Fe^{3+}) \approx 5 \mu_B$, $\mu(Ni^{2+}) \approx 2 \mu_B$).

Activation Energies. Since we are interested in the time dependence of the moment of such a particle, we developed a method to calculate the activation energies associated

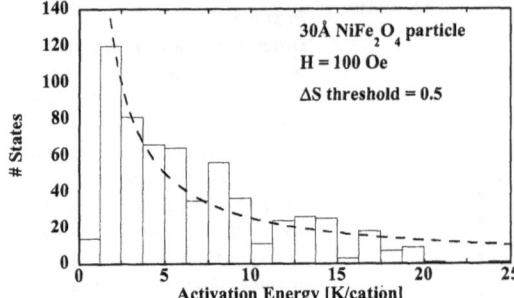

Figure 6.
Calculated activation energy distribution for a 30 Å $NiFe_2O_4$ particle with surface anisotropy of 4 k_B/spin. The dashed curve is a "fit" to $1/E$.

with transitions between surface spin states. States are found by perturbing an initial configuration with a random set of rotations, as described in Section 2. A perturbation energy of 30 K/spin was used, and a set of 600 distinct states were recorded, each time checking that the new state differed by $|\Delta S_i| > 0.02$ from each of the previously recorded states. Each state was tested for stability by applying a perturbation 3 times and checking that each time it converged back to the same state, i.e. $|\Delta S_i|$ less than a threshold value. The perturbation was made small initially, then made incrementally larger until the state was no longer stable under that perturbation. We define the "activation energy" as the perturbation energy required to make the spins converge to a different state, i.e. $|\Delta S_i| > 0.02$.

We find that there is a hierarchy of states in the neighborhood of any starting state. If a state is perturbed by a small energy, it can relax to another state which differs from the original state by a small amount (i.e. $|\Delta S_i|$ is small). However, if the state is perturbed by a large energy, it can relax to another state which differs by a large amount (i.e. $|\Delta S_i|$ is large). We characterized this hierarchy of states by repeating the activation energy calculation using different values of the threshold for $|\Delta S_i|$ in the stability criterion. Results for this calculation on a 30 Å particle with surface anisotropy $k_S = 4k_B$ using the $|\Delta S_i|$ threshold value of 0.5 are shown in Fig. 6. The dashed curve is a fit to $1/E$, which appears appropriate for a limited energy range (2-20 K). We consider the experimental window, as discussed in Section 1, corresponding to these energy barriers. Using the value $f_0 = 3 \times 10^9$ s^{-1} obtained for γ-Fe_2O_3 nanoparticles [49], these barriers correspond to an experimental window of $T = 0.07$ to 0.7 K for relaxation measurements where $t \approx 1000$ s. This implies that thermally-activated relaxation of this system would be temperature-independent at low temperatures, similar to what we found experimentally [41].

A detailed treatment of the relaxation of such a many-state system is a non-trivial problem. The hierarchical distribution of barriers is similar to ones described for other systems with many degrees of freedom such as spin glasses and folding proteins. One can think of the system in terms of a complex "energy landscape", where in the vicinity of each state, there is a "family" of local equilibria which can be accessed with only a small activation energy (a few mK, in this case). The $|\Delta S_i|$ threshold specifies how far away in configuration space the system must be excited before we consider it to be in a new state. Once the $|\Delta S_i|$ threshold is large enough to span the family of local equilibria, we begin to probe the higher activation energies in the distribution.

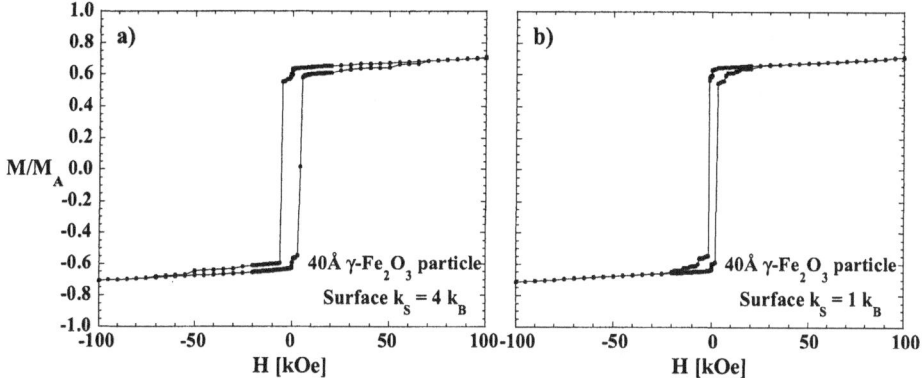

Figure 7. Calculated hysteresis loops for 40 Å γ-Fe$_2$O$_3$ particles, with surface vacancy density SVD = 0.1/0.1/0.1, and broken bond density BBD = 0.8. The particle shown in a) has a surface anisotropy of 4 k_B/spin, whereas b) has a surface anisotropy of 1 k_B/spin. Reproduced with permission from Ref. 6.

3.6 MODELING: γ-Fe$_2$O$_3$

As shown in Fig. 5, we found that the high field irreversibility in NiFe$_2$O$_4$ was primarily due to pairs of antiferromagnetically coupled surface spins which reverse together. It would appear that this type of irreversibility would not occur if all of the cations had the same moment. Such is the case for γ-Fe$_2$O$_3$, which only contains Fe^{3+} ions, and indeed we do not observe high field irreversibility in samples of γ-Fe$_2$O$_3$ nanoparticles, as shown in Fig. 2. However, detailed modeling of spin configurations of γ-Fe$_2$O$_3$ nanoparticles did exhibit similar tendencies for high field irreversibility for the same values of surface anisotropy and roughness. We used the exchange constants of NiFe$_2$O$_4$ which correspond to Fe^{3+} pair interactions:

$$J_{AA} = -21.0, \; J_{AB'} = -28.1, \; J_{B'B'} = -8.6 \tag{13}$$

(using the same notation as in Eq. (12)). Using the Néel model [50] to estimate the corresponding T_C gives a value of 905 K, which is consistent with experimental estimates of the T_C for γ-Fe$_2$O$_3$ in the literature.[51] Since the evidence for vacancy ordering is unclear for small particles [51], we assumed a random distribution of octahedral vacancies. Figure 7a shows a calculated hysteresis loop for a 40 Å γ-Fe$_2$O$_3$ particle having the same roughness and surface anisotropy parameters as those used in Fig. 4. The high field irreversibility is clearly present, in addition to the reduced moment and lack of saturation due to surface spin disorder. The lack of high field irreversibility in the experimental data of Fig. 2 suggests that the surface anisotropy may be less for the chemically prepared specimens. We have calculated the hysteresis loop corresponding to a surface anisotropy of 1 k_B/spin, rather than 4 k_B/spin. This result is shown in Fig. 7b, where we see that the high field irreversibility and large coercivity are no longer present. A lower surface anisotropy could result from different ligands bonded to surface cations, leading to different crystal field splittings. Experimental testing of this hypothesis would require further study.

3.7. SUMMARY

We have observed high field irreversibility in the moment versus field and moment versus temperature of $NiFe_2O_4$ nanoparticles. The onset temperature of this irreversibility is near 50 K. Earlier investigations established that there is spin canting in these particles. The appearance of shifted hysteresis loops lead us to propose that the canted spins are on the particle surfaces, and have multiple stable configurations, one of which is selected by field cooling. We additionally suggested that the open hysteresis loops and time-dependent moment are due to transitions between surface spin configurations, rather than magnetization reversals of whole particles. Our computational model demonstrates the potential for surface spin disorder, arising from reduced coordination and broken exchange bonds between surface spins. Calculation of the energy barrier distribution between surface spin states is consistent with $n(E) \sim 1/E$ which has been shown to produce a thermally-activated temperature-independent viscosity. Thus, a temperature-independent viscosity is not necessarily an indicator of MQT in fine particle systems where spin disorder is present. A model of surface anisotropy is given, based on consideration of crystal field splitting of surface spin states. The combination of surface spin disorder and surface anisotropy accounts for the observed high field irreversibility and gives energy barriers of the correct order of magnitude to explain the low temperature relaxation. For chemically precipitated γ-Fe_2O_3 nanoparticles, we find unsaturated magnetization, consistent with surface spin disorder, but no high field irreversibility. We suggest that the lack of high field irreversibility is due to smaller surface anisotropy in these particles as a result of different ligands bonded to surface cations.

4. Antiferromagnetic Nanoparticles

4.1 BACKGROUND

Néel suggested in 1961 that fine particles of an antiferromagnetic material should exhibit magnetic properties such as superparamagnetism and weak ferromagnetism [52]. Néel attributed the permanent magnetic moment to an uncompensated number of spins on two sublattices. Indeed, large magnetic moments in antiferromagnetic nanoparticles have been observed in materials such as NiO [53-55]; however their origin is not clear. Some investigators attributed these moments to nonstoichiometry, presence of superparamagnetic metallic nickel clusters or Ni^{3+} ions within the NiO lattice [56]. However, a more recent report[57] has shown that these moments are only slightly changed by mild reduction (to eliminate Ni^{3+}) or oxidation (to eliminate Ni metal). Our recent experimental work [58,5] has shown that NiO nanoparticles also exhibit remarkable hysteresis at low temperatures, having coercivities and loop shifts of up to 10 kOe. This behavior is difficult to understand in terms of the 2-sublattice antiferromagnetic ordering which is accepted for bulk NiO. Numerical modeling of spin configurations in these nanoparticles yields 8-, 6-, or 4-sublattice configurations, indicating a new finite size effect, in which the reduced coordination of surface spins causes a fundamental change in the magnetic order throughout the particle. The relatively weak coupling between the sublattices allows a variety of reversal paths for the spins upon cycling the applied field, resulting in large coercivities and loop shifts when bulk and surface anisotropies are included.

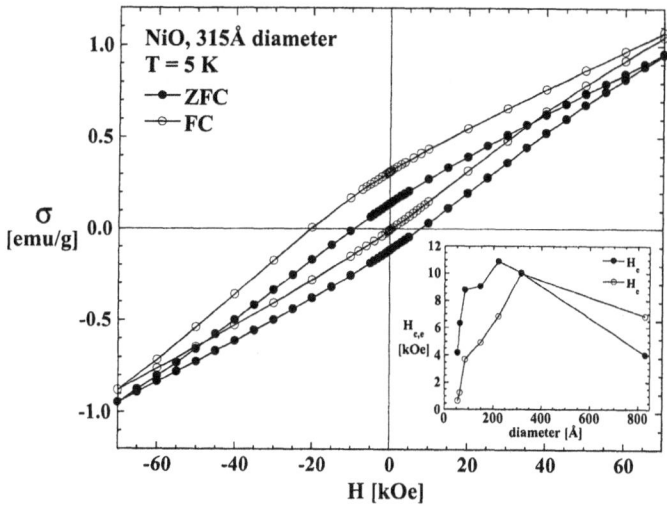

Figure 8. Hysteresis loops at 5 K of 315 Å NiO particles; ZFC and FC from 340 K in 20 kOe. Inset is the coercive field, H_c, and loop shift (or "exchange field", H_e) as functions of particle size. Reproduced with permission from Ref. 5.

4.2 EXPERIMENT

Following the method used by Richardson [53], nickel hydroxide Ni(OH)$_2$ precursor was chemically precipitated by mixing a nickel nitrate Ni(NO$_3$)$_2$·6H$_2$O aqueous solution and a sodium hydroxide NaOH aqueous solution. NiO nanoparticles of various sizes were prepared by calcining portions of the dried gel for 3 hours at various temperatures [58]. X-ray diffraction patterns indicate single phase fcc NiO. The particle size was estimated from both X-ray diffraction line broadening using a modified Debye Scherrer method [59], and BET surface area measurements assuming spherical particles. TEM measurements are consistent with these size determinations, and suggest the possibility of platelet-shaped particles.

Large coercivities and shifted hysteresis loops were observed for all samples after field cooling. Hysteresis loops measured at 5 K for the 315 Å diameter particles both zero field cooled (ZFC) and field cooled (FC) in 20 kOe from 340 to 5 K are shown in Fig. 8. The ZFC coercivity H_c is 10 kOe, the magnetization is increasing even at H = 70 kOe, and the loop is open up to the highest field. The FC loop is slightly broadened and shifted along the applied field direction, with a loop shift (or "exchange field", H_e) of 10 kOe. The dependencies of H_c and H_e on particle diameter are illustrated in the inset of Fig. 8. Very large coercivities and exchange fields (>10 kOe) are obtained for the intermediate sized particles with 220 Å $\leq$ d $\leq$ 315 Å [5].

Recent measurements [60] by Pishko et al. of far-infrared absorption showed two distinct resonances at 33 cm^{-1} for the 57 Å NiO sample, whereas the 435 Å sample showed a single resonance. Resonances in this frequency range are typically spin-wave resonances, hence are intimately tied to the spin structure. Specifically, this result suggests a multi-sublattice state for the smaller particles.

As discussed below, the stability of the 2-sublattice state is connected to the rhombohedral distortion that occurs in bulk NiO below the Néel temperature. Thus, we predict that the most extreme multi-sublattice states correspond to an essentially undistorted cubic structure (as detailed in the following section). Part of our recent neutron diffraction study [61] focused on a full profile refinement of diffraction spectra for the 435 Å NiO sample. The analysis showed a rhombohedral distortion consistent with literature data on bulk NiO at low temperatures. Specifically, we found a contraction of $\sim 1.7 \times 10^{-3}$ along the antiferromagnetic propagation vector. Remarkably, we observed a transition from contraction to *expansion* near 340 K, well below the Néel temperature, which was found to be near 480 K. This expansion ultimately disappeared above the Néel temperature, as expected for a magnetoelastic distortion. We suggest that this transition is an indication of a temperature-driven transition from 2- to multi-sublattice states, as anticipated by the theory [5]. These results may be difficult to reconcile with the spin-wave resonance results, but it is conceivable that the reduced distortion is a precursor to a full-blown multi-sublattice state that exhibits multiple spin-wave resonances (as the 57 Å sample). Further work is required to clarify this point.

4.3 MODELING: NiO

The hysteresis of a collection of 2-sublattice, antiferromagnetic nanoparticles (AFN), having net moments due to uncompensated spins, can be described in terms of a Stoner-Wohlfarth type model [62], in which the spin axis has 2 or more metastable orientations, which depend on the magnetic anisotropy and the applied field. Within this model, *major* hysteresis loops are symmetric since the magnetocrystalline anisotropy has inversion symmetry. If, however, the field is not sufficient to reverse the particle moment (i.e. minor hysteresis loop), one could obtain a shifted loop with no hysteresis. Therefore, a simultaneous loop shift and coercivity can only be described in terms of this model if one attributes it to a broad distribution of reversal fields, both greater and less than the maximum applied field. In order to better understand the magnetic behavior of AFN, we have employed calculations of equilibrium spin configurations as described in Section 2. The anisotropy of bulk NiO was investigated by Hutchings and Samuelsen [63] who used an orthorhombic form for the anisotropy:

$$E_A = D_1 S_z^2 + D_2 S_y^2 \tag{14}$$

where x is the easy axis $\langle 11\bar{2} \rangle$ and z is the hard axis $\langle 111 \rangle$. The structure of NiO is rhombohedral so for the calculations we used instead a six-fold symmetric form, consistent with torque measurements [64]. The anisotropy used in our calculations was:

$$E_A = D_1 \hat{S}_z^2 - (D_2/18) \cos 6\phi_i \sin^6 \theta_i \tag{15}$$

where θ_i and ϕ_i are the conventional spherical coordinates corresponding to the direction of the ionic spin $\mathbf{S}_i$ with ϕ_i referenced to the $\langle 112 \rangle$ direction. The factor of 18 results from matching the leading terms in a small-ϕ_i expansion of our expression to that of Eq. (14). Following the notation of Ref. [63], the exchange and anisotropy parameters used are the following (in units of K):

$$J_1^+ = 15.7, \; J_1^- = 16.1, \; J_2 = -221,$$
$$D_1 = 1.13, \; D_2 = 0.06 \tag{16}$$

where J_1^- is the exchange integral between spins in the same (111) plane (normally ferromagnetically aligned) and J_1^+ is the exchange integral between spins in adjacent (111) planes (normally antiferromagnetically aligned). The only exchange term that makes the 2-sublattice state more stable in bulk NiO is proportional to the small $(J_1^+ - J_1^-)$ splitting. In our model, we find an energetic preference for multi-sublattice states in nanoparticles due to the reduced coordination at surface sites. This preference is enhanced when surface roughness and broken exchange bonds are included. Hence, the multi-sublattice states are ground states for smaller particles.

Since J_1^+ and J_1^- are defined based on a 2-sublattice spin configuration, it is not immediately obvious how to include the $(J_1^+ - J_1^-)$ in the calculations. We must therefore develop a way of defining how the splitting is to be applied to an arbitrary spin configuration, which handles the 2-sublattice configuration as a limiting case. The splitting is associated with the rhombohedral contraction occurring below the Néel temperature, and is believed to be due to the asymmetry in nearest neighbor atomic spacing depending on whether the neighbor is in the contraction plane. The theory of exchange striction [65,66] predicts it to vary as

$$J_1^+ - J_1^- = 2j\bar{S}^2 \tag{17}$$

where $\bar{S}$ is the average spin and $j = -0.26$ K [63]. $\bar{S}$ can be described as an order parameter for the 2-sublattice state. Since we find substantial deviations from the 2-sublattice state, we calculate $\bar{S}$ after each step of the relaxation procedure and rescale J_1 according to Eq. (17). We compute $\bar{S}$ for any given spin configuration as:

$$\bar{S} = \max_{j=1,2,3,4}\{\bar{S}_j\} = \max_j\left\{\left\|\frac{1}{N}\sum_{i=1}^{N}(-1)^{k_{ij}}\vec{\mathbf{S}}_i\right\|\right\} \tag{18}$$

where N is the number of spins, and k_{ij} is defined as:

$$k_{i1} = \vec{\mathbf{P}}_i \cdot (1,1,1) / a$$
$$k_{i2} = \vec{\mathbf{P}}_i \cdot (1,1,-1) / a$$
$$k_{i3} = \vec{\mathbf{P}}_i \cdot (1,-1,1) / a \tag{19}$$
$$k_{i4} = \vec{\mathbf{P}}_i \cdot (-1,1,1) / a$$

where $\vec{\mathbf{P}}_i$ is the position of the i-th spin and a is the cubic lattice parameter. We essentially calculate the order parameter for each possible (111)-type ordering plane, and take the largest value. The magnitude of the uniaxial surface anisotropy was chosen to be 2 K, which is a reasonable value based on EPR determinations of the magnetocrystalline anisotropy of Ni^{2+} ions in bulk oxide crystals with sites of low symmetry [27].

Calculations on spherical particles of different diameters were performed in order to determine the onset of the multi-sublattice spin state. Figure 9 shows the order parameter $\bar{S}$ and the average number of sublattices for 30 different particles in zero applied field as a function of particle diameter. Two curves are plotted for relatively smooth particles with different values of the broken bond density (BBD), and a third

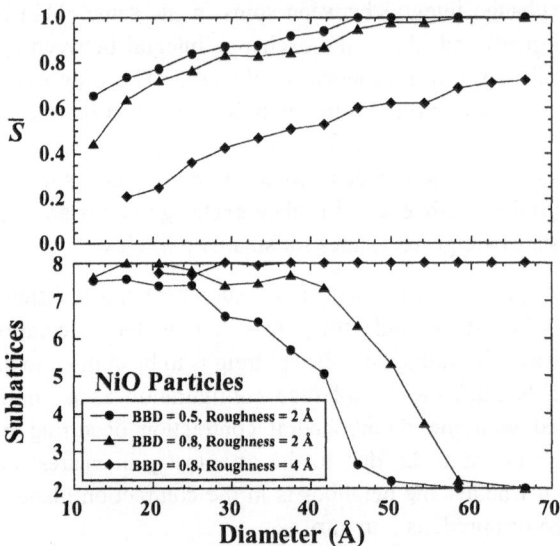

Figure 9. Calculated average order parameter $\bar{S}$ and average number of sublattices for 30 different NiO particles in zero applied field as a function of diameter. The surface broken bond density (BBD), and RMS roughness amplitude are indicated in the legend. Reproduced with permission from Ref. 5.

curve is plotted for a rougher particle. Roughness is created by removing surface cations at random and taking off any asperities afterward, as described in Section 2. The results indicate that the order parameter approaches unity as the particle size increases, while the average number of sublattices is close to 8 for smaller sizes and approaches 2 as the size increases. The size threshold for this behavior is strongly dependent on parameters of the surface.

Hysteresis loops were calculated for both spherical and platelet shaped particles and in both cases we found large coercivities and loop shifts. A simulated field cooling procedure was performed, in which perturbations of 400 K/spin were applied and the spin configuration was allowed to relax in the presence of a 100 kOe field. The perturbation was applied several times, followed by relaxation of the spin configuration each time, to find the lowest energy state. An example of such a calculation is shown in Fig. 10, for a 44 Å diameter, 17 Å thick platelet, with the field applied in the plane of the platelet, which has $\langle 111 \rangle$ orientation. The broken bond density in this case was 0.5. The calculated curve exhibits a large coercivity and loop shift as was found experimentally (e.g. Fig. 8). The inset of Fig. 10 shows the corresponding spin configuration in zero field, which illustrates the complex, multi-sublattice state. We find that the inter-sublattice angles are not fixed, but can change substantially upon cycling the field, giving rise to a variety of reversal paths for the spins. The surface anisotropy and multi-sublattice states are key ingredients to produce simultaneous large coercivities and loop shifts.

Based on our calculated spin configurations, we find that the stability of the 2-sublattice state (versus multi-sublattice states) in *bulk* NiO is directly related to the

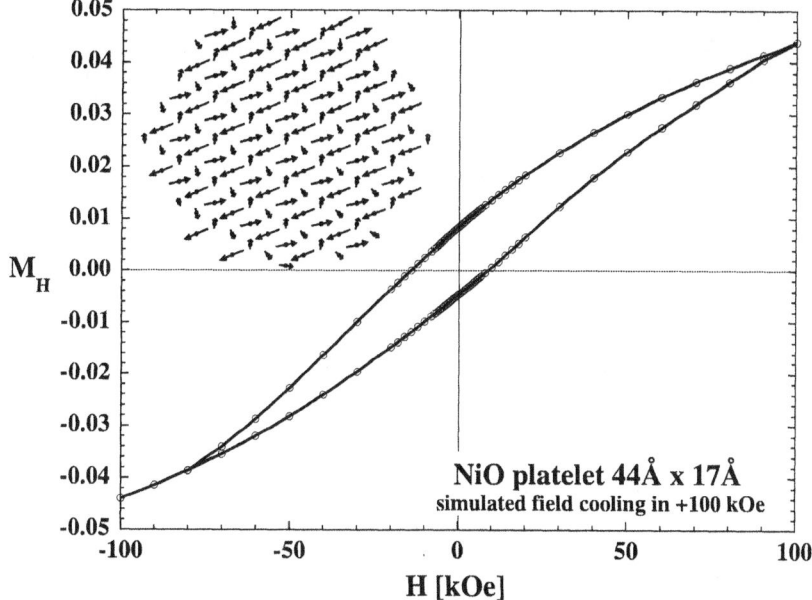

Figure 10. Calculated "field-cooled" hysteresis loop for a 44 Å diameter, 17 Å thick NiO platelet. The field is applied in the plane of the platelet, which has < 111 > orientation. The inset shows the calculated spin configuration for the platelet in zero field. The central cross section of the particle is shown.

$(J_1^+ - J_1^-)$ splitting. Since the rhombohedral contraction is known to diminish with increasing temperature [66], multi-sublattice states should become more prevalent at higher temperatures. We suggest that considerations of multi-sublattice ordering may be useful in describing critical behavior, even in bulk NiO. This idea is supported by recent neutron diffraction studies on relatively large (~435 Å) NiO particles discussed above.

In summary, we have observed large moments in NiO nanoparticles, as well as large coercivities and loop shifts at low temperatures. These observations are consistent with multi-sublattice spin configurations which follow directly from bulk exchange parameters and considerations of low coordination at surface sites. Specifically, we find that the stability of the 2-sublattice state over multi-sublattice states in bulk NiO is directly related to the small exchange term $(J_1^+ - J_1^-)$, and that the low coordination at surface sites tends to make multi-sublattice states more stable. This competition between bulk and surface energies results in the finite size effect. We show that this finite size effect can have a profound effect on low temperature hysteresis properties, giving rise to simultaneous coercivity and loop shift when surface and bulk anisotropies are included.

4.4 MODELING: CoO (TOWARDS A FULLY MAGNETOELASTIC MODEL)

Cobalt monoxide is in the same family of divalent transition metal monoxides as NiO and MnO. All of them are antiferromagnetic and have a cubic, NaCl-type structure

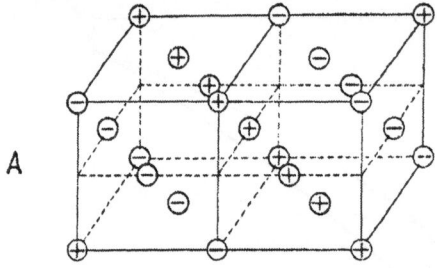

A

Figure 11.
Two different collinear spin arrangements for an
FCC antiferromagnet: (A) The "Type-II" state
accepted for bulk CoO, NiO, and MnO; (B) A
state originally proposed by Y.-Y. Li for bulk
CoO. In the present work, we find the "B" state
as being metastable for CoO nanoparticles.
Reproduced with permission from Li's paper,
Ref. 70.

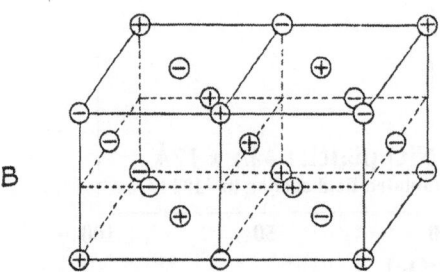

B

above their Néel temperature. In their antiferromagnetically ordered state, they undergo
a magnetoelastic distortion, making the structure slightly rhombohedral in the case of
NiO and MnO, or tetragonal in the case of CoO (more precisely the structure is
monoclinic, due to a slight rhombohedral distortion in addition to the predominant
tetragonal one). Experimental studies of bulk, nanoscale powders of CoO are difficult
due to the tendency to further oxidize to the spinel, Co_3O_4 phase; however, there have
been some recent studies by incorporating such particles in a nanocomposite material or
by reduction in vacuum [67]. CoO thin films have been of significant interest in recent
years in studies of exchange anisotropy. There may be connections between the finite
size effects discussed here and the magnetic behavior of polycrystalline CoO films,
since the dimensions of crystallites are similar to the nanoparticles we consider.

The spin Hamiltonian is similar to that of NiO, with the following exchange and
anisotropy parameters derived from literature data on bulk CoO [68,69]:

$$J_1^+ = -5.8 \text{ K}, \ J_1^- = -6.0 \text{ K},$$

$$J_2^+ = 38.4 \text{ K}, \ J_2^- = 26.0 \text{ K}, \ D_1 = -32.5 \text{ K}$$

where the main difference from NiO is that there is an asymmetry in the next-nearest-
neighbor exchange (J_2), depending on whether the bond is parallel or perpendicular to
the tetragonal c-axis. There is also a relatively large uniaxial anisotropy of 32.5
K/cation (at $T = 30$ K).

In modeling this system, we found a metastable spin state that corresponds to the "B-
state" considered by Li [70], as shown in Fig. 11. This state is consistent with the
complete absence of rhombohedral distortion because of the higher symmetry of the spin
state. The B-state was proposed for bulk CoO by Li, but the bulk state was eventually
shown to have the well-known "Type-II" state [71] as do bulk NiO and MnO.

Nevertheless, in the nanoparticle calculations, we find that the B-state is nearly degenerate with the Type-II state. Interestingly, there appears to be a trend toward structures of higher symmetry in nanoparticles, at least for NiO and CoO.

If we apply the method of computing order parameters $\overline{S}_j$ discussed above, we find that each of the four order parameters are equal to 0.5 for this B-state. Our original method of choosing the maximum one no longer seems appropriate because it amounts to setting:

$$J_1^+ - J_1^- = 2j\overline{S}^2 = 2j(0.5)^2 \quad \text{[for B-state]}$$

when by symmetry arguments, this splitting should be zero. Therefore, we made a refinement to this procedure, in which we apply this splitting to all four order parameters. The net result is that if $\overline{S}_j$ are all equal, these splittings cancel each other out. When one of the order parameters approach 1.0, all the others necessarily approach zero, so this technique gives the proper bulk limit. Following the analysis of Lines and Jones [65], we also can include the following elastic energy term in our rudimentary magnetoelastic Hamiltonian:

$$F_{el} = 2jN\left[(-\overline{S}_0^2 + \overline{S}_1^2 + \overline{S}_2^2 - \overline{S}_3^2)^2\right.$$
$$\left. + (-\overline{S}_0^2 + \overline{S}_1^2 - \overline{S}_2^2 + \overline{S}_3^2)^2 + (-\overline{S}_0^2 - \overline{S}_1^2 + \overline{S}_2^2 + \overline{S}_3^2)^2\right] \quad (20)$$

where j is the same exchange parameter used above and N is the number of cations.

We carried out Monte Carlo calculations of quantities such as the order parameter, magnetic and elastic energies vs. temperature [72,73]. We observed a peak in the order parameter near 160 K, which coincides with a vanishing of the elastic energy (Eq. 20) as temperature increases. This is suggestive of a transition from a Type-II-state to a B-state at higher temperatures, since this term should be zero for the B-state.

5. Remarks

The substantial qualitatitve and quantitative differences between nanoparticle materials within the class of transition metal oxides illustrates the importance of using a realistic structural model in the analysis. We have demonstrated a general framework that takes physically significant input parameters (i.e. crystal structure, exchange parameters) and determines magnetic properties of individual nanoparticles. Parameters of surface structure, moment, anisotropy, and exchange can, in principle, be determined experimentally, but are thus far treated in a phenomenological fashion. The phenomenon of surface spin disorder is given a firm theoretical basis, and its relation to surface anisotropy and coercivity is made clear. Challenges remain in the experimental determination of surface magnetic parameters, and coupling of *ab initio* calculation of such quantities to the spin Hamiltonian approach described here. Magnetoelastic interactions are of particular significance in the antiferromagnetic monoxides, and we have demonstrated a method to include elastic energies as inferred from the magnetic configuration. We have found a tendency for magnetoelastic distortions to diminish in nanoparticles relative to bulk. This approach toward higher symmetry crystal structures may be a generic phenomenon in nanoparticles.

Acknowledgements

High field magnetization data were taken by Ed McNiff at the Francis Bitter National Magnet Laboratory (FBNML). Low temperature relaxation measurements were performed with the assistance of Christopher Seaman at UCSD. Thanks to Elisabeth Tronc of Université Pierre et Marie Curie for providing the γ-Fe_2O_3 nanoparticle samples. Salah Makhlouf, currently of United Arab Emirates University, performed the synthesis and magnetometry on NiO nanoparticles. TEM investigation of NiO nanoparticles was performed by Gregory Fischer and Lea Rudee at UCSD. Thanks to Simon Foner at FBNML for many valuable discussions. This work was supported by the MRSEC Program of the NSF under Award Number DMR-9400439, and DARPA Advanced Materials Research Institute.

References

[1] E. C. Stoner and E. P. Wohlfarth, Phil. Trans. Roy. Soc. London, **A240**, 599 (1948).

[2] C. P. Bean, J. Appl. Phys. **26**, 1381 (1955).

[3] R. H. Kodama, A. E. Berkowitz, E. J. McNiff Jr., S. Foner, Phys Rev. Lett. **77**, 394 (1996).

[4] R. H. Kodama, A. E. Berkowitz, E. J. McNiff Jr., S. Foner, J. Appl. Phys. **81**, 5552 (1997).

[5] R. H. Kodama, Salah A. Makhlouf and A. E. Berkowitz, Phys. Rev. Lett. **79**, 1393 (1997).

[6] R. H. Kodama and A. E. Berkowitz, Phys. Rev. B **59**, 6321 (1999).

[7] R. Street, J. C. Woolley, Proc. Phys. Soc. London A, **62**, 662 (1949).

[8] B. Barbara, C. Paulsen, L. C. Sampaio, M. Uehara, F. Fruchard, J. L. Tholence, and A. Marchand, Magnetic Properties of Fine Particles, eds. J. L. Dormann and D. Fiorani, (Elsevier Science Publishers, 1992) p.235.

[9] W. F. Brown, Rev. Mod. Phys. **17**, 15 (1945).

[10] A. Aharoni, Physica B **306**, 1 (2001).

[11] R. W. De Blois and C. P. Bean, J. Appl. Phys. **30**, 225S (1959).

[12] R. W. De Blois, J. Appl. Phys. **32**, 1561 (1961).

[13] A. Aharoni, J. Appl. Phys. **39**, 5846 & 5850 (1968).

[14] A. Aharoni, J. Appl. Phys. **41**, 2484 (1970).

[15] Yang Wang, G. M. Stocks, D. M. C. Nicholson, W. A. Shelton, V. P. Antropov and B. N. Harmon, J. Appl. Phys. **81**, 3873 (1997).

[16] A. B. Oparin, D. M. C. Nicholson, X.-G. Zhang, W. H. Butler, W. A. Shelton, G. M. Stocks and Yang Wang, J. Appl. Phys. **85**, 4548 (1999).

[17] W. F. Brown, "Micromagnetics" (New York, Interscience Publishers, 1963).

[18] J.-G. Zhu, H. N. Bertram, IEEE Trans. Magn., **24**, 2706 (1988).

[19] D. R. Fredkin, T. R. Koehler, IEEE Trans. Magn., **25**, 3473 (1989).

[20] M. E. Schabes, H. N. Bertram, J. Appl. Phys., **64**, 5832 (1988).

[21] M. Labrune, J. Miltat, IEEE Trans. Magn., **26**, 1521 (1990).

[22] K. Kosavisutte, N. Hayashi, Jpn. J. Appl. Phys., **34**, 5599 (1995).

[23] N. C. Koon, J. Appl. Phys., **79**, 5841 (1996).

[24] W. M. Saslow, N. C. Koon, Phys. Rev. B, **49**, 3386 (1994).

[25] D. P. Pappas, A. P. Popov, A. N. Anisomov, B. V. Reddy, S. N. Khanna, Phys. Rev. Lett., **76**, 4332 (1996).

[26] L. Néel, J. Phys. Rad., **15**, 225 (1954).

[27] A. K. Petrosyan, A. A. Mirzakhanyan, Phys. Stat. Sol. B. **133**, 315 (1986).

[28] E. S. Kirkpatrick, K. A. Müller, R. S. Rubins, Phys. Rev. **135**, A86 (1964).

[29] W. Low, "Paramagnetic Resonance in Solids," (Academic Press, New York, 1960), p. 33.

[30] W. F. Brown, J. Appl. Phys., **33S**, 1308 (1962).

[31] G. F. Hughes, J. Appl. Phys. **54**, 5306 (1983).

[32] H. Goldstein, "Classical Mechanics" (Reading, MA, Addison-Wesley, 1980) p.165.

[33] N. Metropolis, A. Rosenbluth, M. Rosenbluth, A. Teller, E. Teller, J. Chem. Phys. **21**, 1087 (1953).

[34] A. E. Berkowitz, J. A. Lahut, I. S. Jacobs, L. M. Levinson, D. W. Forester, Phys. Rev. Lett. **34**, 594 (1975).

[35] A. E. Berkowitz, J. A. Lahut, C. E. VanBuren, IEEE Trans. Magn. **MAG-16**, 184 (1980).

[36] J. M. D. Coey, Phys. Rev. Lett. **27**, 1140 (1971).

[37] A. E. Berkowitz, F. T. Parker, F. E. Spada, D. Margulies, Studies of Magnetic Properties of Fine Particles, eds. J.L. Dormann and D. Fiorani, (Elsevier Science Publishers, 1992) p.309.

[38] A.H. Morrish, ibid. p.181.

[39] D. Lin, A.C. Nunes, C.F. Majkrzak, A.E. Berkowitz. J. Mag. Magn. Mat. **145**, 343 (1995).

[40] A. H. Morrish, K. Haneda, J. Appl. Phys., **52**, 2496 (1981).

[41] R. H. Kodama, A. E. Berkowitz, C. L. Seaman, M. B. Maple, J. Appl. Phys. **75**, 5639 (1994).

[42] F. Gazeau, E. Dubois, M. Hennion, R. Perzynski and Yu. Raikher, Europhys. Lett., 40, 575 (1997).

[43] A. C. Nunes, E. L. Hall, A. E. Berkowitz, J. Appl. Phys. **63**, 5181 (1988).

[44] J. P. Jolivet, R. Massart, J.-M. Fruchart, Nouv. J. Chim. **7**, 325 (1983).

[45] E. Tronc, J. P. Jolivet, Magnetic Properties of Fine Particles, eds. J. L. Dormann and D. Fiorani, (Elsevier Science Publishers, 1992) p.199.

[46] E. M. Chudnovsky and L. Gunther Phys. Rev. Lett. **60**, 661 (1988).

[47] B. Barbara, C. Paulsen, L. C. Sampaio, M. Uehara, F. Fruchard, J. L. Tholence, and A. Marchand, Magnetic Properties of Fine Particles, eds. J. L. Dormann and D. Fiorani, (Elsevier Science Publishers, 1992) p.235.

[48] A. A. Sidorov, V. P. Suetin, V. G. Pokazan'ev, IAdernyi rezonans v tverdofaznykh soedineniiakh d-metallov, UNTs AN SSSR, Sverdlovsk (1984), pp.21-31.

[49] T. Jonsson, J. Mattsson, P. Nordblad, and P. Svedlindh, J. Mag. Magn. Mat. **168**, 269 (1997).

[50] L. Néel, Ann. Phys. **3**, 137 (1948).

[51] A. H. Morrish, Crystals: Growth, Properties and Applications, v.2, ed. H. C. Freyhardt (Springer-Verlag, 1980), p. 173.

[52] L. Néel, in Low Temp. Phys., edited by C. Dewitt, B. Dreyfus, and P. D. de Gennes (Gordon and Beach, New York, 1962), p.413.

[53] J. T. Richardson and W. O. Milligan, Phys. Rev., **102**, 1289 (1956).

[54] J. Cohen, K. M. Creer, R. Pauthenet and K. Srivastava, J. Phys. Soc. Jpn., **17 Suppl. B-I**, 685 (1962).

[55] W. J. Schuele and V. D. Deetscreek, J. Appl. Phys., **33**, 1136 (1962).

[56] I. S. Jacobs and C. P. Bean, Magnetism Vol. III, eds G. T. Rado and H. Suhl, Academic Press, New York 1963, p.294.

[57] J. T. Richardson, D. I. Yiagas, B. Turk, K. Forster and M. V. Twigg, J. Appl. Phys. **70**, 6977 (1991).

[58] Salah A. Makhlouf, F. T. Parker, F. E. Spada and A. E. Berkowitz, J. Appl. Phys. **81**, 5561 (1997).

[59] G. K. Williamson and W. H. Hall, Acta Metal. **1**, 22 (1953).

[60] V. V. Pishko, S. L. Gnatchenko, V. V. Tsapenko, R. H. Kodama, and Salah A. Makhlouf, J. Appl. Phys. **93**, 7382 (2003).

[61] R. H. Kodama, J. A. Borchers, M. Vedpathak, B. Toby, Salah A. Makhlouf. In preparation.

[62] E. C. Stoner and E. P. Wohlfarth, Phil. Trans. Roy. Soc. London, **A240**, 599 (1948).

[63] M. T. Hutchings and E. J. Samuelsen, Phys. Rev. **B6**, 3447 (1972).

[64] K. Kurosawa, M. Miura, and S. Saito, J. Phys. C: Solid St. Phys. **13**, 1521 (1980).

[65] M. E. Lines and E. D. Jones, Phys. Rev. **139**, A1313 (1965).

[66] L. C. Bartel and B. Morosin, Phys. Rev. **B3**, 1039 (1971).

[67] R. Sappey, E. P. Price, F. Hellman, A. E. Berkowitz and D. J. Smith, "[P26.014] Magnetic and Thermodynamic Features of Antiferromagnetic Nanoparticles in a Metallic Matrix," APS March Meeting 2000.

[68] V. Wagner and D. Hermann-Ronzaud, Neutron Inelastic Scattering 1977. (IAEA, Vienna, 1978) p.135-43.

[69] M. D. Rechtin and B.L. Averbach, Phys. Rev. **B6**, 4294 (1972).

[70] Y. Li, Phys. Rev. **100**, 627 (1955).

[71] P. W. Anderson, Phys. Rev. **79**, 705 (1950).

[72] R. H. Kodama, "Modeling magnetic properties of antiferromagnetic nanoparticles" APS March Meeting (2000).

[73] R. H. Kodama, R. Sappey, A. E. Berkowitz, In preparation.

EXCHANGE COUPLING IN IRON AND IRON/OXIDE NANOGRANULAR SYSTEMS

L. Del Bianco
National Institute for the Physics of Matter (INFM) c/o Department of Physics, University of Bologna, I-40127 Bologna, Italy

A. Hernando
Instituto de Magnetismo Aplicado UCM-Renfe, P.O. Box 155, E-28230 Las Rozas, Madrid, Spain

D. Fiorani
Istituto di Struttura della Materia – CNR, C.P. 10, 00016 Monterotondo Stazione (Roma), Italy

1. Introduction

Nanocrystalline materials are polycrystalline solids with mean grain size D of the order of nanometers (typically, below ~100 nm). In a ferromagnetic nanocrystalline sample, constituted by a single atomic element, the macroscopic magnetic properties are mainly determined by two factors: the fluctuation of the anisotropy easy axis on a scale corresponding to the structural correlation length D and the coincidence of the latter with the exchange correlation length (L), at the nanometric scale. As a consequence of such characteristics, the actual orientation of the atomic magnetic moments results from the competition between the local magnetic anisotropy and the exchange interaction. The effect was nicely described by Herzer through the Random Anisotropy Model (RAM), relating the effective, macroscopic anisotropy of the ferromagnetic sample to the D/L ratio [1].

Engineered nanostructured magnetic materials are generally characterized by the coexistence of two or more phases, magnetically and/or structurally different, modulated on the nanometric scale. In many cases, such systems can be properly described as collections of nanoparticles (or nanograins) dispersed in a matrix. The macroscopic magnetic behavior of these nanocomposite materials is determined by the size, structure and morphology of the constituent phases and also by the type and strength of magnetic interaction between them.

217

When the matrix is diamagnetic or paramagnetic and its thickness is large enough to prevent transmission of exchange coupling between the particles, the sample can be magnetically considered as an assembly of isolated, or weakly dipole-dipole interactive, magnetic elements, eventually showing superparamagnetic relaxation, if they are small enough. Examples are the particulate systems for magneto-recording applications [2] and the nanogranular materials with giant magnetoresistive response [3,4].

On the other hand, when the matrix too is ferromagnetic, the particles and the matrix are exchange coupled and the total system behaves in a collective mode, exhibiting in general new and outstanding macroscopic properties, e.g. enhanced soft magnetic properties, like in Fe-rich nanocrystalline [5] materials (FINEMET-type alloys), as well as enhanced hard magnetic properties, as in NdFeB based nanocomposite (hard spring magnets) [6,7]. Hence, it appears that the magnetic character of the matrix is ultimately responsible for the macroscopic magnetic properties of the composite, nanostructured systems.

Accordingly, modified versions of the RAM have been formulated, accounting for the ability of the matrix to transmit the exchange interaction [8,9]. Indeed, the Herzer model can be applied to such composite systems only under the assumption that the matrix has an infinite exchange correlation length.

Coming back to the nanocrystalline samples constituted by a single atomic element, the reduced grain dimension implies that a large fraction of atoms are located at the interface among the crystallites. In the case of a mean grain size of the order of 10 nm the fraction of atoms located at the grain boundary can be calculated to be as high as 30%, comparable to the number of atoms on regular lattice sites. The atomic structure of grain boundary of nanocrystalline materials has been a very controversial issue. Several investigations have partially modified the early reports of a gas-*like* configuration [10] and have established that the structure and properties of grain boundaries depend on how the material was prepared and on the subsequent thermal history [11,12].

If a different structural arrangement of the atoms at the grain boundary actually exists, a pure nanocrystalline material too may be considered as two-phase systems, constituted by nanocrystallites interconnected by an inter-crystallite matrix.

Hence, it should be expected that the macroscopic magnetic properties of the system are finally determined by the structural features and related magnetic character of the grain boundaries.

1.1   OUTLINE OF THE ARTICLE

In this paper, the mechanisms of inter-phase magnetic coupling in nanostructured materials are discussed. In particular, the magnetic properties of two different systems are reviewed. The first system consists of nanocrystalline Fe samples obtained as bulk by mechanical attrition. Mössbauer spectroscopy allows distinguishing the different physical environment of the atoms in the crystallites and at the grain boundaries, revealing an amorphous-like configuration of the latter. AC and DC susceptibility, remanence and relaxation measurements are consistent with a low-temperature freezing of the crystallite magnetic moments in random directions. The effect is explained in terms of the variation with temperature of the capability of the disordered interface to transmit the exchange interaction to neighboring crystallites.

The second investigated system consists of Fe/Fe oxide granular material obtained by cold-compaction of gas-condensed Fe nanoparticles, surrounded by an Fe oxide layer. The low-temperature magnetic behavior is found to be strongly dependent on the

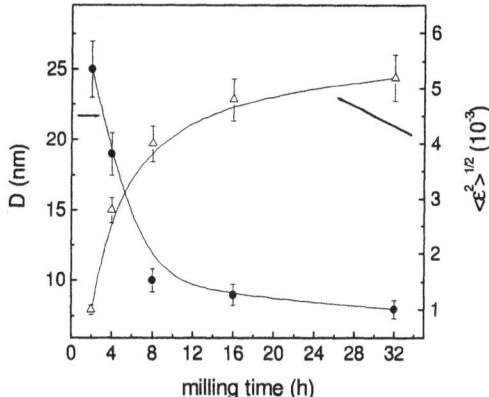

Figure 1. Mean grain size *D* and microstrain $\langle\varepsilon^2\rangle^{1/2}$ as a function of the milling time, as estimated by XRD.

exchange coupling between the metallic crystallites and the disordered (both structurally and magnetically) oxide matrix. Such exchange coupling gives rise to a frozen disordered state of the system at low temperatures and determines the temperature variation of the magnetic properties.

2. Nanocrystalline Fe

Ball-milling is a non-equilibrium processing tool, particularly suitable for the synthesis of various metastable structures: amorphous materials, extended solid solutions, solid solution of immiscible elements [13]. The peculiarity of this technique, in comparison with high-equilibrium synthesis methods, is the possibility of producing nanocrystalline metallic phases presenting metastability on different levels. In fact, the mean grain size is not the only parameter characterizing the structure and hence the physical behavior of pure nanostructured metals. The thermodynamic state of the interfaces, which can be controlled by a tailoring of the synthesis parameter, is also found to affect many physical properties [11,14,15,16].

2.1   X-RAY DIFFRACTION MEASUREMENTS

Pure Fe powder (99.9% purity, 325 mesh) was processed at room temperature for selected times (2, 4, 8, 16 and 32 hours) with a SPEX mixer mill model 8000.

The powder was sealed in a hardened steel vial under an argon atmosphere in order to reduce oxygen contamination. The ball to powder weight ratio was 8/1. Energy dispersive X-ray analysis (EDS) in the transmission electron microscope of the as-milled samples allowed one to exclude the presence of impurities deriving from the milling tools, at least in an amount greater than 0.1 at.% [17].

The average grain size *D* and the root mean squared strain $\langle\varepsilon^2\rangle^{1/2}$ were evaluated by X-ray diffraction (XRD) and through a Warren Averbach analysis method [18].

The results are shown in Figure 1. A strong size reduction and a microstrain increase are experienced at short milling times. Prolonged milling up to 32 hours causes a further small grain size refinement up to the final value of 8 nm whereas the microstrain tends to a constant value of the order of 5×10^{-3}.

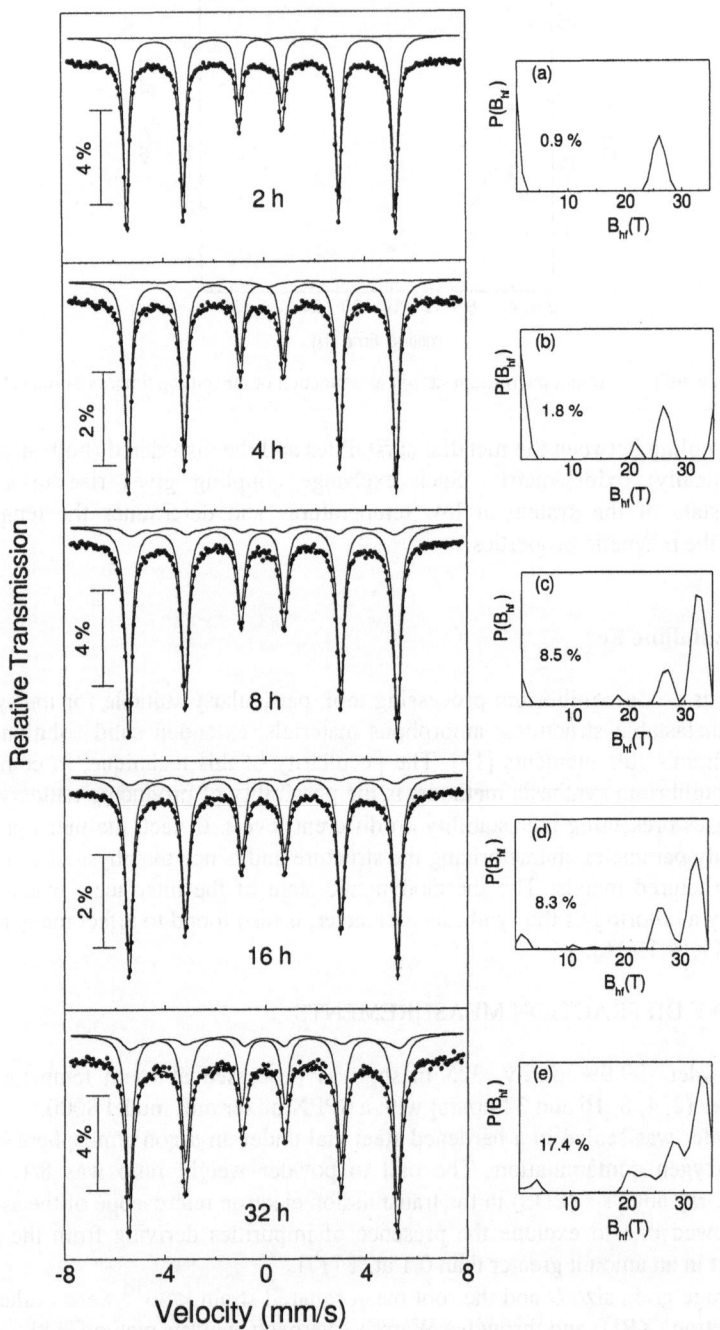

Figure 2. Mössbauer spectra (left) and hyperfine magnetic field distributions (right) of iron powder milled for the indicated time. Percentages are the relative resonant area of the hyperfine field distribution component

2.2 MÖSSBAUER SPECTROSCOPY MEASUREMENTS

Room temperature Mössbauer spectroscopy analysis of the as-milled samples was carried out in a standard transmission geometry using a Co source in a Rh matrix [19].

A progressive increase of the linewidth with the milling time was observed (from 0.25 mm/s up to 0.36 mm/s in the 32 h sample). In agreement with previous works on nanocrystalline iron [20,21], we considered the presence of two different components in the Mössbauer spectra: a sextet plus a distribution of hyperfine magnetic fields (HFF). The fitted spectra and relative HFF distributions are displayed in Figure 2. The hyperfine parameters of the sextet are characteristic of bcc iron (hyperfine field B_{hf} = 33.0 T and isomer shift relative to standard iron IS = 0.00 m/s) and hence this component is due to the crystalline grains. By going with the mechanical attrition process, the HFF distribution profile changes and, together with the peak at 26 T, already visible after short milling time (Fig. 2a,b), further contributions are present for subsequent milling (Fig. 2c-e). The presence of a distribution of hyperfine field is representative of a disordered, amorphous-like structure [22]. In our case, it has been associated with the interfacial region between the nanocrystallites. The relative resonant areas of the HFF distributions (which are assumed to be proportional to the interface volume fraction) are also reported in Figure 2: they do not include the small contributions (of the order of 1%) visible at fields lower than 5 T, attributed to iron oxide. The resonant area of the distribution increases with the milling time, and therefore with reducing the mean grain size. The peak at 26 T in the HFF distributions reveals the presence of atoms less coordinated, with respect to the ones in the crystallites. As this peak is visible also in the spectrum of the 2 h sample, it should be concluded that, in the adopted experimental conditions, it is possible entering the nanometer regime already after the very first hours of the milling process, as confirmed by XRD (Fig. 1).

According to the Mössbauer results, the interface volume fraction, which is less than 1% after 2 hours of milling, rises to 8.5% after 8 hours as a consequence of the observed strong reduction of the grain size. At this stage, the crystallites are very near their ultimate dimensions and no further increase in the amount of interface is experienced passing from 8 hours to 16 hours of milling. The progressive grain size refinement is not the only possible explanation for the observed increase of the component of the Mössbauer spectra associated with the interface. In fact, for prolonged milling, the interfacial region is supposed to become more disordered [23], with a reduction of the degree of coincidence at the grain boundaries. The final result is an enhancement of the interfacial region, namely of the resonant area linked to the hyperfine field distribution, despite the almost negligible decrease of the average grain size. The high interface volume fraction (17.4%) in the 32 h powder can be explained just in this way; moreover it should be considered that a strong reduction of the width of the grain size distribution is observed at this stage of the milling treatment. The presence of different contributions, besides the peak at 26 T, in the hyperfine field distributions for milling time longer than 2-4 hours, is consistent with the increasing disorder of the interfacial region since it is indicative of a distribution of interatomic spacings at the grain boundaries, becoming particularly wide during the final stages of the milling process. In the case of the 32 h powder, where a dynamically stable microstructure is present and no significant modifications are induced by further mechanical attrition, the hyperfine magnetic fields are found to be broadly distributed between 20 and 35 T (Fig. 2e). In summary, the grain boundary structure of as-milled samples presents a wide distribution of nearest-neighbor

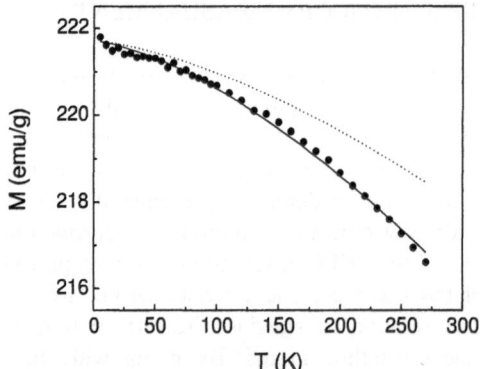

Figure 3. Experimental magnetization curve (applied field H = 50 kOe) of as-milled Fe (circles) and $T^{3/2}$ law fitting (solid line). The dotted line is the $T^{3/2}$ law curve for coarse-grained Fe.

distances together with a reduction of short-range order and may be considered close to the amorphous configuration.

After annealing the as-prepared sample for 1 h at 970 K in flowing Ar, only the bcc Fe sextet is observed, which implies a full recovery of the Mössbauer spectrum of Fe. It is to be noted that the thermal treatment causes a growth of the nanograins up to such a dimension (hundreds of nanometers) that the interface volume fraction is no more comparable with that of the crystals.

2.3   MAGNETIZATION MEASUREMENTS

We studied the low temperature magnetic properties of a ball-milled Fe sample with mean grain size $D \sim 10$ nm and microstrain of about 4×10^{-3}. Magnetic measurements were carried out by a commercial SQUID magnetometer in the 5-250 K temperature range. The thermal dependence of the magnetization has been measured under an applied magnetic field $H = 50$ kOe. Figure 3 shows the experimental curve, which has been fitted to the $T^{3/2}$ Bloch law. For comparison, the curve for bulk coarse-grained Fe, with Bloch's constant of 3.3×10^{-6} $K^{-3/2}$, has been also reported. Nanocrystalline Fe exhibits a stronger dependence of magnetization on temperature (the fitting to the $T^{3/2}$ law provides a Bloch's constant value of 4.94×10^{-6} $K^{-3/2}$). This strong temperature dependence can be explained considering that the reduced atomic coordination at the interface makes the spins at the grain boundary more susceptible to thermal excitation.

Similar effects were observed in ultrafine Fe particles and connected to larger fluctuation of surface moments compared to that at the interior [24]. The value of magnetization at $T = 5$ K is about 221 emu/g, very close to the saturation magnetization of bulk Fe. This provides a further evidence of the excellent purity of the investigated sample.

The temperature dependence of zero-field-cooled (M_{ZFC}) and field-cooled (M_{FC}) magnetization was measured at different values of the applied magnetic field H. In particular, M_{ZFC} was measured on heating up to 250 K, whereas M_{FC} was recorded during the subsequent cooling [25].

The susceptibility M_{ZFC-FC}/H curves are reported in Figure 4. Unlike M_{FC}, for low fields ($H = 60, 100, 400$ Oe), M_{ZFC} decreases monotonously with reducing T down to a temperature T^* below which it exhibits a rapid fall. T^* decreases with increasing field (~

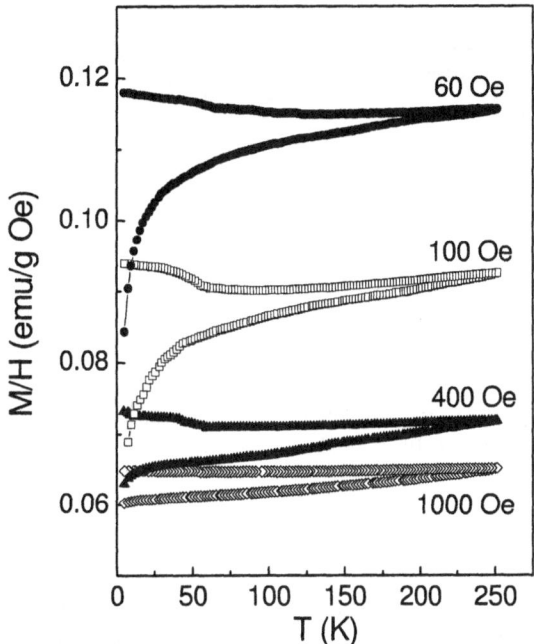

Figure 4. M/H as a function of temperature measured at different magnetic fields after zero-field-cooling (lower branch of each displayed curve) and field-cooling (upper branch).

70 K for $H = 60$ Oe and ~ 30 K for $H = 400$ Oe) and, for $H = 1000$ Oe, the fall is substantially absent. The strong decrease of the low field ZFC susceptibility and the concomitant increase of irreversibility (namely the difference between the M_{FC} and M_{ZFC} curves) with decreasing temperature are consistent with the passage from a ferromagnetic state to a low-temperature disordered regime, where a collective freezing of the crystallite magnetic moments occurs [26, 27].

Since T^* is not precisely defined, it is not possible to check the analytical $T^*(H)$ relationship, to be compared to that expected for reentrant spin-glass-like systems [28,29]. Nevertheless, it is evident that the applied magnetic field favors the ordered regime, moving the onset of the freezing process to lower temperatures and tending to erase the glassy state, which seems indeed the case for $H = 1000$ Oe.

The high-temperature, ferromagnetic regime is also confirmed by Mössbauer spectroscopy measurements on this sample, showing a hyperfine magnetic pattern, without any evidence of presence of a paramagnetic component in the range from 77 K to 300 K, in agreement with the results in Figure 2.

2.4   AC SUSCEPTIBILITY MEASUREMENTS

Support for such a description comes from the analysis of the temperature dependence of the real χ' (in-phase) and imaginary χ'' (out-of-phase) components of the AC magnetic susceptibility. The measurements were performed by a LakeShore 7000 system using the mutual-inductance technique and the data were collected on warming from 13 to 300 K after zero-field-cooling of the sample. The calibration was performed with a $Gd_2(SO_4)_3 \cdot 8H_2O$ paramagnetic standard having the same shape and size as the investigated

L. Del Bianco et al.

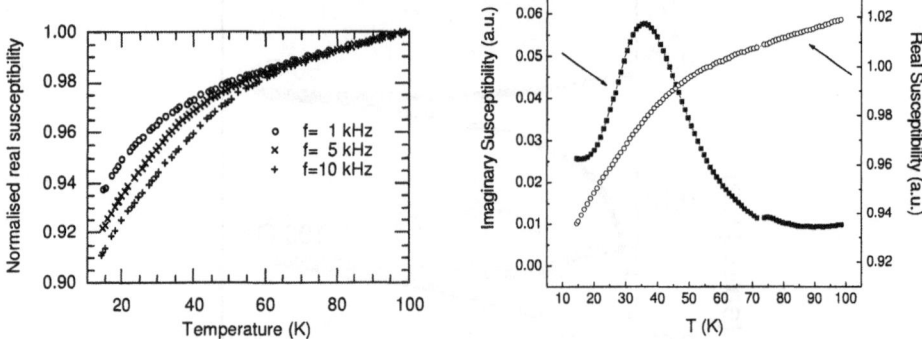

Figure 5. Left: real part of the AC susceptibility χ' as a function of temperature, measured for three different frequencies (1 kHz, 5 kHz and 10 kHz) of the driving field (field amplitude $H_0 = 100$ A/m). The data are normalized to the value recorded at $T=100$ K. Right: real and imaginary components of the AC susceptibility as a function of temperature at the frequency of 7500 Hz.

samples. Figure 5 shows the results obtained with a driving field of amplitude $H_0 = 100$ A/m, oscillating at different frequencies in the 1-10 kHz range. With decreasing temperature below about $T = 70$ K, the in-phase susceptibility component χ' shows a strong frequency dependence, which reflects a marked slowing down of the dynamics (Fig. 5,left). Moreover, χ' shows a rapid decrease below a frequency-dependent temperature which is accompanied by a well-defined peak in the imaginary component χ" (Fig. 5, right), confirming the presence of relaxation processes.

2.5   MAGNETIC RELAXATION MEASUREMENTS

To gain more information on the dynamical magnetic behavior, magnetization relaxation measurements were performed at different temperatures ($T_m = 20$ K, 30 K, 70 K, 250 K) by a SQUID magnetometer. The measurements were performed as follows: first the sample was cooled in zero field down to T_m; at T_m, 10 s elapsed (waiting time, t_w) before the application of the magnetic field ($H = 50$ Oe) and then the time variation of M_{ZFC} was recorded. At the end of this first measurement, the field was removed and the temperature was raised to 300 K, lowered again at the same T_m as before and, after $t_w=1000$ s, the magnetic field was applied and the time variation of M_{ZFC} recorded again. In Figure 6, time-relaxation measurements of M_{ZFC} at $T_m= 20$ K and 70 K are shown. The magnetization relaxes with time and, at 20 K, is clearly dependent on the waiting time, i.e. the M_{ZFC} value is lower in the measurement carried out after waiting for 1000 seconds (Fig. 6a). The waiting-time effect is reduced at $T_m = 30$ K and disappears at $T_m = 70$ K (Fig. 6b). The waiting-time dependence supports the existence of a collective frozen magnetic state at low temperature. Indeed, the aging effect is usually observed in disordered spin systems governed by a correlated dynamics and it is the consequence of the onset, with decreasing temperature, of a very complex multi-valley phase space, with many quasi-degenerate energy minima [30]. It is also noteworthy that the M_{ZFC} was found to relax with time in the whole investigated temperature range.

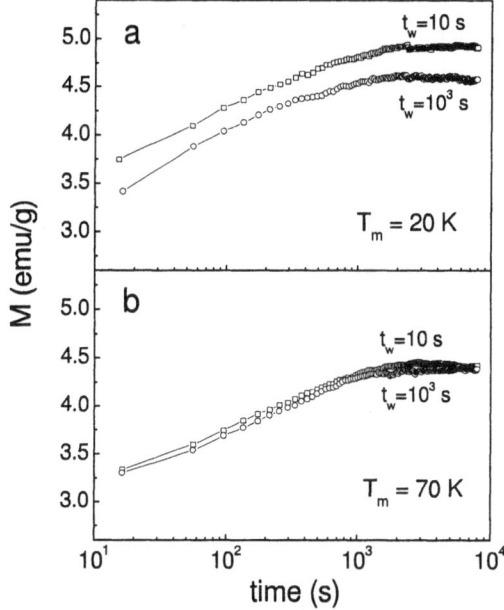

Figure 6. Time dependence of the zero-field-cooled magnetization, measured at $T_m = 20$ K (a) and $T_m = 70$ K (b) for two different waiting times (10 s and 10^3 s).

2.6 DISCUSSION

In pure Fe, the structural disorder and the wide distribution of interatomic spacings at the interface (Section 2.2) are expected to result in a distribution of anisotropy directions and a distribution, in magnitude and even in sign, of exchange interactions.

In analogy to what has been proposed for surface spins of small ferrimagnetic particles [31] or for boundary spins in antiferromagnetic ball-milled Fe-Rh [32], the combination of topological disorder and competing magnetic interactions can lead to a spin-glass-like character of the interface region. Indeed, it is worth noting that amorphous Fe is not accessible to experiments, but extrapolations of the magnetic behavior of amorphous Fe-Zr alloys to pure amorphous Fe [33] as well as ab-initio band-structure calculations [34] predict for amorphous iron a spin-glass or a speromagnetic behavior.

On account of the previous considerations, the experimental results here reported may be well interpreted in the following terms. At sufficiently high temperature, the magnetic domains extend over many grains since the crystallites are ferromagnetically coupled by exchange interaction, which dominates on the single crystallite anisotropy energy, leading to an averaging out of the effective anisotropy (indeed, the coercivity at room temperature is very low, $H_C \sim 3$ Oe). According to the Random Anisotropy Model (RAM) for a two-phase system [8], this is expected to occur when the crystallite dimension is smaller than the ferromagnetic exchange length of the grain (L_G, which is around 20 nm for iron) and the grain boundary thickness (δ) is smaller than the ferromagnetic exchange length of the interface (L_i). At any temperature, L_i may be supposed up to one order of magnitude shorter than L_G. In fact, it is assumed that, due to

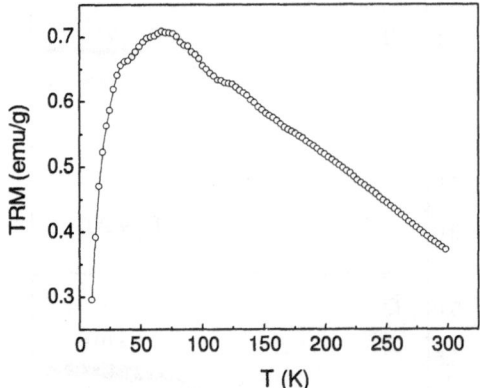

Figure 7. Thermoremanent magnetization (TRM) as a function of temperature.

the topological fluctuations in interatomic distances, the effective exchange stiffness parameter is smaller at the boundary than at the crystal core, whereas the anisotropy may be enhanced, because of the lack of symmetry.

Substantially, the condition $L_i > \delta$ implies that the spins at the grain boundary are oriented in the direction imposed by the surrounding environment: it is as if the boundary spins were polarized by the core of the ferromagnetic grains. However, the existence of irreversibility (difference between M_{FC} and M_{ZFC} curves in Fig. 4) and of relaxation phenomena even at 250 K indicates that the material is not a homogeneous ferromagnet. There should be some small crystallites, which, maybe because of a less effective exchange coupling through the interface, do not join any of the large multi-grain magnetic domains and behave as blocked superparamagnetic particles.

With reducing temperature, the anisotropy at the grain boundary grows more and more and the links holding the boundary spins aligned to the magnetization vectors of the surrounding crystallites are progressively severed. Competing magnetic interactions stabilize a spin-glass-like state at the grain boundary. Since the boundary spins are frozen in random directions, the exchange interaction cannot be transmitted across the interfaces.

In the context of the RAM, this is expressed as follows: with reducing temperature, the local anisotropy at the interface increases more rapidly than the exchange stiffness parameter [35] and increases so much that L_i becomes smaller than δ. Therefore the crystallites are uncoupled, but they are expected to strongly interact through dipole-dipole interactions, which can be ferromagnetic as well as antiferromagnetic. The consequent mixing of interactions, which compete with the magnetocrystalline anisotropy, is responsible for the random freezing of the grain moments.

Therefore, the freezing of the boundary spins implies that the whole system acts as a spin-cluster glass, which is fully coherent with the experimental results.

This explanation is also supported by the temperature dependence of the thermoremanent magnetization (TRM) (Fig. 7). The measurement was performed, by SQUID, as follows: first the sample was cooled down from $T = 300$ K to 5 K in a field of 25 kOe; at 5 K, the field was removed and the remanence was measured for increasing values of temperature, up to 300 K. It is observed that, with decreasing temperature, TRM first increases, as expected for an assembly of ferromagnetic

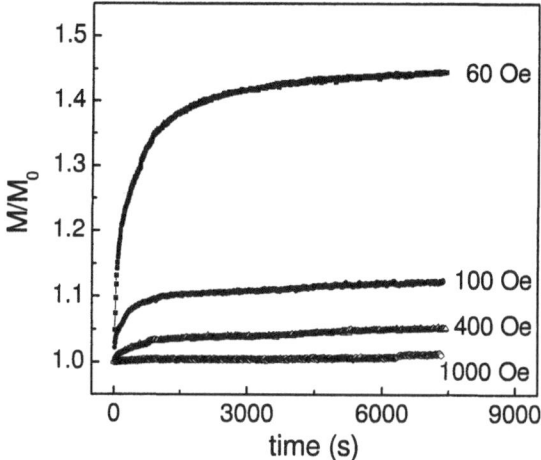

Figure 8. Time dependence of the zero-field-cooled magnetization for different applied magnetic fields at $T = 25$ K. M_0 is the initial magnetization value.

domains, but below 75 K, rapidly decreases. This is fully consistent with the onset and development of a collective frozen magnetic state with grain moments randomly oriented.

It must be pointed out that the magnetization measured at $T = 5$ K for an applied magnetic field of 50 kOe was about 220 emu/g (Fig. 3), namely very close to the saturation magnetization of bulk polycrystalline iron. This means that such magnetic field is sufficiently strong to fully align the randomly locked spins.

The erasing of the low-temperature glassy state, following the application of a sufficiently strong external field, is also confirmed by magnetic relaxation measurements performed at $T = 25$ K in various fields. The results are shown in Figure 8, where M is normalized to its initial value M_0. A progressive reduction of the relaxation is experienced by increasing the magnetic field and the effect is totally annihilated in the measurement carried out at $H = 1000$ Oe, in agreement with the results displayed in Figure 4.

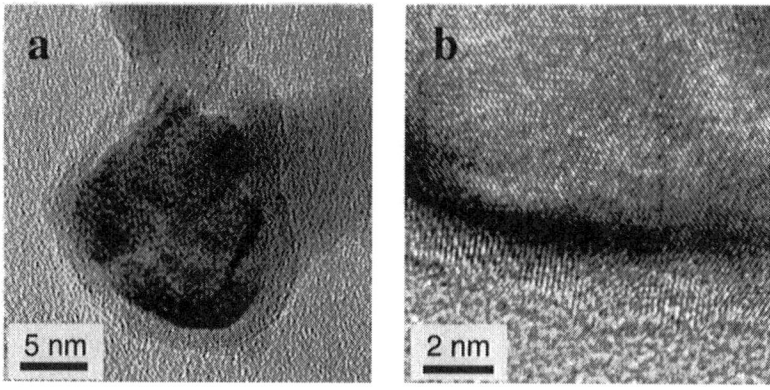

Figure 9. (a) TEM micrograph of an oxygen passivated Fe nanoparticle. (b) Detail showing the oxide layer around the metallic core.

3. Fe/Fe oxide nanogranular system

3.1 SYNTHESIS AND STRUCTURAL PROPERTIES

Fe nanoparticles were produced by inert gas condensation (IGC) method [36]: 99.98% purity Fe was evaporated in a tungsten boat located in a chamber filled with He at a pressure of 133 Pa. Aerosol iron nanoparticles accumulated on a metallic rotating drum cooled by liquid N_2; then they were exposed to a mixture of 133 Pa O_2 and 1200 Pa He for 12 hours to achieve passivation [37].

Figure 9a shows the transmission electron microscopy (TEM) micrograph of a typical Fe particle obtained by IGC and oxygen passivation (in this case, the nanoparticles were collected on a carbon-coated copper grid put inside the evaporation chamber during the synthesis process). The dark core corresponds to metallic iron whereas the surrounding light gray layer is the oxide phase. The thickness of such layer is ~ 2 nm, independently of the Fe core size. The TEM contrast of the oxide phase resembles that of an amorphous structure and the presence of crystallites in this small region is hardly detectable (Fig. 9b). The interface region between metallic core and oxide phase appears as a rather sharp boundary due to Fresnel fringes (Fig. 9b).

The as-prepared particles are not always distinguishable as singular and well-separated entities, as shown in Figure 9a. More generally, the products of the evaporation and passivation procedures are agglomerates of several nanoparticles.

Nanoparticles with increasing mean size have been obtained by increasing the current crossing the tungsten boat, and hence the vapor temperature, during three evaporation processes. After the passivation, the particles were scraped from the cold finger and pressed in high vacuum with a uniaxial pressure of 1.5 GPa, to obtain the pellets labeled D6, D10 and D15, which were analyzed by XRD. The results are shown in Figure 10.

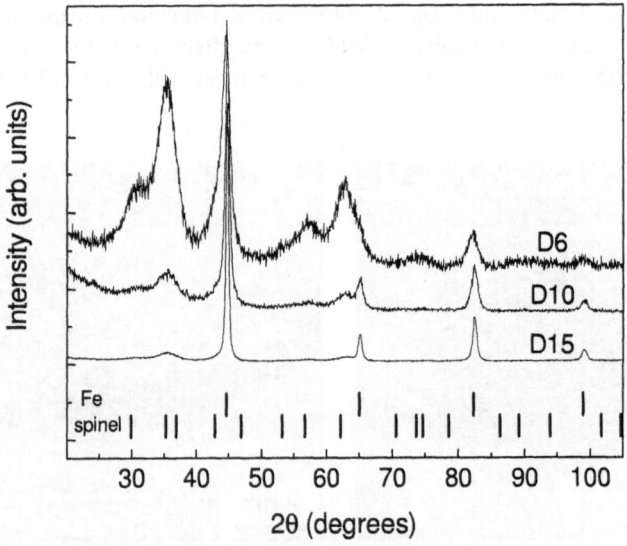

Figure 10. XRD spectra for samples D6, D10 and D15.

Two components are detected: the peaks of bcc Fe and the broadened and overlapping peaks of the oxide phase. The position and the integrated intensity of the latter correspond to the values usually given for magnetite (Fe_3O_4) and maghemite (γ-Fe_2O_3), both having spinel structure. However, the peak broadening does not allow us to distinguish between them.

The XRD patterns have been analyzed following the Rietveld method [38] to determine the volume averaged grain size D of the iron phase (providing a measure of the mean size of the Fe cores) and its weight fraction (a spinel structure for the oxide phase was assumed). In samples D6, D10 and D15, $D = 5.6$ nm (Fe weight fraction $x_{Fe} = 18 \pm 3\%$), $D = 9.6$ nm ($x_{Fe} = 28 \pm 3\%$) and $D = 15.2$ nm ($x_{Fe} = 59 \pm 3\%$) respectively (the error on D is 10%).

For the oxide lattice parameter, the Rietveld analysis provides an intermediate value between those of magnetite and maghemite ($8.399 \div 8.335$ Å). Indeed, we cannot exclude that the oxide phase is a mixture of both Fe_3O_4 and γ-Fe_2O_3, as suggested in various research works on similar samples [39, 40]. The mean grain size of the oxide phase is ~ 2 nm and the microstrain of the order of 10^{-2}, consistent with a high structural disorder. The compacted sample can indeed be modeled as a collection of Fe particles embedded in a poorly crystallized oxide matrix.

3.2   MAGNETOTHERMAL BEHAVIOR

Samples for magnetic characterization were obtained by the subsequent fragmentation of the compacted pellets in powder of micrometric size. Each micrometric fragment consisted of a dense agglomerate of nanoparticles.

Hysteresis loops were measured by a commercial SQUID magnetometer in the temperature range 5 K $\leq T \leq 300$ K [37].

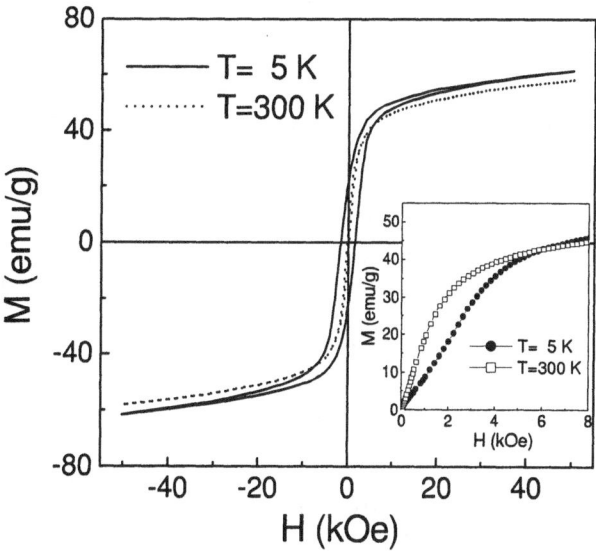

Figure 11. Hysteresis loops at $T = 5$ K, 300 K. Inset: initial magnetization curves as a function of magnetic field at $T = 5$ K, 300 K.

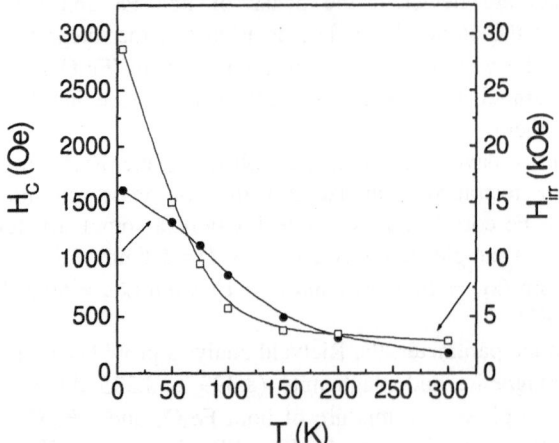

Figure 12. Coercivity (H_C) and irreversibility field (H_{irr}) as a function of temperature.

For the D6 sample, the loops at $T = 5$, 300 K are shown in Figure 11 (in the inset, the initial magnetization curves are displayed). At $T = 5$ K, the loop presents a linear variation in the high field region. The magnetization value at $H = 50$ kOe (61.4 emu/g) is well below the calculated one (110 emu/g), considering the Fe and oxide contents estimated by XRD. Such behavior is consistent with a non-collinear spin arrangement and recalls that found in ferrimagnetic particles (γ-Fe$_2$O$_3$ [41, 42], NiFe$_2$O$_4$ [31]) where the coexistence of topological disorder and frustration of magnetic interactions at the surface resulted in spin-canting and spin-glass-like behavior. In our sample too, the spins of the structurally disordered oxide matrix could be frozen in a spin-glass-like state at low temperature. Moreover, it should be expected that at very low temperature, the interplay between matrix-particle exchange coupling and particle-particle dipolar interactions results in a frozen disordered magnetic state for the whole system.

This hypothesis is supported by the high irreversibility field ($H_{irr} \sim 28$ kOe) at $T=5$ K (magnetic field value above which the two branches of the loop in the first quadrant merge together in a single curve). Both H_{irr} and the coercivity H_C strongly increase with reducing T starting from $T \sim 150$ K (Fig. 12).

The zero-field-cooled (M_{ZFC}), field-cooled (M_{FC}) magnetization vs. T were measured at different applied magnetic field (100 Oe $\leq H_{appl} \leq 20$ kOe; M_{ZFC} was measured on warming and M_{FC} during the subsequent cooling). Some results are shown in Figure 13a. For $H_{appl} < 1000$ Oe, magnetic irreversibility persists up to temperatures close to 300 K. We define T_{irr} as the temperature where the difference between M_{FC} and M_{ZFC}, normalized to its maximum value at $T = 5$ K, becomes smaller than 3%.

With increasing H_{appl}, T_{irr} reduces, the shape of the M_{ZFC} curves exhibits a strong field-dependence, suggesting blocking/freezing phenomena at field-dependent temperatures. It is worth noticing that magnetic irreversibility persists at low temperature even at $H_{appl} = 20$ kOe. The results indicate that the hypothesized spin-glass-like state is not observed to evolve in a pure paramagnetic or superparamagnetic regime with increasing T up to 300 K.

This is confirmed by the Mössbauer spectroscopy analysis. At $T = 4.2$ K, the Mössbauer spectrum can be well fitted using the characteristic six-line pattern of bcc-Fe, given by the metallic particles, plus two wide sextets for the oxide component,

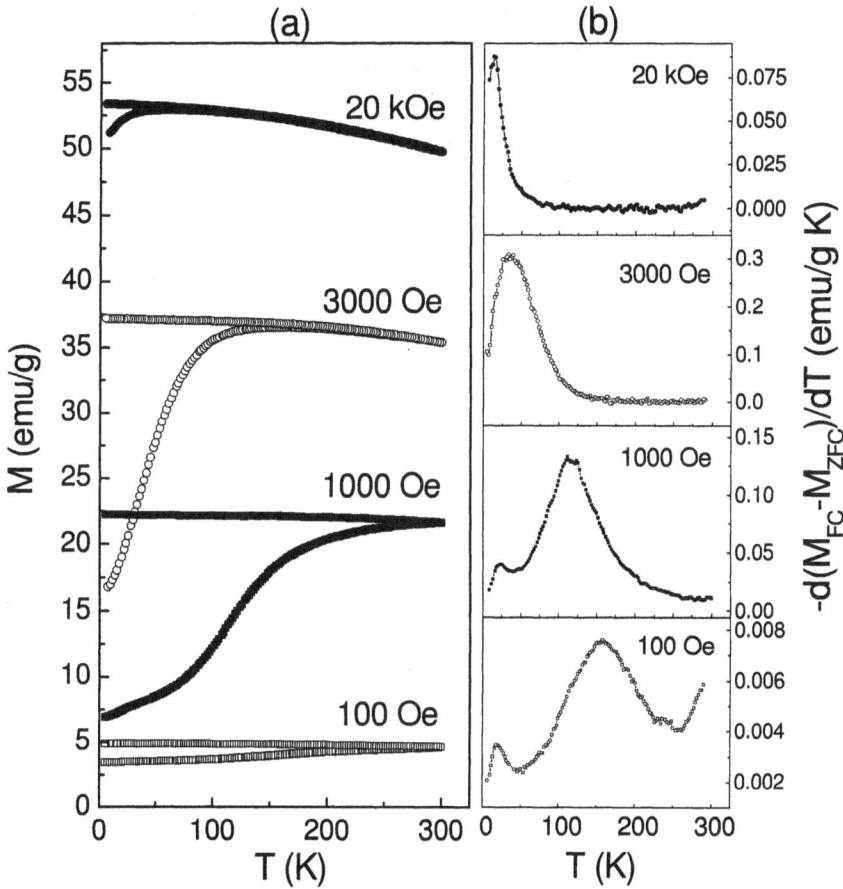

Figure 13. (a) Zero-field-cooled (ZFC, lower branch of each curve) and field-cooled (FC, upper branch) magnetization vs. T for different H_{appl}. (b) Temperature derivative $[-d(M_{FC}-M_{ZFC})/dT]$ of the difference between FC and ZFC magnetization.

consistent with structural disorder. Typical Mössbauer spectra at $T = 77$ K (a) and 300 K (b) are shown in Figure 14. No significant paramagnetic component is observed in the Mössbauer spectra up to room temperature. At $T = 300$ K, the two oxide sextets are replaced by a broad relaxing component, superimposed to the bcc-Fe sextet.

This behavior was also observed by other authors on similar samples [39,40].

In Figure 13b, the temperature derivative curves $[-d(M_{FC}-M_{ZFC})/dT]$ are shown. It is worth reminding that, in the presence of independent relaxation phenomena (i.e. non-interacting particles), the temperature derivative of the remanent magnetization actually reflects the effective distribution of anisotropy energy barriers of the system [43]. Such derivative curve has the same trend as the temperature derivative of the difference between M_{FC} and M_{ZFC}. Although in the present case the relaxation processes cannot be considered independent, qualitative information can be drawn from such analysis.

A common feature of these curves is the presence of a weakly field-dependent peak at $T_1 \sim 20$ K, consistent with the presence of a low-temperature, frozen magnetic state for the whole system. This is also confirmed by the lack of magnetic relaxation below T

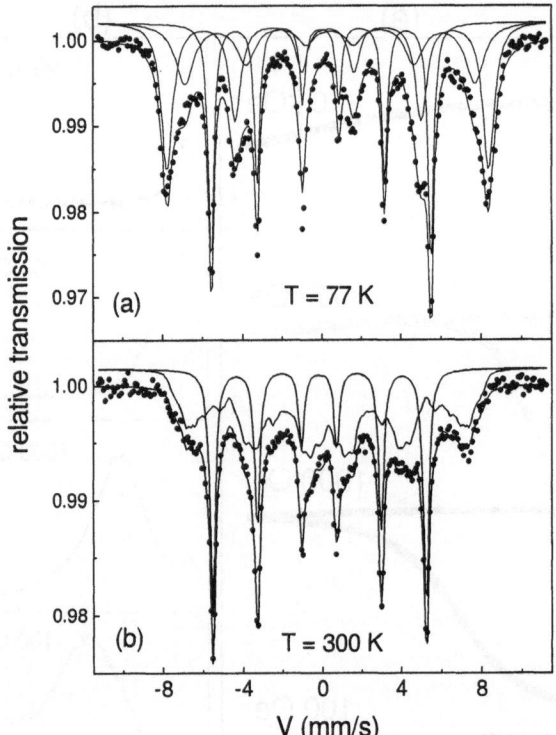

Figure 14. Mössbauer spectra at $T = 77$ K (a) and $T = 300$ K (b) measured on an Fe/Fe oxide sample.

$= 25$ K, as shown in Figure 15, where the time dependence of M_{ZFC} is reported at different temperatures ($H_{appl} = 100$ Oe; note that the maximum relaxation is measured at $T = 150$ K). A second large peak is observed at a temperature T_2 ($T_2 \sim 150$ K for $H_{appl} = 100$ Oe): it becomes narrower and shifts to lower temperature with increasing field.

In a simplified picture, we can describe our system as constituted by two components, strongly coupled at the interface: a non-relaxing, (quasi-static) component (the Fe particles); a relaxing, magnetically disordered component (given by regions of exchange-interacting spins of the oxide matrix). Below T_1, the oxide region moments do not relax and are frozen in the spin-glass-like (or *cluster-glass-like*) state. On increasing the temperature above T_1, such moments become progressively unfrozen, according to the distribution of effective anisotropy energy barriers, determined by their size and strength of the magnetic interaction with the surrounding.

The large peak in the $[-d(M_{FC}-M_{ZFC})/dT]$ curve at $H_{appl} = 100$ Oe (Fig. 13b) reflects the width of such distribution. Once the net moments of the oxide magnetic regions become able to thermally fluctuate, they tend to be polarized by the Fe particle moments, which thus prevent (or shift to higher temperature) the passage into the superparamagnetic regime. As a matter of fact, at $T = 300$ K the virgin M vs. H curve is ferromagnetic-like and it is clearly S-shaped at $T = 5$ K, as found for frustrated systems [44] (Fig. 11, inset). The polarizing action exerted by the Fe particle moments is boosted by the magnetic field, favoring the formation of a ferromagnetic network throughout the sample.

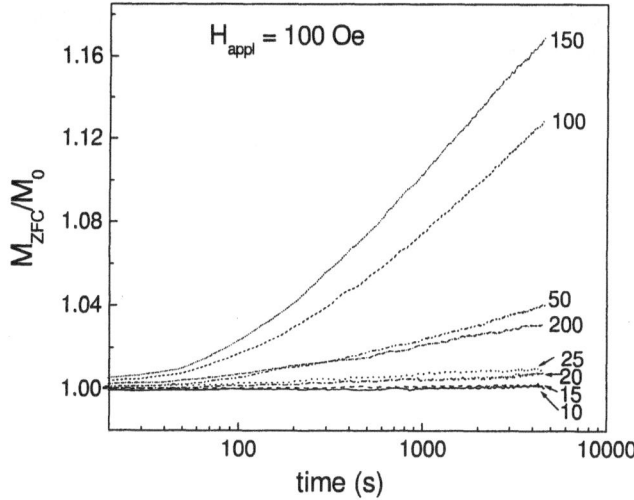

Figure 15. M_{ZFC} vs. time for $H_{appl} = 100$ Oe at the indicated temperatures. M_0 is the magnetization at the beginning of the measurements.

The thermal evolution of H_C (Fig. 12) is coherent with the above picture: at very low temperature the frozen oxide matrix exerts a strong pinning action on the Fe particle moments, whereas at $T = 300$ K the oxide matrix is polarized by the Fe particles. Its contribution to H_C can be neglected compared to that of the Fe particles.

Actually, at $T = 300$ K, the value of H_C is in agreement with that estimated by the Stoner Wohlfarth model for randomly oriented, spherical single-domain Fe particles, $H_C = 0.64 \cdot K_{Fe}/M_S \sim 180$ Oe (K_{Fe}: magnetocrystalline anisotropy of bcc Fe; M_S: Fe saturation magnetization) [45]. Such high value of H_C compared to that of compacted ferromagnetic nanoparticles [46] rules out the possibility that the Fe particle moments are ferromagnetically aligned by exchange interaction through the oxide matrix and support the hypothesis that the particles are predominantly magnetic single-domain.

Hence, the oxide matrix does not transmit the exchange interaction to neighboring Fe particles, which is consistent with the relaxing behavior of the oxide moments, as revealed by the Mössbauer analysis (Fig. 14).

3.3   EXCHANGE BIAS EFFECT

The exchange interaction at the interface between a ferromagnet (FM) and an antiferromagnet (AFM) may create a preferential direction in the spin orientation of the FM. The effect occurs when the FM-AFM sample is cooled through the Néel temperature (T_N) of the AFM, the Curie temperature of FM (T_C) being larger than T_N.

Below T_N, the spins of the AFM couple to those of the FM so as to minimize the interface exchange interaction. In turn, because of the AFM-FM exchange interaction, a single stable configuration for the FM spins is induced and a unidirectional anisotropy appears (exchange anisotropy). When the sample is cooled through T_N in a static magnetic field (H_{cool}), the exchange anisotropy manifests itself in the form of a shift of the hysteresis loop towards negative H values (exchange bias). The value of the shift is conventionally taken as an estimate of the exchange field H_{ex}.

In addition to AFM-FM systems, the exchange bias may also be observed in samples involving a ferrimagnetic (FI) (FI/AFM, FM/FI) [47] or a spin-glass (SG) phase (FI/SG, AFM/SG). Good examples of this last class of materials are the non-interacting $NiFe_2O_4$ [31] and NiO [48] nanoparticles, which are composed of an ordered core (ferrimagnetic or antiferromagnetic, respectively) and a structurally disordered surface region, showing spin-glass-like properties.

In this context, we have studied the exchange bias phenomenon and its dependence on temperature and on the cooling field in the nanogranular Fe/Fe oxide system.

3.3.1 Temperature dependence

Field-cooled (FC) hysteresis loops were measured at different temperatures in the range 5 - 250 K on samples D6, D10 and D15 for H_{cool} = 20 kOe. In the FC procedure, the sample was cooled down from T = 250 K, under H_{cool}. Once the measuring temperature was reached, the field was set at H = 50 kOe and the measurement of a hysteresis loop started. At T = 5 K, a shift of the loop towards the negative field has been observed for the three samples. The shift is usually quantified through the positive exchange field parameter $H_{ex} = -(H_{right}+H_{left})/2$, H_{right} and H_{left} being the points where the loop intersects the field axis. In Figure 16, H_{ex} has been reported as a function of temperature for the three samples. H_{ex} increases with decreasing D, revealing that exchange anisotropy effects at the interface between particles and matrix are more important when the surface to volume ratio of the particles increases. It is worth noting that H_{ex} appears below T = 150 K, i.e. in correspondence with the freezing of most of the moments of the oxide magnetic regions, as indicated by the value of T_2 in the $[-d(M_{FC}-M_{ZFC})/dT]$ curve at H_{appl} = 100 Oe (Fig. 13b); with reducing temperature below T = 150 K, H_{ex} shows a marked increase because of the progressive freezing of an increasing number of oxide region moments. To have an insight into the exchange bias phenomenon, a useful starting point is the Meiklejohn and Bean model, which predicts for FM/AFM systems the relation [49]

$$H_{ex} \cong J_{int} / M_{FM} \cdot t_{FM} \qquad (1)$$

where J_{int} is the exchange constant across the FM/AFM interface per unit area and M_{FM} and t_{FM} are the magnetization and the thickness of the FM layer, respectively. It should be noted that important parameters are not considered in Equation (1), such as the AFM layer anisotropy and thickness, the non-collinearity of the AFM-FM spins, the formation of domains in the AFM and FM layers [50]. For real granular systems, the particle random orientation, the distribution of particle sizes and shapes, the difficulty to characterize the microstructure of the interface and, in our case, the spin-glass-like nature of the oxide matrix make it difficult to extract quantitative information. Nevertheless, the observed increase of H_{ex} with decreasing D at T = 5 K (Fig. 16) constitutes an interesting result, resembling the $1/t_{FM}$ dependence of Equation (1). Moreover, a necessary condition for the observation of exchange anisotropy is [49]

$$K_{AFM} \cdot t_{AFM} \geq J_{int} \qquad (2)$$

where K_{AFM} and t_{AFM} are the anisotropy and the thickness of the AFM layer, respectively.

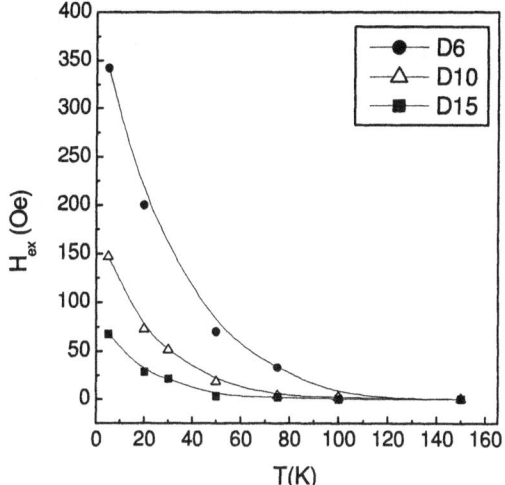

Figure 16. Exchange field (H_{ex}) as a function of temperature for samples D6, D10 and D15.

In our case, if one assumes for M_{FM} the value of bulk Fe (1714 emu/cm^3), for t_{FM} the mean particle size D and for t_{AFM} the double of the oxide shell thickness ($\sim$ 4 nm, independently of D), Equation (1) yields $J_{int} = 0.35$ ergs/cm^2 for sample D6 and then $K_{AFM} \geq 9\times10^5$ ergs/cm^3 from Equation (2); for D15, $K_{AFM} \geq 4\times10^5$ ergs/cm^3. Hence, K_{AFM}, namely the oxide matrix anisotropy, should be at least one order of magnitude larger than the magnetocrystalline anisotropy of bulk γ-Fe$_2$O$_3$ (4.6×10^4 ergs/cm^3) and

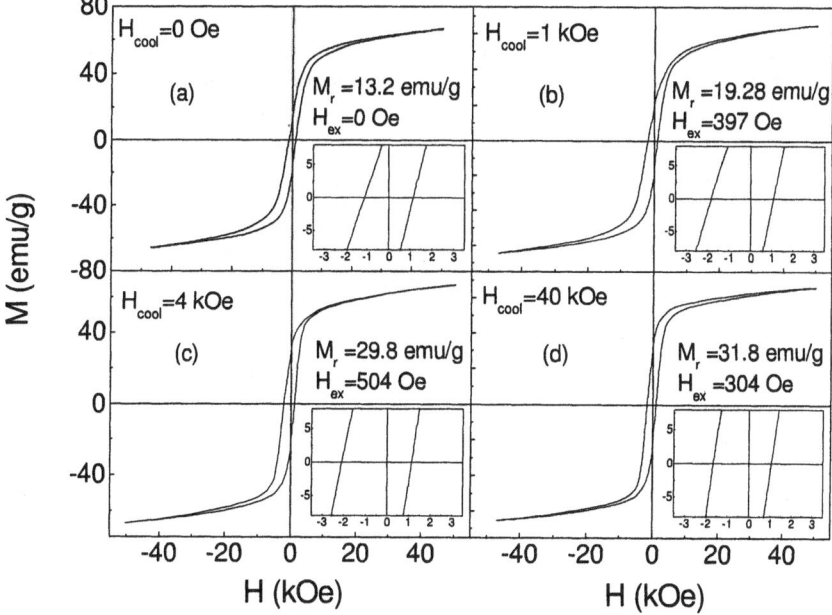

Figure 17. (a) Zero-field-cooled hysteresis loop measured at $T = 5$ K; (b-d) field-cooled hysteresis loops at $T = 5$ K for different cooling field (H_{cool}). The remanent magnetization M_r and the coercivity H_C are reported. Insets: enlarged views of the central region of the loops.

definitely larger than that of Fe_3O_4 (1×10^5 ergs/cm^3). Such estimation is close to the value found in γ-Fe_2O_3 nanoparticles (7×10^5 ergs/cm^3) and associated with surface and finite size effects [51,29].

3.3.2   Cooling field dependence

For this study, we considered a sample with $D \sim 6$ nm and Fe weight fraction $\sim 20\%$.

Hysteresis loops were measured at $T = 5$ K, for different values of $0 \le H_{cool} \le 50$ kOe. In Figure 17, the loops corresponding to $H_{cool} = 0, 1, 4, 40$ kOe are shown and the values of the remanent magnetization (M_r) and H_{ex} are reported. Both M_r and H_{ex} increase up to $H_{cool} = 4$ kOe. For higher H_{cool}, M_r reaches a plateau value of $\sim$30 emu/g whereas H_{ex} decreases monotonously down to $\sim$ 250 Oe after cooling in a field of 50 kOe. A strong influence of the cooling field on the exchange bias phenomenon may be expected in the case of the ordered/SG systems, due to the multivalley energy structure and multiple equivalent spin configurations of the SG phase.

According to the value of the cooling field, applied above T_f (freezing temperature of SG), the magnetization of the ordered phase tends to align more and more in the field direction. As the temperature is lowered across T_f, a spin configuration of the SG phase will be selected through the exchange interaction with the ordered component. Thus, depending on H_{cool}, the degeneracy of the SG state can be reduced (actually, strong enough magnetic fields can destroy the SG state entirely).

On the other hand, for a FM/AFM uniaxial system, only two energetically equivalent spin configurations exist for the AFM and, following the field-cooling process, the exchange interaction with the FM phase selects one of the two.

Hence, the results in Figure 17 can be explained considering that different frozen magnetic configurations of the system may be selected through the variation of H_{cool}.

As previously stated, the exchange anisotropy between the metallic particle and the oxide matrix sets in when the temperature is lowered across $\sim$ 150 K. In the demagnetized sample, as the Fe particle moments point in random directions, the exchange anisotropy is averaged out and hence in the ZFC loop measurement at $T = 5$ K no shift is observed (Fig. 17a). Following the field-cooling process in a low H_{cool} ($<$ 4 kOe), an enhancement in the alignment degree of the Fe moments at $T = 5$ K is produced, consistent with the observed increase in M_r. Therefore, the different contributions to the exchange anisotropy sum together and a net exchange bias effect can be observed. In particular, the maximum $H_{ex} = 504$ Oe is measured for $H_{cool} = 4$ kOe (Fig. 17c).

For $H_{cool} > 4$ kOe, H_{ex} decreases. It should be considered that, with increasing H_{cool}, the magnetic coupling between the field and the oxide matrix moments increases as well, tending to orient them along the field direction. At $T = 5$ K, for high enough fields, such coupling may compete with the mix of magnetic interactions, proper of the system, overcoming the exchange coupling at the interface between Fe particles and oxide matrix. The system freezes in a configuration in which the energy associated with such internal interactions is not minimized. The matrix-particle interface energy becomes higher and higher with increasing the cooling field and hence H_{ex} decreases (Fig. 17d).

4. Conclusions

Nanostructured systems are ideal samples for the investigation of basic aspects such as surface and interface effects, magnetic interactions and influence of nanoscale fluctuations of the local magnetic properties on the macroscopic behavior.

Both the materials we have investigated – nanocrystalline Fe and Fe/Fe oxide granular samples – can be described as a collection of Fe nanocrystallites, spatially separated by an inter-crystallite phase, represented by the grain boundaries and by the oxide matrix, respectively. In these materials, such inter-crystallite phase presents a peculiar characteristic consisting of a high structural disorder. The combination of structural disorder and related frustration of magnetic interactions results in spin disorder and spin-glass-like properties, similarly to what was observed for surface spins of ferrimagnetic oxide particles [31,41,42]. This conclusion holds not just for the oxide matrix, where the magnetic coupling is carried out by the superexchange interaction, but also for the grain boundary phase of nanocrystalline Fe, where the distribution of interatomic spacings generates frustration of the exchange interaction.

Hence, a spin disordered configuration of the boundary region (surface or interface) is possible in nanostructured systems, independently of the type of the magnetic interaction which is involved.

The magnetic character of the inter-crystallite phase determines the macroscopic magnetic properties of the system. With varying temperature, the magnetic coupling between the Fe crystallites and the matrix gives rise to different magnetic behaviors, the difference being essentially determined by the change in the matrix effective anisotropy.

In both materials, at high temperature the matrix effective anisotropy is low and the polarizing action of the Fe crystallite moments on the matrix moments dominates.

This implies an enhancement of the degree of collinearity of the magnetic moments of the system. In particular, in nanocrystalline Fe, a long-ranged ferromagnetic order is stabilized since the polarized grain boundary phase transmits the exchange coupling to neighboring grains.

With reducing temperature a progressive freezing of the inter-crystallite phase moments occurs along random directions. Below a critical temperature the mix of crystallites-matrix exchange coupling and crystallite-crystallite dipolar interaction – which competes with the magnetocrystalline anisotropy – results in a frozen state for the whole system.

Acknowledgments

This work was partially supported by the Italian Ministry of Education, University and Research (MIUR) under project FIRB 'Microsystems based on novel magnetic materials structured on the nanoscale.

Professor E. Bonetti (University of Bologna, Italy) and Dr. A.M. Testa (ISM-CNR, Roma, Italy) are acknowledged for their valuable support and helpful discussion.

References

[1] G. Herzer, Scr. Metall. Mater. **33**, 1741 (1995)
[2] J.C. Lodder, Thin Solid Films **281-282**, 474 (1996)
[3] J.Q. Xiao, J.S. Jiang, C.L. Chien, Phys. Rev. Lett. **68**, 3749 (1992)
[4] A.E. Berkowitz, J.R. Mitchell, M.J. Carey, A.P. Young, S. Zhang, F.E. Spada, F.T. Parker, A. Hutten, G. Thomas, Phys. Rev. Lett. **68**, 3745 (1992)

[5] Y. Yoshizawa, S. Oguma, K. Yamauchi, J. Appl. Phys. **64**, 6044 (1988)

[6] H. Kronmüller, R. Fischer, M. Seeger, A. Zern, J. Phys. D: Appl. Phys. **29**, 2274 (1996)

[7] A. Manaf, R.A. Buckley, H.A. Davies, M. Leonowicz, J. Magn. Magn. Mater. **101**, 360 (1991)

[8] J. Arcas, A. Hernando, J. M. Barandiarán, C. Prados, M. Vázquez, P. Marín, A. Neuweiler, Phys. Rev. B **58**, 5193 (1998)

[9] J.F. Loffler, H.B. Braun, W. Wagner, Phys. Rev. Lett. **85**, 1990 (2000)

[10] R. Birringer, H. Gleiter, H.P. Klein, P. Marquardt, Phys. Lett. **102A**, 365 (1984)

[11] J. Löffler, J. Weissmüller, Phys. Rev. B **52**, 7076 (1995)

[12] A. Tschöpe, R. Birringer, J. Appl. Phys. **71**, 5391 (1992)

[13] C.C. Koch, Nanostruct. Mater. **2**, 109 (1993)

[14] K. Reimann, R. Würshum, J. Appl. Phys. **81**, 7186 (1997)

[15] A. Di Cicco, M. Berrettoni, S. Stizza, E. Bonetti, G. Cocco, Phys. Rev. B **50**, 12386 (1994)

[16] E. Bonetti, L. Del Bianco, L. Pasquini, E. Sampaolesi, Nanostruct. Mater. **10**, 741 (1998)

[17] L. Del Bianco, A. Hernando, E. Bonetti, C. Ballesteros, Phys. Rev. B **59**, 14788 (1999)

[18] B.E. Warren, *X-Ray Diffraction* (Dover Press, Cambridge, 1990)

[19] L. Del Bianco, A. Hernando, E. Bonetti, E. Navarro, Phys. Rev. B **56**, 8894 (1997)

[20] U. Herr, J. Jing, R. Birringer, U. Gonser, H. Gleiter, Appl. Phys. Lett. **50**, 472 (1987)

[21] S.J. Campbell, J. Chadwick, R.J. Pollard, H. Gleiter, U. Gonser, Physica B **205**, 72 (1995)

[22] See for instance: A.K. Bhatnagar, R. Jagannathan in *Metallic Glasses-Production, Properties and Applications*, T.R. Anantharaman Ed. (Trans. Tech Publications, Chur, Switzerland, 1984), p. 89

[23] H.J. Fecht, E. Hellstern, Z. Fu, L. Jonhson, Metall. Trans. A **21**, 2333 (1990)

[24] Dajie Zhang, K.J. Klabunde, C.M. Sorensen, G.C. Hadjipanayis, Phys. Rev. B **58**, 14167 (1998)

[25] E. Bonetti, L. Del Bianco, D. Fiorani, D. Rinaldi, R. Caciuffo, A. Hernando, Phys. Rev. Lett. **83**, 2829 (1999)

[26] B.R. Coles, B.V.B. Sarkissian, R.H. Taylor, Philos. Mag. B **37**, 489 (1978)

[27] S.N. Kaul, S. Srinath, J. Phys.:Condens. Matter **10**, 11067 (1998)

[28] V. Cannella, J.A. Mydosh, Phys. Rev. B **6**, 4220 (1972)

[29] B. Martinez, X. Obradors, Ll. Balcells, A. Rouanet, C. Monty, Phys. Rev. Lett. **80**, 181 (1998)

[30] T. Jonsson, J. Mattsson, C. Djurberg, F. A. Khan, P. Nordblad, P. Svedlindh, Phys. Rev. Lett. **75**, 4138 (1995)

[31] R.H. Kodama, A.E. Berkowitz, E.J. McNiff, S. Foner, Phys. Rev. Lett. **77**, 394 (1996)

[32] A. Hernando, E. Navarro, M. Multigner, A. R. Yavari, D. Fiorani, M. Rosenberg, G. Filoti, R. Caciuffo, Phys. Rev. B **58**, 5181 (1998)

[33] D. H. Ryan, J. M. D. Coey, E. Batalla, Z. Altounian, J. O. Ström–Olsen, Phys. Rev. B **35**, 8630 (1987)

[34] M. Liebs, M. Fähnle, Phys. Rev. B **53**, 14012 (1996)

[35] A. Hernando, P. Marín, M. Vázquez, J. M. Barandiarán, G. Herzer, Phys. Rev. B **58**, 366 (1998)

[36] R. Birringer, H. Gleiter, H.P. Klein, P. Marquardt, Phys. Lett. **102A**, 365 (1984)

[37] L. Del Bianco, D. Fiorani, A.M. Testa, E. Bonetti, L. Savini, S. Signoretti, Phys. Rev. B **66**, 174418 (2002)

[38] H.M. Rietveld, Acta Crystallogr. **20**, 508 (1966)

[39] K. Haneda, A.H. Morrish, Nature **282**, 186 (1979)

[40] S. Gangopadhyay, G.C. Hadjipanayis, B. Dale, C.M. Sorensen, K.J. Klabunde, V. Papaefthymiou, A. Kostikas, Phys. Rev. B **45**, 9778 (1992)

[41] J.M. Coey, Phys. Rev. Lett. **27**, 1140 (1971)

[42] E. Tronc, A. Ezzir, R. Cherkaoui, C. Chanéac, M. Nogués, H. Kachkachi, D. Fiorani, A.M. Testa, J.M. Grenèche, J.P. Jolivet, J. Magn. Magn. Mater. **221**, 63 (2000)

[43] K. O'Grady, R.W. Chantrell, in *Magnetic Properties of Fine Particles*, J.L. Dormann and D. Fiorani Eds. (North-Holland, Amsterdam, 1992), p. 93

[44] P. Zhang, F. Zuo, F.K. Urban III, A. Khabari, P. Griffiths, A. Hosseini-Tehrani, J. Magn. Magn. Mater. **225**, 337 (2001)

[45] E.C. Stoner, E.P. Wohlfarth, Proc. Phys. Soc. A **240**, 599 (1948)

[46] J.F. Loffler, J.P. Meier, B. Doudin, J.P. Ansermet, W. Wagner, Phys. Rev. B **57**, 2915 (1998)

[47] P.J. van der Zaag, R.M. Wolf, A.R. Ball, C. Bordel, L.F. Feiner, R. Jungblut, J. Magn. Magn. Mater. **148**, 346 (1995)

[48] R.H. Kodama, S.A. Makhlouf, A.E. Berkowitz, Phys. Rev. Lett. **79**, 1393 (1997)

[49] W.H. Meiklejohn, J. Appl. Phys. **33**, 1328 (1962)

[50] Review: J. Nogués, I.K. Schuller, J. Magn. Magn. Mater. **192**, 203 (1999)

[51] B. Martinez, A. Roig, X. Obradors, E. Molins, A. Rouanet, C. Monty, J. Appl. Phys. **79**, 2580 (1996)

SURFACE AND INTERPARTICLE EFFECTS IN AMORPHOUS MAGNETIC NANOPARTICLES

R.D. ZYSLER, E. DE BIASI, C.A. RAMOS
Centro Atómico Bariloche,8400 S.C. de Bariloche, RN, Argentina

D. FIORANI
ISM-CNR, Area della Ricerca di Roma, C.P. 10, I-00016 Monterotondo (Rome), Italy

H. ROMERO
Facultad de Ciencias, Departamento de Física, Universidad de Los Andes, Mérida, 5101, Venezuela

1. Introduction

The knowledge of the magnetic properties of nanometer particles is central in basic research [1, 2] and technological applications, e.g. in high-density magnetic storage media [3, 4], hard magnets [5], and biomedicine [6]. Finite size effects induce a magnetic behaviour which may strongly differ in several aspects from those observed on conventional bulk materials. Surface effects dominate the magnetic properties of the smallest particles since decreasing the particle size ratio of the surface spins to the total number of spins increases, e.g. in a particle of diameter ~3 nm, about 70% of atoms lie on the surface, which is structurally and magnetically disordered. Consequently, the picture of a single-domain magnetic particle where all spins are pointing into the same direction, leading to coherent relaxation process, is no longer valid when one considers the effect of misaligned spins on the surface on the global magnetic properties of the particle. Defects, missing bonds and then the decrease of the average coordination number determine a weakening of the exchange interactions between surface atoms. This contributes to reduction of the magnetic transition temperature with decreasing particle size, as shown by experiments [7, 8] and Monte Carlo (MC) calculations [9]. Moreover, the symmetry breaking at the surface results in a surface anisotropy (SA), which can represent the dominant contribution to the total particle anisotropy for small enough particles. In most cases SA is strong enough to compete with the exchange energy that favours full alignment of particle spins. Actually, it is expected that the magnetization vector will point along the bulk axis (the *particle* anisotropy axis) in the core of the particle, and it will then gradually turn into a different direction when it approaches the surface. As a consequence of the combination of finite-size and surface effects, the profile of the magnetization is not uniform across the particle and the magnetization of the surface layer is smaller than that corresponding to the core spins. This effect has been reported in several nanoparticle systems. High-field magnetization measurements on γ-Fe$_2$O$_3$ [10] and Co [11] nanoparticles have shown that the magnetization is strongly influenced by surface effects, depending on the particle size. Also nickel ferrite particles show high-field open hysteresis loops, due to the large surface anisotropy, and high-field magnetic relaxation, due to the progressive overcoming of energy barriers of a spin-glass-like surface state [12, 13]. MC calculations were carried out simulating maghemite particles described by a Heisenberg model, including internal dipolar interactions and surface exchange different from those of the core value [10, 14]. Other MC simulations with an Ising model for spherical particles of maghemite show that broken bonds at the surface may originate a spin-glass phase [15]. These results indicated that magnetic disorder at the surface induces a reduction of the saturation magnetization and facilitates the thermal demagnetization of the particle at zero field. For 6 nm γ-Fe$_2$O$_3$ particles, the surface spin canting was considered responsible for the non-saturation of the magnetization even at a magnetic field of 50 kOe [16]. Interactions between surface and core spin structure (exchange anisotropy) should also be taken into account. Furthermore, these two contributions can be in competition making it difficult to model the particle magnetic behaviour and experimentally separate their contribution.

Interparticle interactions (dipole-dipole and exchange interactions, if the particles are in close contact) can also affect the surface spin orientation. The behaviour of interacting superparamagnetic moments with volume distribution, disordered arrangement and random axis orientation is a non-trivial problem. For weak interaction, the magnetization is still superparamagnetic and is well described with the Dormann-Bessais-Fiorani model [1,17]. For strong interparticle interactions, the individual energy barriers can no longer be identified and only the energy of the particle assembly is significant. A collective state is established with magnetic properties recalling those of spin glasses. Surface magnetic properties themselves may change under the effect of interparticle interactions inducing local deviations of surface spins which will depend on the type of interaction, particle arrangement and surface microstructure.

In all the reported examples of magnetic nanoparticles, the bulk samples are crystalline and then the magnetocrystalline anisotropy term represents a substantial contribution to the total anisotropy energy of the particle. This anisotropy term has a large influence on the actual magnetic order in the particles and on the effective surface anisotropy.

The magnetic behaviour of amorphous magnetic nanoparticles has not been much investigated so far. In such particles, shape anisotropy and surface anisotropy are expected to play the major role. Recently, experiments and Monte Carlo (MC) simulations have been performed on amorphous ferromagnetic nanoparticles [18-21]. In this work, amorphous nanoparticles have a core-shell structure, and we consider the disordered shell spins to contribute to a large proportion of the total magnetic response, making them attractive to study (the core-shell intraparticle interaction as well as interparticle effects present in concentrated systems).

In order to gain a deeper insight into surface properties and interparticle interaction effects, we have investigated the magnetic properties of ~3 nm ferromagnetic amorphous nanoparticles (Fe-Ni-B and Co-Ni-B alloys) in a very diluted dispersion in a non-magnetic matrix, where actually particles are non-interacting, and in a powder sample, where particles are strongly interacting. The results of Monte Carlo simulations, assuming a *core-shell* model, agree with observed magnetic behaviour confirming that it is strongly affected by the disordered shell spins.

We will present first a description of the core-shell model, then the experimental results of magnetization measurements, and finally the results of MC simulations.

2. The core-shell model

In this model we consider each single-domain particle as a system of interacting classical spins in the presence of an applied magnetic field. In this case, we can write the Hamiltonian in the form

$$H = - \sum_{<i,a>} J_{i,i+a} \, \mathbf{S}_i \cdot \mathbf{S}_{i+a} - \sum_i K_i (\hat{\mathbf{S}}_i \cdot \hat{\mathbf{n}}_i)^2 - \mu \sum_i \hat{\mathbf{S}}_i \cdot \mathbf{H} \tag{1}$$

where the first term is the isotropic ferromagnetic ($J>0$) Heisenberg exchange interaction between nearest neighbour spins, the second term is a uniaxial anisotropy contribution (the unit vector $\hat{\mathbf{n}}_i$ denotes the easy direction) and the last corresponds to the Zeeman energy (μ is the magnetic moment per atom). All terms are summed over all N_{total} spins of the particle.

We propose that the particle is composed of a core containing N_{core} spins, and a shell surrounding it that contains N_{shell} spins. All spins are identical, subject to the same exchange interactions and anisotropy within the core and differing from the shell values. This assumption is based on the fact that the shell atoms can be expected to have missing bonds or other defects that lead to a reduced exchange interaction as compared to the core exchange. Furthermore, we consider a uniaxial anisotropy in the core (along our z direction) and a single-site anisotropy on the spins of the shell external surface. With these assumptions, the total energy of the system is given by

$$E = - \sum_{<i,a>}^{N_{core}} J_{core} \, \mathbf{S}_i \cdot \mathbf{S}_{i+a} - \sum_i^{N_{core}} K_{core} (\hat{\mathbf{S}}_i \cdot \hat{z})^2 - \mu \sum_i^{N_{core}} \hat{\mathbf{S}}_i \cdot \mathbf{H}$$
$$- \sum_{<i,a>}^{N_{shell}} J_{shell} \, \mathbf{S}_i \cdot \mathbf{S}_{i+a} - \sum_i^{N_{surf}} K_{surf} (\hat{\mathbf{S}}_i \cdot \hat{\mathbf{n}}_{surf})^2 - \mu \sum_i^{N_{shell}} \hat{\mathbf{S}}_i \cdot \mathbf{H} \tag{2}$$

where J_{core} is the exchange constant between core spins and J_{shell} is the corresponding one to shell-shell and core-shell spins, K_{core} and K_{surf} are the core and surface anisotropy per spin, and $\hat{\mathbf{n}}_{surf}$ indicates a unit vector perpendicular to the surface. The core-shell exchange acts on the third term only when considering the shell spins that have a core spin as a first neighbour. As we only consider amorphous nanoparticles, K_{core} comes from the shape anisotropy, and K_{surf} is originated by the spin-orbit coupling of each surface ion. Considering that the temperature range of our experiments is well below the bulk ferromagnetic order temperature, we assume that the core of each particle is ordered as a ferromagnetic single domain. In this condition, the ordered core behaves as a superparamagnetic moment [22] and we can consider the spins of the core as a single moment $\mu_{core} = N_{core} \, \mu$. Therefore, the first term in Eq. (2) is constant and may be neglected, allowing the energy to be rewritten as follows:

$$E = -K_{core} (\hat{\mu}_{core} \cdot \hat{z})^2 - \mu_{core} \cdot \mathbf{H} - \sum_{<i,a>}^{N_{shell}} J_{shell} \, \mathbf{S}_i \cdot \mathbf{S}_{i+a}$$
$$- \sum_i^{N_{surf}} K_{surf} (\hat{\mathbf{S}}_i \cdot \hat{\mathbf{n}}_{surf})^2 - \mu \sum_i^{N_{shell}} \hat{\mathbf{S}}_i \cdot \mathbf{H} \tag{3}$$

In the following, results from MC simulations are reported and compared with results of experimental measurements. The first two terms of Eq. (3) give origin to the core behaviour and reproduce the superparamagnetic and blocked regimes where the maximum in the ZFC magnetization is given by the anisotropy energy barrier K_{core}. The exchange term between the surface and core spins leads to a ferromagnetic order of the surface. As this ferromagnetic order is achieved, small correlated spin regions grow and the core-shell interaction tends to align them parallel to the core. These ordering effects compete with the surface anisotropy term (fourth term in Eq. (3)), which, we propose, tends to align the spins perpendicular (parallel) to the surface for $K_{surf} > 0$ ($K_{surf} < 0$). Then, the surface anisotropy competes with the ferromagnetic order among the shell spins and it is necessary to decrease the temperature enough so that the short-range order occurs. Consequently, the total magnetization of each particle consists of two components, i.e., the core magnetic moment, which has a low temperature-dependence, and the shell magnetization, which at low temperature contributes to an increase of the saturation magnetization. This model explains the reduced saturation magnetization with respect to the bulk value and the existence of a non-saturated component of the magnetization at high field observed in some systems [10, 14, 15, 18, 19].

3. Sample characteristics

Amorphous nanoparticles of $(Fe_{0.26}Ni_{0.74})_{50}B_{50}$ (S1) and $(Co_{0.25}Ni_{0.75})_{65}B_{35}$ (S2) were obtained by reduction of aqueous solutions of metallic salts of $FeSO_4$ ($CoSO_4$) and $NiCl_2$ with a $NaBH_4$ solution [19,23]. The relative Fe/Ni and Co/Ni composition were determined by energy disperse spectroscopy microanalysis (EDS) and the boron concentration by atomic absorption analysis. X-ray powder diffraction measurements show some broad reflections [20, 23], as expected for amorphous compounds. The

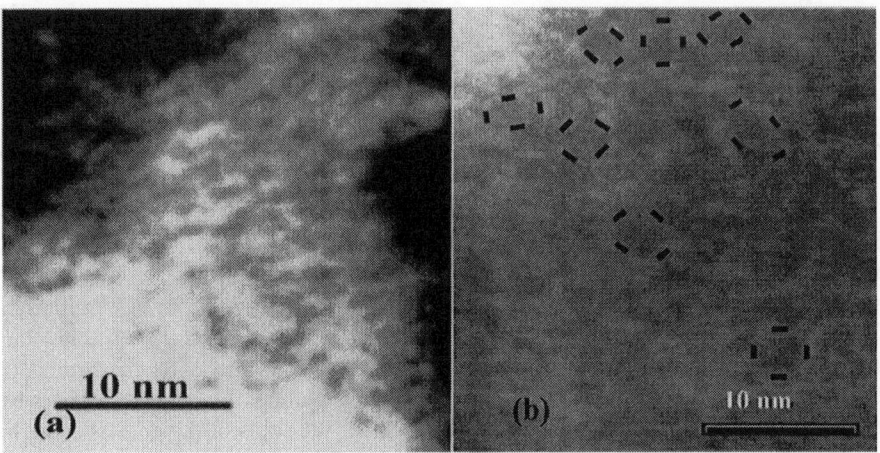

Figure 1. TEM images for S1 (a) and S2 (b) samples.

particle size distribution, determined by TEM and light-scattering experiments, is log-normal with a mean diameter $<\phi> = 3$ nm and a dispersion $\sigma = 0.3$ [23, 19] (see Fig. 1). In order to study the magnetic properties of non-interacting nanoparticle systems, diluted dispersions of these particles in a polymer (polyvinyl-pyrrolidone) were prepared (samples S1D and S2D). The dispersion resulted in an S1D particles/polymer ratio of 0.3% by weight and leads to an estimated mean center-to-center interparticle distance $d \sim 35$ nm. In S2D the dispersion resulted in a particles/polymer ratio of 4% by weight and the estimated mean interparticle distance is $d \sim 16$ nm. For both samples, the interparticle distance is large enough that dipole-dipole interparticle interactions are negligible.

4. Non-interacting nanoparticles

4.1 MAGNETIZATION MEASUREMENTS

The magnetic properties were investigated in the temperature range 2 – 300 K and up to 50 kOe using a commercial SQUID magnetometer.

Magnetization measurements as a function of temperature for the dispersed samples, performed according to the standard zero field cooling (ZFC) and field cooling (FC) procedures, are reported in Fig. 2 [18, 19]. The data show an irreversibility on the magnetization below T_{irr} (T_{irr} (S1D) = 85 K and T_{irr} (S2D) = 200 K), typical of single-domain particles characterized by a superparamagnetic regime at high temperatures and the blocked regime at $T < T_{max}$ (T_{max} (S1D) = 55 K; T_{max} (S2D) = 61 K), i.e. the temperature of the maximum in M_{ZFC}. The observed behaviour reveals the progressive

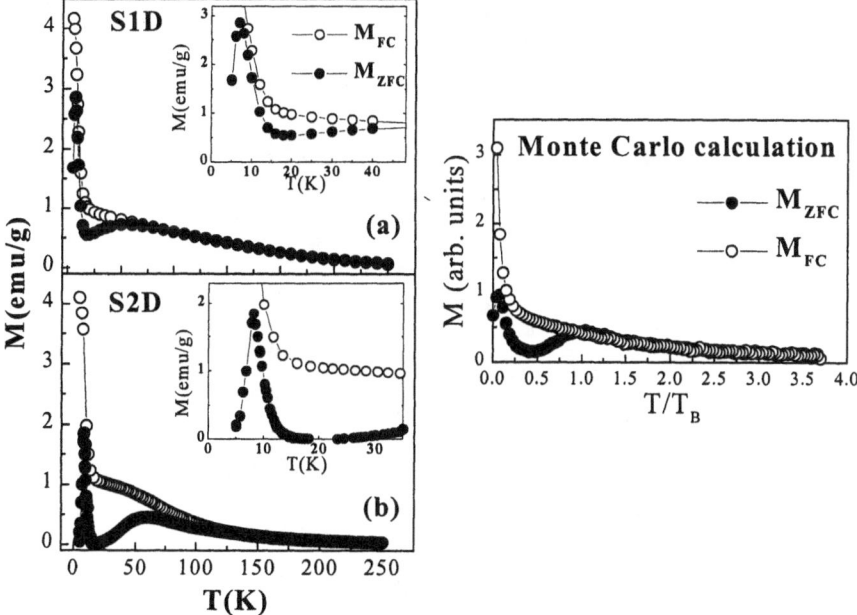

Figure 2. Temperature dependence of the magnetization measured at 50 Oe under zero-field-cooling (solid circles) and field-cooling conditions (open circles) for (a) $(Fe_{0.26}Ni_{0.74})_{50}B_{50}$ and (b) $(Co_{0.25}Ni_{0.75})_{65}B_{35}$. In the inset, low temperature magnetization is reported, showing a second sharp maximum in M_{ZFC}. (c) Monte Carlo simulation of the $M_{ZFC}(T)$ and $M_{FC}(T)$ experiment.

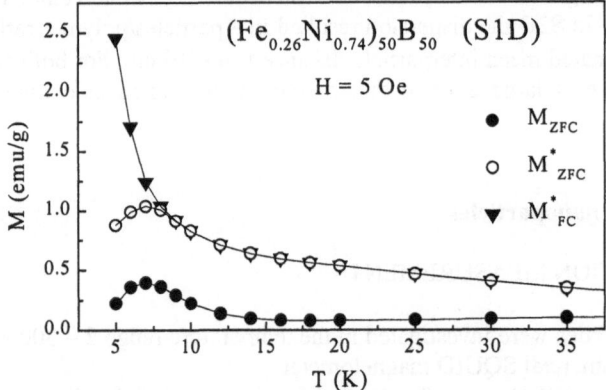

Figure 3. ZFC and FC magnetization for S1D sample as a function of temperature after cooling the sample down to 8.5 K applying 50 Oe and then recording the M^*_{ZFC} and M^*_{FC} applying 5 Oe (see text).

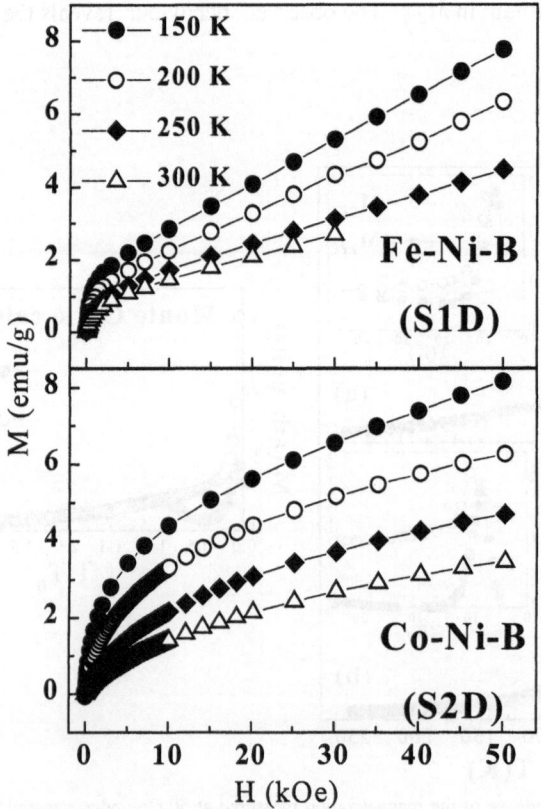

Figure 4. Magnetization versus applied field at some temperatures above T_{irr} (superparamagnetic regime) for S1D sample and S2D sample. The lines are only a guide for the eye.

blocking of the superparamagnetic particle moments, with a distribution of relaxation times related to the size and anisotropy axis direction distributions. In general, for an assembly of non-identical, non-interacting, magnetic nanoparticles, the low field $M(T)$ depends on the type of anisotropy energy barrier distribution function, which governs the relationship between T_{max} and the average blocking temperature $<T_B>$ [1, 24]. For a particle of volume V, T_B is defined as the temperature at which the relaxation time, described by the Néel-Brown expression, $\tau = \tau_0 \exp(E_B/k_BT)$ where $E_B = KV$ (for uniaxial symmetry), becomes equal to the measuring time t_m. T_{irr} corresponds to the highest blocking temperature, i.e. to that of particles with highest energy barrier.

At very low temperatures, a second sharp maximum in M_{ZFC} is observed in both samples at T_S (T_S (S1D) = 7 K and T_S (S2D) = 9 K) (see inset of Fig. 2). In this case, the FC curve does not superimpose the ZFC curve at temperatures above T_S because the superparamagnetic component associated with this behaviour has an additional background due to the FC magnetization of the blocking at higher temperature. In order to observe clearly the low temperature irreversibility we have measured the sample S1D with the following procedure: (a) the sample was cooled under an applied field of 50 Oe down to 8.5 K; (b) the magnetic field was removed and the sample was cooled in zero magnetic field down to 5 K, and (c) the magnetization curve, M^*_{ZFC} was recorded applying a magnetic field of 5 G. The M^*_{FC} curve was obtained in the same way but cooling the sample applying a field of 50 Oe below 8.5 K. Using this procedure both M^*_{ZFC} and M^*_{FC} have the same background added. Figure 3

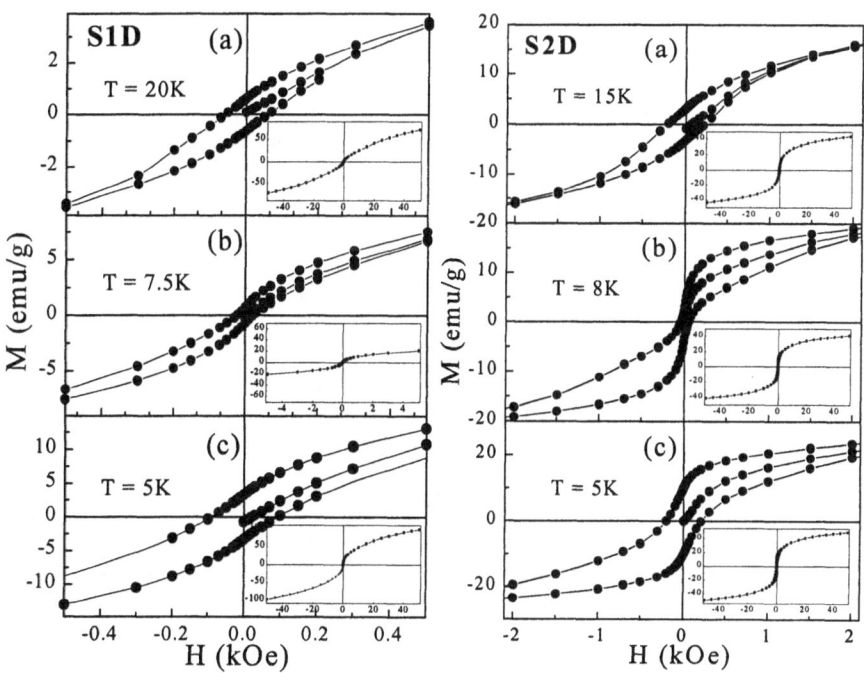

Figure 5. Magnetization cycles measured at $T < T_{irr}$ for the S1D sample. The inset shows the complete cycle.

Figure 6. Magnetization cycles measured at $T < T_{irr}$ for the S2D sample. The inset shows the complete cycle.

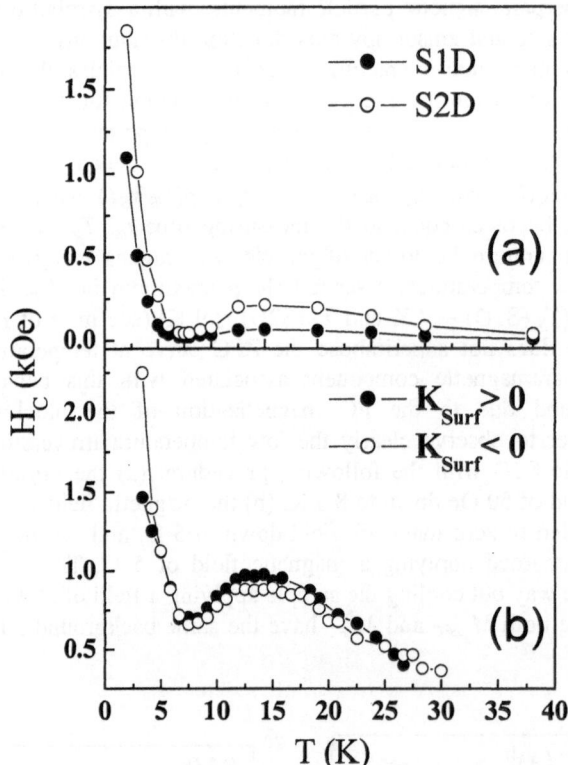

Figure 7. H_C as a function of temperature. (a) Experimental results for both samples S1D and S2D (open circles and solid circles respectively). The lines are only a guide for the eye. (b) Coercive field vs. temperature obtained from the simulated hysteresis loops at different temperatures with $K_{surf} > 0$ (solid circles) and $K_{surf} < 0$ (open circles).

shows the measured magnetization curves at low temperatures where the superposition of M^*_{ZFC} and M^*_{FC} above T_S is observed revealing a *reversible* behaviour of this contribution.

These results suggest, within the core-shell model, that the broad maximum of ZFC at T_{max} is associated with the blocking of core particle moments, whereas the sharp maximum at T_S is related to the freezing of surface spins. With decreasing temperature, surface spin fluctuations slow down and short range correlations among them develop progressively in magnetic correlated spins regions of growing size.

Magnetization measurements were performed as a function of the magnetic field at different temperatures after ZFC. For $T > T_{irr}$, $M(H)$ does not present hysteresis, as expected in the unblocked regime, whereas for $T < T_{irr}$ a hysteretic behaviour occurs, due to the growing fraction of blocked particle moments with decreasing temperature. As is shown in Figure 4, the magnetization of the sample at $T > T_{irr}$ clearly consists of two components, $M(H) = M_{sp}(H) + \chi H$, where M_{sp} is a superparamagnetic one which follows a Langevin-like function. Below T_{irr}, magnetization measurements as a function of the magnetic field show hysteretic behaviour, without saturation of the magnetization (Figs. 5 and 6). A change of hysteresis loop shape, more evident for the sample S2D, is

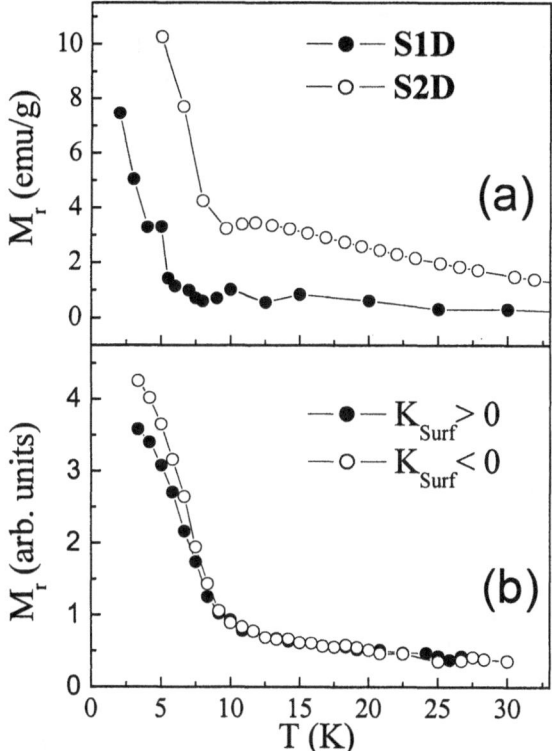

Figure 8. Remanent magnetization $M_r(T)$. (a) Experimental results for samples S1D (solid circles) and S2D (open circles). The data were recorded after cooling the sample in a field of 50 kOe from room temperature down to 5 K, then switching off the field and measuring M_r with increasing temperature (circles). The lines are only a guide for the eye. (b) $M_r(T)$ obtained from the simulated hysteresis loops at different temperatures with $K_{surf} > 0$ (solid circles) and $K_{surf} < 0$ (open circles).

observed close to T_S: the cycle narrows at low fields (the demagnetizing and remagnetizing curves vary rapidly in the low field region). With decreasing temperature, the coercive field H_c increases up to a maximum, then decreases down to a minimum and finally strongly increases again at low temperatures, below T_S (Fig. 7a). The remanent magnetization, M_r, also shows a rapid increase below T_S (Fig. 8a). The temperature dependence of the remanent magnetization, $M_r(T)$, is reported in Fig. 8a M_r has been measured in two ways: i) the sample was cooled under a field of 50 kOe from room temperature (RT) down to 5 K, then the field was switched off and M_r was measured heating up to RT; ii) from the hysteresis loops at different temperatures.

Also, the $M(H)$ curves suggest the existence of two contributions to the magnetization. In Fig. 9, we plotted the magnetization at $H = 50$ kOe as a function of temperature for both samples, S1D and S2D. High temperature behaviour, determined by the core contribution, is weakly temperature dependent. At low temperature, the large magnetization is determined by the shell contribution. Moreover, the hysteresis loops for $T \leq T_S$ indicate that the magnetization comes from the superposition of two contributions: one from the spins of the ferromagnetic particle core, which tends to saturate at low fields, and the other from the magnetically disordered surface frozen spins, which does not saturate up to 50 kOe. The latter is responsible for the change of

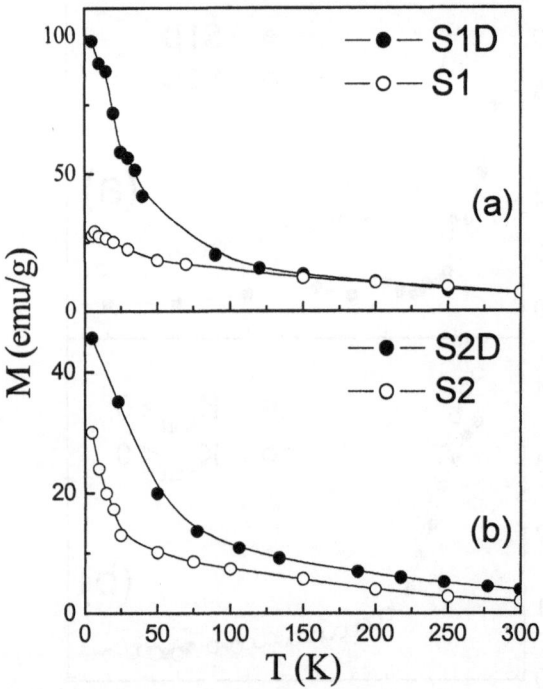

Figure 9. Magnetization at *H* = 50 kOe vs. *T* for dispersed (solid circles) and powder (open circles) samples: (a) Fe-Ni-B and (b) Co-Ni-B.

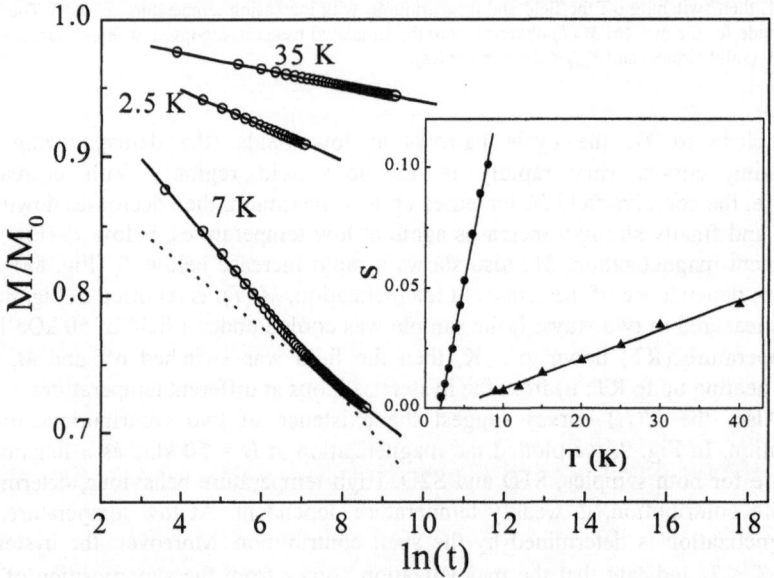

Figure 10. Magnetization relaxation as a function of time for S1D sample. The solid lines result from the fitting of the magnetization data with the $M(t,T) = C - S(T)\ln(t-t_0)$ function. The inset plots the viscosity of the logarithmic decay of the magnetization, *S*, as a function of temperature.

the shape of hysteresis loop at T close to T_S and for the change in the temperature dependence of H_c and M_r for $T \leq T_S$. At $T \sim 10$ K, the hysteresis curves show a narrowing at low fields, where a sharp variation of the magnetization is observed in the demagnetizing and remagnetizing branches (Figs. 5b and 6b). This corresponds to a decrease of H_c and is responsible for the minimum in the temperature dependence of H_c. On the other hand, above 20 K the coercive field varies as $T^{1/2}$ according to the Stoner-Wohlfarth model [25] for an assembly of particle moments with uniaxial anisotropy and a random distribution of anisotropy axes. This indicates that above this temperature the magnetic behaviour is governed by the particle core contribution to the magnetization. Deviations from the model observed below 20 K are due to the progressive blocking of surface spin correlated region effective moments.

The magnetization relaxation phenomena shows evidence of intraparticle structure. The time decay of the magnetization was recorded as follows: the sample was first heated to 100 K and a +100 Oe field was applied. It was then cooled to the measurement temperature (in the range 2 K - 40 K). Then the field was removed and the zero field magnetization was recorded for a few hours, starting a few seconds after removal of the magnetic field. The observed magnetization relaxation is described by a logarithmic function: $M(t,T) = C - S(T)\ln(t-t_0)$, where C is a constant and S is the so-called magnetic viscosity [26]. Different relaxation behaviours have been observed in different temperature ranges, above and below 9 K (see Figure 10). Above and well below 9 K, the $M(t)$ curves can be satisfactorily described and fitted by a single logarithmic function. Just below 9 K, two regimes of the logarithmic magnetization relaxation are observed, at short and long times. In the inset of Fig. 10, showing the slope S of the magnetization logarithmic decay as a function of temperature (the short-time slope is reported for the intermediate relaxation region), the two relaxation regimes are clearly distinguished. The core-shell model explains the observed magnetization relaxation behaviour assuming that at low temperature, below T_S, the shell spins are ordered (frozen) and coupled with the core magnetic moment, whereas at intermediate temperatures, between T_S and T_{max}, the shell spins form correlated regions interacting between them and the core spins. We will come back to this point after describing the MC simulation results.

4.2 MONTE CARLO SIMULATIONS OF THE MAGNETIZATION

4.2.1 Numerical method

Taking into account that our simulations were intended to qualitatively reproduce the experimental results in amorphous nanoparticles, we have chosen the particles' ellipsoidal shape because this is a simple shape that can originate a uniaxial anisotropy and it is easy to simulate in our Monte Carlo calculations. Then, we assume that our system consists of an assembly of non-interacting ellipsoidal particles with identical sizes of $7a$ major axis (in z direction) and $4.5a$ minor axis, where a is the cell parameter. Each single-domain particle is considered as a system of interacting classical spins on a simple cubic lattice in the presence of an applied magnetic field, where the total energy may be described by Eq. (3). In this case, there is a shell width of $\sim 3a$ and a skin of $\sim 1a$ on which the surface anisotropy acts (Fig. 11).

In our simulations we start from a random orientation of the ferromagnetic core and shell spins. At this point, the classical Monte Carlo method based on the Metropolis

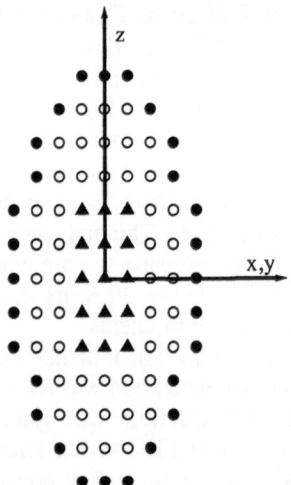

Figure 11. Cross section in the x-z plane of the particle where the core spins (triangles) and shell spins (circles) are indicated. Solid circles indicate the spins in which the surface anisotropy determines the spin orientation.

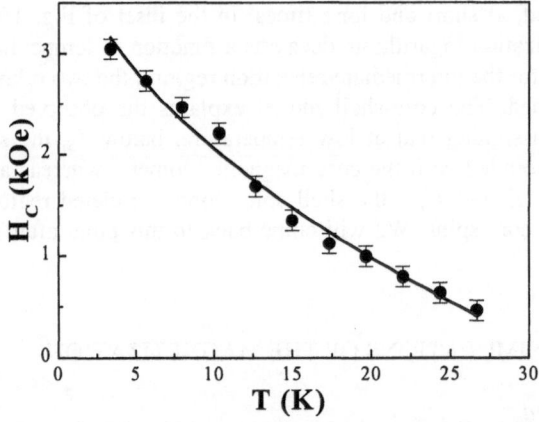

Figure 12. Coercive field vs. temperature obtained from the simulated hysteresis loops at different temperatures by simulating only the core magnetization curves (symbols). The line is a fitting of the data with the $H_C(T) = H_C(0) (1-(T/T_B)^{1/2})$ function.

algorithm is used. In order to obtain the new spin configuration, in each MC step a trial orientation of each spin (or the total moment of the core) is generated. To get the new orientation of the spins we used the method described in Ref. 18. The new orientation is picked randomly from the interval $\theta_0 \pm \delta$ and $\varphi_0 \pm 2\delta$, where θ_0 and φ_0 are the initial angles of the spins [27]. Subsequently, if the total energy of the configuration picked is smaller than the original one, the new orientation of the magnetic moment is taken as a more stable configuration. Otherwise, if the total energy of the picked configuration is greater than the original one, the attempted orientation is accepted with a probability exp($\Delta E/k_B T$), where ΔE is the energy difference between attempted and present orientations. The maximal jump angle (δ) allows us to simulate the existence of the energy barrier in order to obtain the non-equilibrium behaviour (blocked regime). Varying δ is possible to modify the range of acceptance in order to optimize the simulation. Each magnetization value is taken after 1000 MC steps, averaging over the last 400 configurations. In order to reduce the simulation noise, we performed an average over 200 particles for the high temperature cycles and 50 particles for the lower temperature cycles.

Magnetization as a function of temperature has been calculated under ZFC and FC conditions. The M_{ZFC} curve was simulated by cooling the system in zero field (i.e., the simulation starts from a random configuration of the shell spins and the core orientation at high temperatures), and after attaining the lowest temperature, a fixed magnetic field is applied calculating the magnetization while the particles were warmed up. The M_{FC} curve was achieved calculating the magnetization from temperatures above T_B down to the lowest temperature with the magnetic field at fixed value.

Magnetization curves at different temperatures have been calculated. The simulation also begins from a random configuration of the shell spins and the core orientation at high temperatures (at $T > T_B(core)$, the magnetization has a superparamagnetic behaviour and the results are independent of the initial state) and subsequently the calculation follows a ZFC experiment. This process allows the ordering among shell spins driven by the exchange interaction. At a given temperature, the magnetic field was applied and increased progressively and the magnetization was calculated at different fixed fields building the $M(H)$ curve. We have determined the optimum sweep rate of the magnetic field (the magnetic field step) in order to ensure that the blocked magnetic moment "follows" the applied field in the correct form, i.e., the magnetization calculated decreasing field after saturation must decrease, the cycles

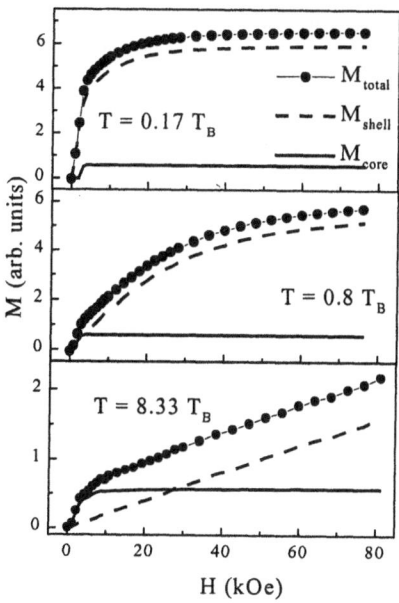

Figure 13. Net moment of the core (M_{core}), shell (M_{shell}) and the total magnetization versus magnetic fields at different temperatures calculated by Monte Carlo simulation under ZFC condition and $K_{surf} > 0$.

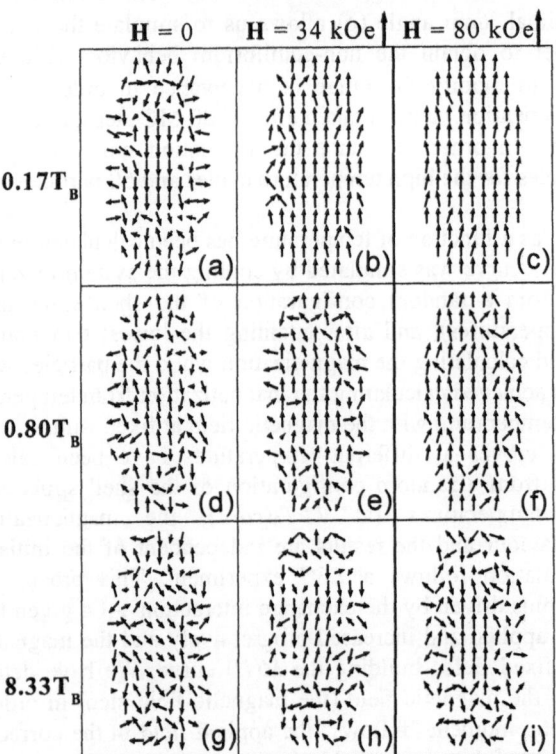

Figure 14. Cross section of the MC calculated spin configurations, after cooling the sample in zero field and $K_{surf} > 0$, in the x-z plane of the particle at three temperatures. (a)-(c) $T < T_S$, (d)-(f) $T_S < T < T_{max}$, (g)-(i) $T_{max} < T$. Note that at $H = 0$ (a, d and g figures) the obtained core orientations were approximately equally distributed in the positive and negative directions.

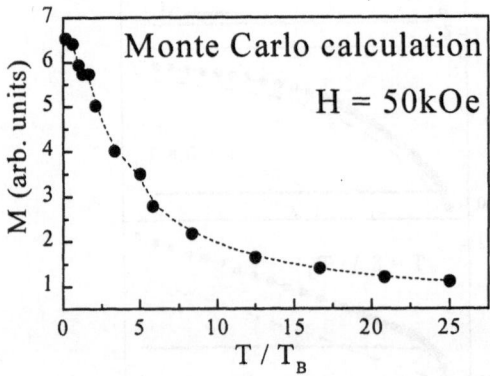

Figure 15. Computed thermal variation of the total magnetization taken at $H = 5$ kOe from the simulation under the same conditions as Figs. 13 and 14.

must be closed and symmetric and the positive and negative saturation magnetization must be equal.

The validity of our Monte Carlo method has been tested by simulating only the core magnetization curves as a function of temperature and magnetic field retrieving the Stoner-Wohlfarth expected behaviour [25] in the $M(T)$ ZFC-FC curves and the coercive field follows a $T^{1/2}$ dependence (Fig. 12).

4.2.2 Simulation results

The simulations were performed for a system of *non-interacting* ellipsoidal particles of $\sim$7a major radius and $\sim$4.5a minor radius where a is the cell parameter ($c/a = 1.5$). Particles with this size have 45 atoms in the core and 476 atoms in the shell where 218 atoms lay on the external surface. The adopted size does not correspond to that of the particles described in the experimental results, but it illustrates the general features we have proposed in our model. In the simulations we assume that the surface anisotropy term tends to align the spins perpendicular to the surface in the Fe-based particles and parallel to the surface in the Co-based particles, as reported in the literature for Fe [28, 29] and Co [30] particles, respectively. Extrapolation of literature results to our samples would imply $K_{surf} > 0$ for the sample S1D and $K_{surf} < 0$ for the sample S2D. The constants used in the simulation were $K_{sup}/K_{core} = 6.3$ and $K_{sup}/J_{shell} = 6$. The set of parameters K_{sup}/K_{core} and K_{sup}/J_{shell} were chosen so as to reproduce the experimental data.

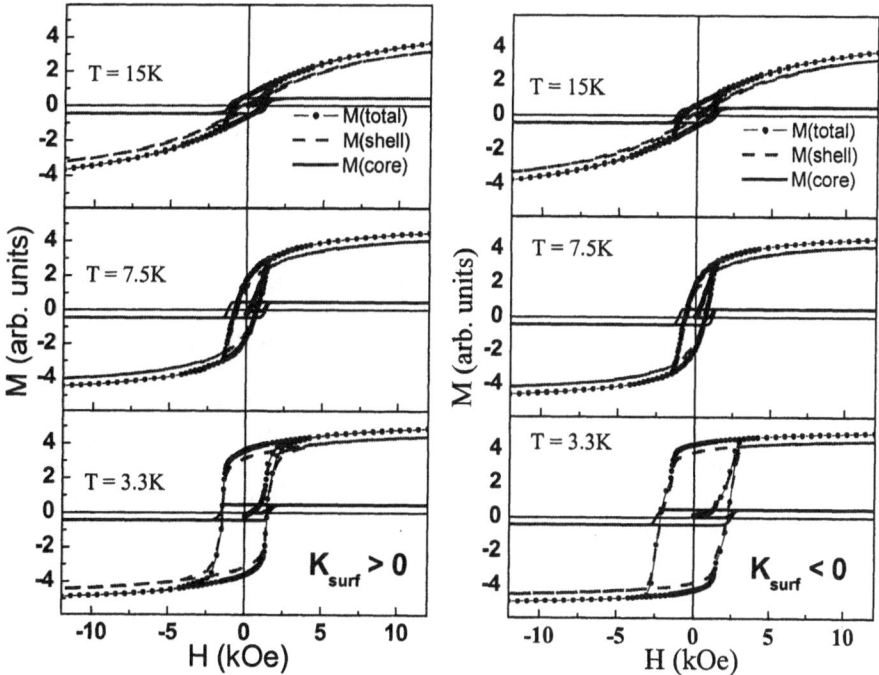

Figure 16. Monte Carlo calculations of hysteresis cycles at different temperatures for $K_{surf} > 0$ (see Eq. (1)) where the net moment of the core (M_{core}), shell (M_{shell}), and the total magnetization are indicated.

Figure 17. Same as Fig. 16 for $K_{surf} < 0$.

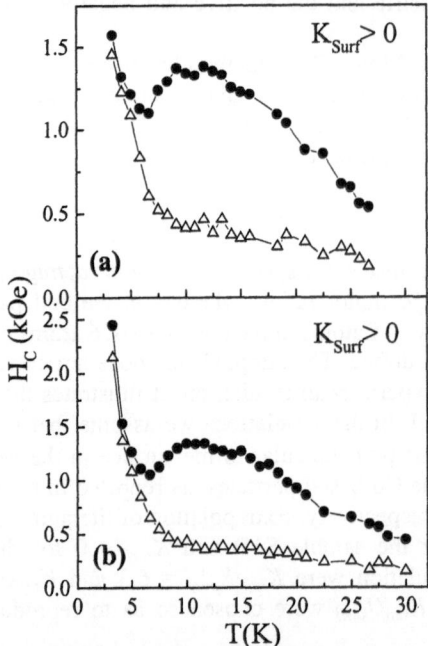

Figure 18. Core (solid circles) and shell (open triangles) contribution of the coercive field vs. temperature obtained from the simulated hysteresis loops at different temperatures. (a) $K_{surf} > 0$ and (b) $K_{surf} < 0$.

We plot in Figure 13 the calculated core moment (M_{core}), shell component, (M_{shell}) and the total magnetization versus magnetic field at different temperatures after cooling the sample in zero field for $K_{surf} > 0$. The $M(H)$ curves at high temperatures for $K_{surf} < 0$ have the same behaviour. Unlike the M_{core}, which saturates at relatively low applied field at all temperatures, M_{shell} needs large magnetic fields in order to achieve saturation and leads to a linear M_{shell} behaviour at high temperatures. The M_{shell} behaviour is more clearly displayed in Figure 14 where we have drawn cross sections of the calculated spin configurations in the x-z plane of the particle. We note that decreasing temperature, at $H = 0$, some of the shell spins are aligned forming ferromagnetic regions not necessarily in the same direction of the core magnetization, and others without any apparent magnetic order. This disorder comes from the competition between the exchange energy that tends to align the spins parallel to each other and the surface anisotropy that tends to align the spins perpendicular to the surface. When a magnetic field is applied, the correlated regions of the shell and the single spins tend to align along the external field direction. The temperature dependence of the ferromagnetic regions yields the observed behaviour of M_{shell} at high applied fields, i.e. at high temperatures the shell component consists of paramagnetic spins with a linear in-field magnetization behaviour, whereas with decreasing temperatures superparamagnetic spin correlated regions are formed and grow in size, with a Langevin behaviour.

Figure 15 presents the MC simulation of $M(T)$ at $H = 50$ kOe ($M(T) = M_{core}(T) + M_{shell}(T)$). For $T/T_B > 10$, $M(T)$ shows a weak temperature dependence, whereas   for $T/T_B < 10$, $M(T)$ varies rapidly, M_{shell} contributing   significantly to the total magnetization, with the same behaviour as the magnetization measured at this field.

T=15K   T=7.5K   T=3.3K

H=0
ZFC

H=4.8kOe

H=12kOe

H=0
after
saturation

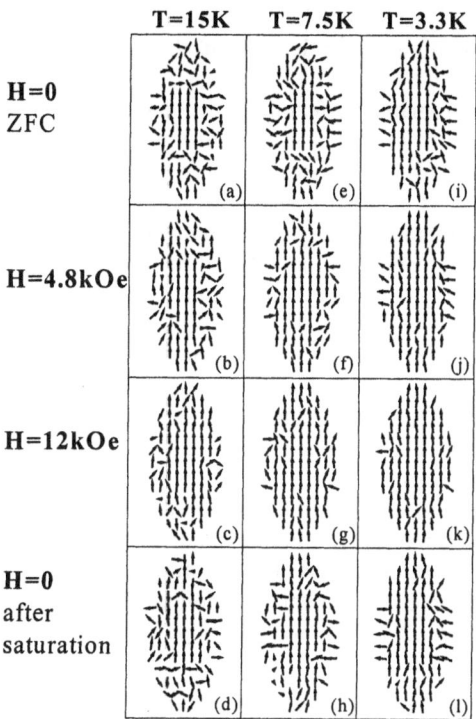

Figure 19. Cross section of the Monte Carlo calculated spin configurations with $K_{surf} > 0$ (surface anisotropy perpendicular to the particle surface), after cooling the sample in zero field, in the x-z plane of the particle at three temperatures and magnetic fields: (a-d) $T_{max} < T$, (e-h) $T_S < T < T_{max}$, and (i-l) $T < T_S$.

The competition between J_{shell} (which has a low value relative to J_{core}) and surface anisotropy leads to short-range ordering of the surface spins only at low temperature. Just below T_{max}, shell spins are paramagnetic while the core is blocked, and with decreasing temperature they form ferromagnetic regions whose moment is blocked according to their size and local environment. Such correlated regions, roughly behaving as single domains, interact between them and the core spins. The spin correlated region moments are not necessarily parallel to the core magnetization direction (Figs. 14a, 14b, 14c). The low-temperature maximum observed in the M_{ZFC} is due to the freezing of interacting ferromagnetic regions. In Fig. 2c is plotted Monte Carlo simulations of the ZFC-FC which can be compared with the experimental results. In these simulations, the significant features of the data are reproduced, confirming the validity of our model.

Hysteresis loops were calculated for $T<T_{irr}$ for positive and negative K_{surf} in order to reproduce the expected behaviour of S1D and S2D samples, respectively. The calculated net moment of the core (M_{core}), shell component (M_{shell}), and the total magnetization versus magnetic field at different temperatures after cooling the sample in zero field are plotted in Figures 16 and 17. Unlike the M_{core}, which saturates at relatively low applied field at all temperatures, M_{shell} needs large magnetic fields in order to achieve saturation and leads to a linear M vs. H behaviour at high fields at low temperature. As the temperature is decreased, the $M(H)$ behaviour shows a noticeable

anomaly at low fields, in close agreement with the experimental data. The results of these simulations give temperature dependences of H_c and M_r similar to the experimental ones (Figs. 7 and 8).

As mentioned above, a remarkable feature is the change of the loop shape and a loop distortion (sharp variation at low field in the demagnetizing and remagnetizing branches) occurring at $T \sim 10$ K. Distorted hysteresis loops were described as due to the addition of superparamagnetic (Langevin behaviour) and blocked magnetization (hysteretic behaviour) contributions [31]; the two components of the magnetization do not interact between them and the magnetization loops of the irreversible component are regular. On the other hand, in our Monte Carlo simulations (using the energy defined in Eq. (3)) not only does the total hysteresis loop become distorted with decreasing temperature but it is noteworthy that the core component too shows irregular branches in the loop at temperature at which H_c shows a minimum (Figs. 16, 17 and 18). This would indicate that the observed low temperature magnetic behaviour is due to the *interaction* between surface and core components, i.e. between disordered surface correlated spin and ferromagnetically ordered core spins (actually the ferromagnetic order in the core is perturbed because of the size confinement). With decreasing temperature, the ferromagnetic regions grow and interact among them and with the high magnetic core moment via exchange. As long as the ferromagnetic regions moments are unfrozen, they basically follow the core moment direction under the orienting action of an applied field (Figs. 19a-19d, and 20a-20d) and the shape of the magnetization cycles

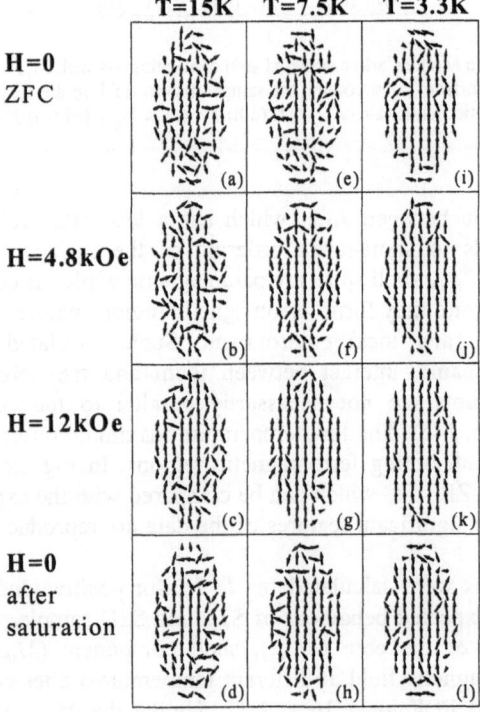

Figure 20. Cross section of the Monte Carlo calculated spin configurations with $K_{surf} < 0$ (surface anisotropy parallel to the particle surface), after cooling the sample in zero field, in the x-z plane of the particle at three temperatures and magnetic fields: (a-d) $T_{max} < T$, (e-h) $T_S < T < T_{max}$, and (i-l) $T < T_S$.

is regular (Figs. 5a, 6a, 16a and 17a). At lower temperature the random freezing of the surface regions of correlated spins tends to demagnetize the sample and the magnetization goes to zero at a lower reverse field producing the observed decrease of H_c (Figs. 19e-19h and 20e-20h). This should also be responsible for the small decrease of M_r. At very low temperatures, below T_S, the "anomalous" effects disappear and the shape of the hysteresis cycle becomes regular again (Figs. 5c, 6c, 16c and 17c), i.e. without sharp variation at low fields in the demagnetizing and remagnetizing branches. Both the coercive field and the remanence rapidly increase as the temperature decreases (Figs. 7a, 8a). This should be due to the growth of surface ferromagnetic regions in an extended frozen disordered surface spin state, exchange coupled to the ferromagnetic core. Because of this coupling, the freezing of the surface spins actually exerts a strong pinning action on the core moment, making its reversal more difficult and thus strongly increasing the effective anisotropy of the whole particle (Figs. 19i-19l, and 20i-20l).

In principle, the decrease in H_c, appearing at ~7 K and 9 K for S1D and S2D respectively, could be explained as simple addition of a Stoner-Wohlfarth cycle for the blocked particles and a superparamagnetic behaviour of the "loose" particles. However, this simple model [31] would not explain the large increase of H_c (and M_r) at lower temperatures. On the other hand, our Monte Carlo calculations, assuming a core-shell particle model for the particles, reproduce correctly the two-peak behaviour observed in Fig. 2 using the same parameters for calculating the $M(T)$ response [18]. This agreement reinforces the validity of the proposed model.

In the Monte Carlo simulations we assumed different signs for the surface anisotropy: $K > 0$ and $K < 0$ that correspond to easy axis perpendicular to the surface and parallel to it, respectively. In Fig. 8, we observe that the sign of K is relevant only below the temperature at which the surface ferromagnetic regions begin to "freeze" (~9 K). At lower temperatures, M_r is larger for the $K < 0$ case, which can be understood as an effective easy axis along the elongated particle direction. The opposite occurs for the $K > 0$ case. The difference $M_r(K < 0) > M_r(K > 0)$ at low temperatures is outside the calculation statistical error. Although the S1D and S2D data may not be directly comparable with each other, the *calculated* $M_r(T)$ seem to correlate favourably with $K > 0$ (S1D) and $K < 0$ (S2D) data respectively.

Furthermore, the proposed model explains the observed magnetization relaxation behaviour. For a single particle the time decay of the magnetization is predicted to be exponential, $M(t) = M(0) \exp(-t/\tau)$ [22, 1]. When an assembly of non-interacting particles with a volume distribution is present, a distribution of energy barriers does exist and the time dependence of the magnetization results forms the integration over the volume distribution. In the case of a narrow volume distribution, the logarithmic decay of the magnetization is obtained with a linear dependence of the viscosity as a function of temperature, $S = \beta M_0 T$ [26], as observed in our

measurements (Fig. 10), where M_0 is the equilibrium magnetization of the system and β is a constant in the range of $1s \leq t \leq 10^4 s$. In our magnetic relaxation measurements, the viscosity behaves linearly with temperature (inset of Fig. 10) and it is clear that two different regimes exist. This change of $S(T)$ behaviour is associated with the freezing of the shell spin correlated regions. Their moments, as proposed in our model, interact with the core of each independent nanoparticle. This core-shell interaction is supported by the fact that the $M(t)$ curves at $T < 9$ K cannot be fitted by a simple superposition of two logarithmic functions. Only at shorter times (where core and shell ferromagnetic regions relax together) and at long times (where only the core moment relaxes) is a logarithmic fit possible. At intermediate times, the core-shell interaction makes the relaxation more complex.

5. Interacting nanoparticles

In the presence of strong interparticle interactions, the magnetic behaviour of an assembly of nanoparticles is considerably more complicated with respect to the non-interacting particle case. For very small particles, the coexistence of significant surface effects and interparticle interactions adds an additional complexity to the problem [10, 13, 15, 16, 28, 29, 32]. Surface and interparticle interaction effects are correlated, with a high degree of interplay when particles are in close contact and exchange interparticle interaction are established, directly involving surface atoms. Due to the complexity of the problem, it is not straightforward to perform numerical calculations for interacting nanoparticles with *core-shell* structure, because of the requirement of long machine-time. Consequently, in this section we only show experimental data of the interacting samples and a phenomenological treatment is considered.

      In order to study the effect of interparticle interactions, powder samples of both compositions (S1 and S2) were used. The temperature dependence of the M_{ZFC} and M_{FC} is reported in Figure 21. The higher T_{max} and the lower T_{irr} values with respect to those for dispersed particles are due to interparticle interactions, since the particles are in close contact. Indeed, their effect is to increase the effective anisotropy energy barrier, then T_{max}, and to reduce the apparent T_B distribution. The flattening of the M_{FC} for S1 sample is also clear evidence of the presence of strong interactions, larger than in S2 sample. Unlike the diluted samples, there is no low-temperature maximum of M_{ZFC}. For S2 sample, at very low temperatures, a slow increasing of the magnetization is observed.

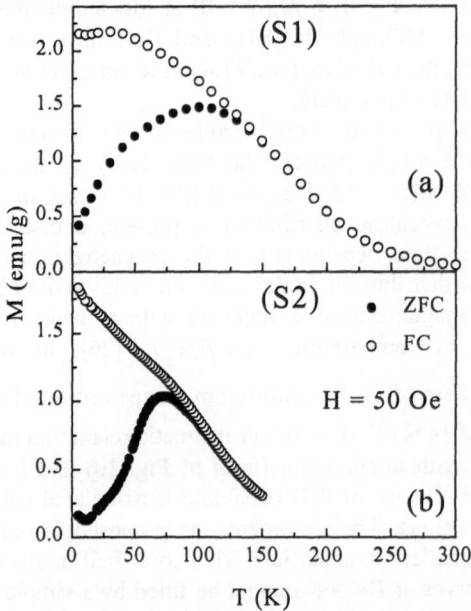

Figure 21. $M(T)$ measured at 50 Oe under ZFC (solid circles) and FC conditions (open circles) for the powder samples: (a) S1 and (b) S2.

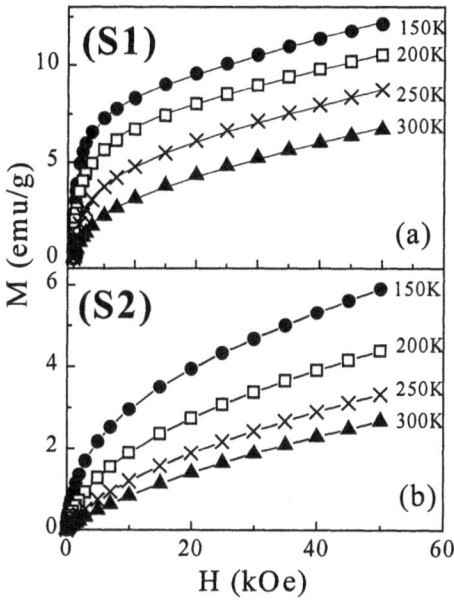

Figure 22. $M(H)$ at $T > T_{irr}$ (superparamagnetic regime) for S1 sample and S2 sample. The lines are only a guide for the eye.

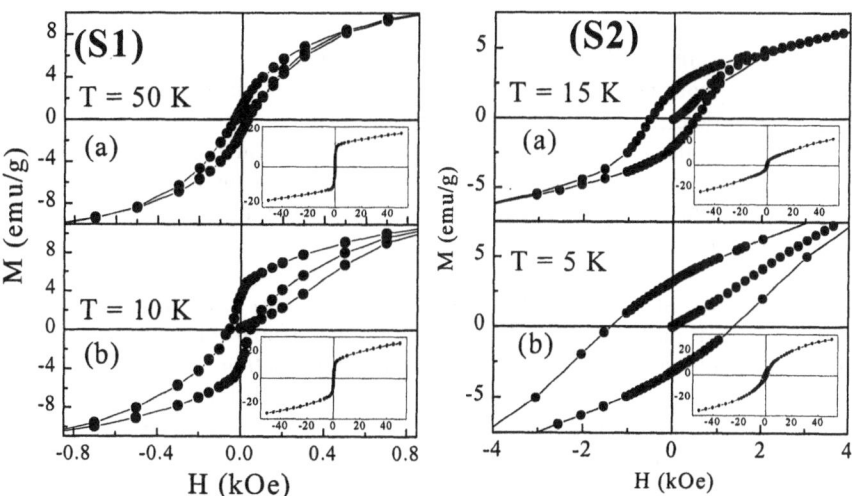

Figure 23. Magnetization cycles measured at $T < T_{irr}$ for the S1 sample. The inset shows the complete cycle.

Figure 24. Magnetization cycles measured at $T < T_{irr}$ for the S2 sample. The inset shows the complete cycle.

Magnetization curves as a function of the magnetic field also show differences with respect to the non-interacting samples. For $T > T_{irr}$, $M(H)$ curves also suggest the existence of two contributions to the magnetization (Fig. 22), and the magnetization values measured at 50 kOe are lower than the dispersed samples case (Fig. 9), suggesting frustration between magnetic interactions. The hysteresis loops at low temperatures (Figs. 23 and 24), well below T_{max}, have different behaviour for each sample revealing the different degree of the interparticle interactions. S2 sample shows a regular loop shape and H_c decreases monotonously with increasing temperatures. On the other hand, S1 sample shows an anomalous behaviour at low temperature, i.e. the virgin magnetization curve lies *below* the remagnetizing one (Fig. 23b). This was observed also in other interacting systems [33-35]. Above 50 K, the cycle becomes regular (Fig. 23.a), i.e. with the virgin magnetization curve within the loop. Additionally, H_c shows a maximum at 10 K, decreasing rapidly at low temperatures. This behaviour illustrates the interplay between surface effects and strong interparticle interactions, where the interactions contribute to "frustrate" the magnetic order in the surface. Also, as in S1 sample, the surface anisotropy is perpendicular to the particle surface, the magnetic order in the *shell* spins is basically frustrated (see Fig. 19) and is strongly affected by the interactions. In this case, the growth of surface ferromagnetic regions, oriented along their anisotropy axes, is actually frustrated because of the presence of strong interparticle interactions, modifying the surface spin orientation. The low magnetization at high fields with respect to the non-interacting samples is also evidence of the frustration in the *shell* spins and, due to the local orientation of the surface anisotropy, the magnetization of sample S1 is strongly affected by interparticle interactions.

6. Conclusions

Magnetization measurements on ~3 nm $(Fe_{0.26}Ni_{0.74})_{50}B_{50}$ and $(Co_{0.25}Ni_{0.75})_{65}B_{35}$ ferromagnetic amorphous non-interacting nanoparticles indicate that surface effects play a dominant role in determining the low temperature magnetic behaviour. Anomalous features are observed: the ZFC magnetization presents two maxima, a large one at T_{max} (~60 K), and a sharp one at low temperatures, T_S (7-9 K); the magnetization loops show a change of shape at low temperature, giving rise to a minimum of H_c, followed by a rapid increase, as also shown by M_r.

We proposed a *core-shell* model, consisting of a magnetically ordered core and a "disordered" surface shell. Using this model, Monte Carlo simulations based on Metropolis algorithm performed in non-interacting ellipsoidal particles reproduce satisfactorily the significant features observed in the magnetization measurements. The M_{ZFC} and M_{FC} curves were simulated using this procedure and the results were in excellent agreement with the experimental data. The calculated hysteresis loops too reproduce satisfactorily the experimentally observed ones. As a matter of fact, the curves for $H_c(T)$ and $M_r(T)$ resulting from the simulated cycles agree with the experimental data.

The results indicate that above ~20 K the magnetic behaviour is governed by the particle core contribution to the magnetization, whereas below this temperature surface effects become increasingly more important. Below ~20 K, the slowing down of surface spin fluctuations gives rise to a progressive blocking of surface spin correlated region moments of increasing size. Below T_S, exchange interactions among them and with the core moment give rise to a disordered frozen surface state. This behaviour is responsible for the observed anomalies in the loop shape and the consequent temperature dependence of M_r and H_c and for the large increase of the effective anisotropy at very low temperature, as revealed by the large increase of H_c and the tendency to non-saturation of the magnetization.

Interparticle interactions affect the magnetic order of the surface layer. In this case, the growth of surface ferromagnetic regions, oriented along their anisotropy axes, is actually frustrated because of the presence of strong interparticle interactions, modifying the surface spin orientations.

References

[1] Dormann, J.L., Fiorani, D., Tronc, E., *Adv. Chem. Phys.* **98**, 283-494 (1997).
[2] Chudnovsky, E.M., Tejada, J. (1994) *Macroscopic Quantum Tunneling of the Magnetic Moment*, Cambridge University Press, Cambridge.
[3] Hadjipanayis, G.C., Prinz, G.A. (1991) *Science and Technology of Nanostructurated Materials*, Plenum Press, New York.
[4] Chantrell, R.W., O'Grady, K. (1994) in Gerber, R. et al. (eds), *Applied Magnetism*, Kluwer Academic Publishers, The Netherlands, p. 113.
[5] Hadjipanayis, G.C., *J. Magn. Magn. Mater.* **200**, 373-391 (1999).
[6] Pankhust, Q.A., Connolly, J., Jones, S.K., Dobson, J., *J. Phys. D: Appl. Phys.* **36**, R167-R181(2003).
[7] Sako, S., Ohshima, K., Sakai, M., Bandow, S, *Surf. Rev. Lett.* **3**, 109-113 (1996).
[8] Sako, S., Ohshima, K., *J. Phys. Soc. Jpn.* **64**, 944-950 (1995).
[9] Merikoski, J. Timonen, J., Mannien, M., Jena, P., *Phys. Rev. Lett.* **66**, 938-941 (1991).
[10] Kachkachi, H., Ezzir, A., Noguès, M., Tronc, E., *Eur. Phys. J. B* **14**, 681-689 (2000).
[11] Chen, J.P. et al., *Phys. Rev. B* **51**,11527-11532 (1995).
[12] Kodama, R.H., Berkowitz, A.E., McNiff, E.J., Foner, S., *Phys. Rev. Lett.* **77**, 394-397 (1996).
[13] Kodama, R.H., Berkowitz, A.E., *Phys. Rev. B* **59**, 6321-6336 (1999); and references therein.
[14] Kachkachi, H., Ezzir, A., Noguès, M., Tronc, E., Garanin, D.A., *J. Magn. Magn. Mater.* **221**, 158-163 (2000).
[15] Iglesias, O., Labarta, A., *Phys. Rev. B* **63**, 184416 (2001).
[16] Coey, J.M.D., *Phys. Rev. Lett.* **27**, 1140-1142 (1971).
[17] Dormann, J.L., Bessais, L., Fiorani, D., *J. Phys. C* **21**, 2015-2034(1988).
[18] De Biasi, E., Ramos, C.A., Zysler, R.D., Romero, H., *Phys. Rev. B* **65**, 144416 (2002).
[19] Zysler, R.D., Romero, H., Ramos, C.A., De Biasi, E., Fiorani, D., *J. Magn. Magn. Mater.* **266**, 233-242 (2003).
[20] Vargas, J., Ramos, C., Zysler, R.D., Romero, H., *Physica B* **320**, 178-180 (2002).
[21] De Biasi, E., Zysler, R.D., Ramos, C.A., Romero, H., Fiorani, D. (submitted to *Phys. Rev. B*).
[22] Néel, L., *Ann. Geophys.* **5**, 99-136 (1949).
[23] Zysler, R.D., Ramos, C.A., Romero, H., Ortega, A., *J. Mater. Sci:* **36 (9)**, 2291-2294 (2001).
[24] Gittleman, J.I., Abeles, B., Bozowski, S., *Phys. Rev. B* **9**, 3891-3897 (1974).
[25] Stoner, E.C., Wohlfarth, E.P., *Philos. Trans. Roy. Soc. London A* **240**, 599-642 (1948).
[26] Vincent, E., Hammann, J., Prené, P., Tronc, E., *J. Phys. I France* **4**, 273-282 (1994).
[27] García-Otero, J., Porto, M., Rivas, J., Bunde, A., *Phys. Rev. Lett.* **84**, 167-170 (2000).
[28] Dimitrov, D.A., Wysin, G.M., *Phys. Rev. B* **50**, 3077-3084 (1994).
[29] Dimitrov, D.A., Wysin, G.M., *Phys. Rev. B* **51**, 11947-11950 (1995).
[30] Gómez-Abal, R., Llois, A.M., *Phys. Rev. B* **65**, 155426 (2002).
[31] Tauxe, L., Mullender, T.A.T., Pick, T., *J. Geophys. Res.* **101**, 571-583.
[32] Batlle, X., Labarta, A., *J. Phys. D: Appl. Phys.* **35**, R15-R42 (2002).
[33] Tronc, E., Ezzir, A., Cherkaoui, R., Chanéac, C., Noguès, M., Kachkachi, H., Fiorani, D., Testa, A.M., Grenèche, J.M., Jolivet, J.P., *J. Magn. Magn. Mater.* **221**, 63-79 (2000).
[34] Fiorani, D., Testa, A.M., Lucari, F., D'Orazio, F., Romero, H., *Physica B* **320**, 122-126, (2002).
[35] Tronc, E., Fiorani, D., Noguès, M., Testa, A.M., Lucari, F., D'Orazio, F., Grenèche, J.M., Wernsdorer, W., Galvez, N., Chanéac, C., Mailly, D., Jolivet, J.P., *J. Magn. Magn. Mater.* **262**, 6-14 (2003).

Magnetic anisotropy and magnetization reversal studied in individual nanoparticles

Wolfgang Wernsdorfer
Lab. L. Néel - CNRS, BP166,
38042 Grenoble Cedex 9, France

1. Introduction

Since the pioneering work of Néel in the late 1940s, nanometer-sized magnetic particles have generated growing interest for their novel behavior and enormous potential in technological applications. In particular, it was recognized that the ferromagnetic state, with a given orientation of the particle moment, has a remanent magnetization if the particle is small enough. This was the starting point of huge permanent magnets and magnetic recording industries. However, despite intense activity during the last few decades, the difficulties in making nanoparticles of good enough quality has slowed the advancement of this field. As a consequence, for 50 years, these applications concentrated above and then near the micrometer scale. In the last decade, this has no longer been the case because of the emergence of new fabrication techniques that have led to the possibility of making small objects with the required structural and chemical qualities. In order to study these objects new techniques were developed such as magnetic force microscopy, magnetometry based on micro-Hall probes, or micro-SQUIDs. This led to a new understanding of the magnetic behavior of nanoparticles, which is now very important for the development of new fundamental theories of magnetism and in modeling new magnetic materials for permanent magnets or high density recording.

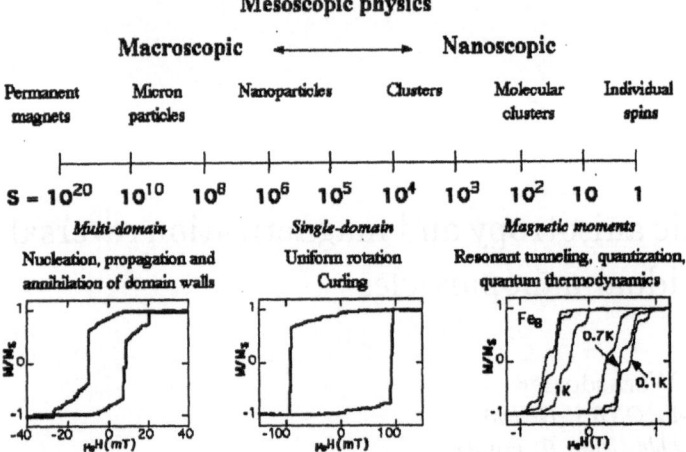

Figure 1. Scale of size that goes from macroscopic down to nanoscopic sizes. The unit of this scale is the number of magnetic moments in a magnetic system (roughly corresponding to the number of atoms). The hysteresis loops are typical examples of magnetization reversal via nucleation, propagation, and annihilation of domain walls (*left*), via uniform rotation (*middle*), and via quantum tunneling (*right*).

In order to put this chapter into perspective, let us consider Fig. 1, which presents a scale of size ranging from macroscopic down to nanoscopic sizes. The unit of this scale is the number of magnetic moments in a magnetic system. At macroscopic sizes, a magnetic system is described by magnetic domains [1] that are separated by domain walls. Magnetization reversal occurs via nucleation, propagation, and annihilation of domain walls (see the hysteresis loop on the left in Fig. 1 which was measured on an individual elliptic CoZr particle of 1 μm × 0.8 μm and a thickness of 50 nm). Shape and width of domain walls depend on the material of the magnetic system, on its size, shape and surface, and on its temperature [2].

When the system size is of the order of magnitude of the domain wall width or the exchange length, the formation of domain walls requires too much energy. Therefore, the magnetization remains in the so-called single-domain state. Hence, the magnetization might reverse by uniform rotation, curling or other nonuniform modes (see hysteresis loop in the middle of Fig. 1).

For system sizes well below the domain wall width or the exchange length, one must take into account explicitly the magnetic moments (spins) and their couplings. The theoretical description is complicated by the particle's boundaries.

At the smallest size (below which one must consider individual atoms and spins) there are either free clusters made of several atoms [3, 4] or molecular clusters which are macromolecules with a central complex containing magnetic atoms. In the last case, measurements on the Mn_{12} acetate and Fe_8 molecular clusters showed that the physics can be described by a collective moment of spin $S = 10$ [5]. By means of simple hysteresis loop measurements, the quantum character of these molecules showed up in well-defined steps which are due to resonance quantum tunneling between energy levels (see hysteresis loop on the right in Fig. 1) [6, 7, 8, 9, 10].

In the following sections, we review the most important theories and experimental results concerning the magnetization reversal of single-domain particles and clusters. Special emphasis is laid on single-particle measurements avoiding complications due to distributions of particle size, shape, and so on. We mainly discuss the low-temperature regime in order to avoid spin excitations.

2. Single-particle measurement techniques

This section reviews commonly used single-particle measuring techniques avoiding complications due to distributions of particle size, shape, and so on, which are always present in particle assemblies [11]. Special emphasis is laid on the micro-SQUID technique which allowed the most detailed studies at low temperatures.

2.1. OVERVIEW OF SINGLE-PARTICLE MEASUREMENT TECHNIQUES

The dream of measuring the magnetization reversal of an individual magnetic particle goes back to the pioneering work of Néel [12, 13]. The first realization was published by Morrish and Yu in 1956 [14]. These authors employed a quartz-fiber torsion balance to perform magnetic measurements on individual micrometer-sized γ-Fe_2O_3 particles. With their technique, they wanted to avoid the complication of particle assemblies which are due to different orientations of the particle's easy axis of magnetization and particle–particle dipolar interaction. They aimed to show the existence of a single-domain state in a magnetic particle. Later on, other groups tried to study single particles, but the experimental precision did not allow a detailed study. A first breakthrough came via the work of Knowles [15] who developed a simple optical method for measuring the switching field, defined as the minimum applied field required to reverse the magnetization of a particle. However, the work of Knowles failed to provide quantitative information on well-defined particles. More recently, insights into the magnetic properties of individual and isolated particles were obtained with the help of electron holography [16], vibrating reed magnetometry [17], Lorentz microscopy [18, 19], magneto-optical Kerr effect [20], and magnetic force microscopy [21, 22]. Recently, magnetic nanostructures have been studied by the technique of magnetic linear dichroism in the angular distribution of photoelectrons or by photoemission electron microscopy [23, 24]. In addition to magnetic domain observations, element-specific information is available via the characteristic absorption levels or threshold photoemission [1]. Among all mentioned techniques, most of the studies have been carried out using magnetic force microscopy at room temperature. This technique has an excellent spatial resolution but time-dependent measurements are difficult due to the sample–tip interaction.

Only a few groups were able to study the magnetization reversal of individual nanoparticles or nanowires at low temperatures. The first magnetization measurements of individual single-domain nanoparticles and nanowires at very low temperatures were presented by Wernsdorfer et al. [25]. The detector (a Nb microbridge-DC-SQUID) and the studied particles were fabricated using electron-beam lithography. By measuring the electrical resistance of isolated Ni wires with diameters between 20 and 40 nm, Giordano and Hong studied the motion of magnetic domain walls [26, 27]. Other low-temperature techniques that may be adapted to single-particle measurements are Hall probe magnetometry [28, 29, 30], magnetometry based on magnetoresistance [31, 32, 33] or spin-dependent tunneling with Coulomb blockade [34, 35]. At the time of writing, the micro-SQUID technique allows the most detailed study of the magnetization reversal of nanometer-sized particles [36, 37, 38, 39, 40, 41, 42]. The following section reviews the basic ideas of the micro-SQUID technique.

2.2. MICRO-SQUID MAGNETOMETRY

The Superconducting Quantum Interference Device (SQUID) has been used very successfully for magnetometry and voltage or current measurements in the fields of medicine, metrology and science [43, 44]. SQUIDs are mostly fabricated from a $Nb-AlO_x-Nb$ trilayer, several

---

[1] We refer to the literature concerning other domain observation techniques [1].

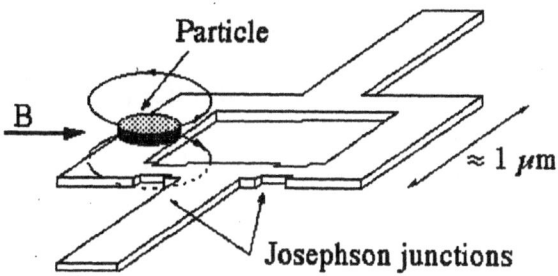

Figure 2. Drawing of a planar Nb micro-bridge-DC-SQUID on which a ferromagnetic particle is placed. The SQUID detects the flux through its loop produced by the sample magnetization. Due to the close proximity between sample and SQUID a very efficient and direct flux coupling is achieved.

hundreds of nanometers thick. The two Josephson junctions are planar tunnel junctions with an area of at least $0.5~\mu m^2$. In order to avoid flux pinning in the superconducting film the SQUID is placed in a magnetically shielded environment. The sample's flux is transferred via a superconducting pickup coil to the input coil of the SQUID. Such a device is widely used as the signal can be measured by simple lock-in techniques. However, this kind of SQUID is not well suited to measuring the magnetization of single submicron-sized samples as the separation of SQUID and pickup coil leads to a relatively small coupling factor. A much better coupling factor can be achieved by coupling the sample directly with the SQUID loop. In this arrangement, the main difficulty arises from the fact that the magnetic field applied to the sample is also applied to the SQUID. The lack of sensitivity to a high field applied in the SQUID plane, and the desired low temperature range led to the development of the micro-bridge-DC-SQUID technique [45] which allows us to apply several teslas in the plane of the SQUID without dramatically reducing the SQUID's sensitivity.

The planar Nb micro-bridge-DC-SQUID can be constructed by using standard electron beam lithography, and the magnetic particle is directly placed on the SQUID loop (Fig. 2) [25]. The SQUID detects the flux through its loop produced by the sample magnetization. For hysteresis loop measurements, the external field is applied in the plane of the SQUID, so that the SQUID is only sensitive to the flux induced by the stray field of the sample magnetization. Due to the close proximity between sample and SQUID, magnetization reversals corresponding to $10^3~\mu_B$ can be detected, i.e., the magnetic moment of a Co nanoparticle with a diameter of 2–3 nm [41, 42].

3. Mechanisms of magnetization reversal at zero kelvin

As already briefly discussed in the introduction, for a sufficiently small magnetic sample it is energetically unfavorable to form a stable magnetic domain wall. The specimen then behaves as a single magnetic domain. For the smallest single-domain particles, the magnetization is expected to reverse by uniform rotation of magnetization. For somewhat larger ones, nonuniform reversal modes are more likely–for example, the curling reversal mode [2]. For larger particles, magnetization reversal occurs via a domain wall nucleation process starting in a rather small volume of the particle. For even larger particles, the nucleated domain wall can be stable for certain fields. The magnetization reversal happens then via nucleation and annihilation processes. In these sections we neglect temperature and quantum effects.

The following section discusses in detail the uniform rotation mode that is used in many theories, in particular in Néel, Brown, and Coffey's theory of magnetization reversal by thermal activation (Section 4) and in the theory of macroscopic quantum tunneling of magnetization.

3.1. MAGNETIZATION REVERSAL BY UNIFORM ROTATION (STONER–WOHLFARTH MODEL)

The model of uniform rotation of magnetization, developed by Stoner and Wohlfarth [46] and Néel [47], is the simplest classical model describing magnetization reversal. One considers a particle of an ideal magnetic material where exchange energy holds all spins tightly parallel to each other, and the magnetization magnitude does not depend on space. In this case the exchange energy is constant, and it plays no role in the energy minimization. Consequently, there is competition only between the anisotropy energy of the particle and the effect of the applied field. The original model of Stoner and Wohlfarth assumed only uniaxial shape anisotropy with one anisotropy constant—that is, one second-order term. This is sufficient to describe highly symmetric cases like a prolate spheroid of revolution or an infinite cylinder. However, real systems are often quite complex, and the anisotropy is a sum of mainly shape (magnetostatic), magnetocrystalline, magnetoelastic, and surface anisotropy. One additional complication arises because the different contributions of anisotropies are often aligned in an arbitrary way one with respect to each other. All these facts motivated a generalization of the Stoner–Wohlfarth model for an arbitrary effective anisotropy which was done by Thiaville [48, 49].

One supposes that the exchange interaction in the cluster couples all the spins strongly together to form a giant spin whose direction is described by the unit vector $\vec{m}$. The only degrees of freedom of the particle's magnetization are the two angles of orientation of $\vec{m}$. The reversal of the magnetization is described by the potential energy

$$E(\vec{m}, \vec{H}) = E_0(\vec{m}) - \mu_0 V M_\mathrm{s} \vec{m}.\vec{H} \tag{1}$$

where V and M_s are the magnetic volume and the saturation magnetization of the particle respectively, $\vec{H}$ is the external magnetic field, and $E_0(\vec{m})$ is the magnetic anisotropy energy which is given by

$$E_0(\vec{m}) = E_\mathrm{shape}(\vec{m}) + E_\mathrm{MC}(\vec{m}) + E_\mathrm{surface}(\vec{m}) + E_\mathrm{ME}(\vec{m}) \tag{2}$$

E_shape is the magnetostatic energy related to the cluster shape. E_MC is the magnetocrystalline anisotropy (MC) arising from the coupling of the magnetization with the crystalline lattice, similar as in bulk. E_surface is due to the symmetry breaking and surface strains. In addition, if the particle experiences an external stress, the volumic relaxation inside the particle induces a magnetoelastic (ME) anisotropy energy E_ME.

3.1.1. Shape anisotropy

Shape anisotropy E_shape comes from the anisotropy of the demagnetizing field arising from long range dipolar interaction in the particle. The summation of dipolar interaction over the whole nanoparticle approximated as a giant spin gives after a short calculation [50]:

$$E_\mathrm{shape}(\vec{m}) = -\frac{3\mu_{at}^2}{2} \sum_{i \neq j} \frac{(\vec{m}.\vec{r_{ij}})^2}{\| \vec{r_{ij}} \|^5} \tag{3}$$

where $\vec{r_{ij}}$ is the vector connecting any two atoms on sites i and j carrying a magnetic moment μ_at. Shape anisotropy is a quadratic form of the magnetization and thus can be diagonalized. If we note $\hat{x}$, $\hat{y}$ and $\hat{z}$ the eigenvectors, shape anisotropy can be written:

$$E_\mathrm{shape}(\vec{m})/v = K_1^\mathrm{shape} m_z^2 + K_2^\mathrm{shape} m_x^2 \tag{4}$$

where $K_1^\mathrm{shape} < 0$ and $K_2^\mathrm{shape} > 0$ are shape anisotropy constants. They

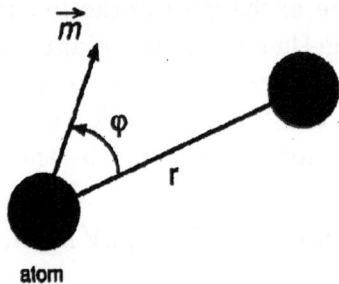

atom

Figure 3. Schematic drawing illustrating the Néel pair model. r is the distance between two magnetic atoms (black dots). $\vec{m}$ is the local magnetization and φ is the angle between the atomic bond and $\vec{m}$.

depend on the atomic moment as μ_{at}^2 and the particle geometry. In this expression we have replaced m_x^2 by $1 - m_y^2 - m_z^2$ since $\| \vec{m} \|^2 = 1$. The z direction along which the particle is the longest is called the easy magnetization axis and the x direction along which the particle is the shortest the hard axis. Along y, the particle has an intermediate length. x, y, z are not necessarily connected to the crystal axes.

In order to quantify the particle shape, one usually defines three characteristic lengths called gyration radii [51] r_x, r_y and r_z as:

$$
r_x = \sqrt{\frac{\sum_{i=1}^{N_{tot}} (G\vec{M_i}.\hat{x})^2}{N_{tot}}}
$$

$$
r_y = \sqrt{\frac{\sum_{i=1}^{N_{tot}} (G\vec{M_i}.\hat{y})^2}{N_{tot}}}
$$

$$
r_z = \sqrt{\frac{\sum_{i=1}^{N_{tot}} (G\vec{M_i}.\hat{z})^2}{N_{tot}}}
$$

(5)

where N_{tot} is the number of atoms in the particle, G is its center of mass and M_i is the position of atom i. $\hat{x}$, $\hat{y}$ and $\hat{z}$ are unit vectors along x, y and z axes respectively. In the continuous limit in which particles are considered as ellipsoids, r_x, r_y and r_z are similar to the half-axes a, b and c. The cluster aspect ratio is then defined as r_z/r_x which is the maximum of all possible ratios obtained with the gyration radii[52].

3.1.2. *Magnetocrystalline anisotropy*

Magnetocrystalline anisotropy E_{MC} arises from the coupling of the magnetization with the crystal lattice through spin-orbit interaction. In an infinite cubic crystal, this anisotropy can be written:

$$E_{\mathrm{MC}}(\vec{m})/v = K_4^{\mathrm{MC}}(m_{x'}^2 m_{y'}^2 + m_{x'}^2 m_{z'}^2 + m_{y'}^2 m_{z'}^2) \qquad (6)$$
$$+ K_6^{\mathrm{MC}} m_{x'}^2 m_{y'}^2 m_{z'}^2$$

where x', y' and z' are [100], [010] and [001] crystal axes. It begins with fourth order terms since x', y' and z' must be equivalent directions. K_4^{MC} and K_6^{MC} are magnetocrystalline anisotropy constants. Higher order terms are commonly neglected. For bcc iron (respectively fcc cobalt), K_4^{MC} is positive (respectively negative) and x', y', z' are easy (respectively hard) directions.

3.1.3. Surface magnetocrystalline anisotropy

At the cluster surface, the crystal symmetry is broken which involves second order terms. This contribution is called surface magnetocrystalline anisotropy E_{surface} and may be either in plane or out of plane. In order to estimate it, we use a phenomenological model based on the Néel anisotropy model [53, 50]. In Néel's original work, the magnetic pair interaction energy between atoms is given by (Fig. 3):

$$E = L(\vec{m}.\vec{e})^2 = L \cos^2 \varphi \qquad (7)$$

where $\vec{m}$ is a unit vector pointing along the magnetization direction. $\vec{e}$ is an interatomic vector and L the Néel constant. This expression satisfies the cylindrical symmetry. L depends on the interatomic distance r according to the following expression:

$$L(r) = L(r_0) + \left(\frac{dL}{dr}\right)_{r_0} r_0 \eta \qquad (8)$$

where r_0 is the bulk unstrained bond length and η the bond strain.
For fcc crystals:

$$L(r_0) = -\frac{3}{4}\lambda_{100}(c_{11} - c_{12}) + \frac{3}{4}\lambda_{111}c_{44} \qquad (9)$$
$$\left(\frac{dL}{dr}\right)_{r_0} r_0 = \frac{3}{2}\lambda_{100}(c_{11} - c_{12}) - \frac{9}{2}\lambda_{111}c_{44}$$

Table I. Magnetic anisotropy energy as a function of the atom position in a truncated octahedron. For (100) and (111) facets, $\vec{n}$ is a unit vector perpendicular to the facet. For an edge, $\vec{n}$ is along its axis. $\vec{m}$ is a unit vector pointing along the local magnetization.

bulk	(100) facet	(111) facet	(100)-(111) edge	(111)-(111) edge	apex
$4L$	$-L(\vec{m}.\vec{n})^2$	$-1.5L(\vec{m}.\vec{n})^2$	$L(\vec{m}.\vec{n})^2$	$L(\vec{m}.\vec{n})^2$	$2L$

and for bcc crystals:

$$L(r_0) = -\frac{9}{16}\lambda_{100}(c_{11} - c_{12}) \tag{10}$$

$$\left(\frac{dL}{dr}\right)_{r_0} r_0 = \frac{9}{16}\lambda_{100}(c_{11} - c_{12}) - \frac{27}{8}\lambda_{111}c_{44}$$

where $(\lambda_{100}, \lambda_{111})$ are magnetostriction constants and (c_{11}, c_{12}, c_{44}) elastic constants. Using this model, magnetic anisotropy can be calculated

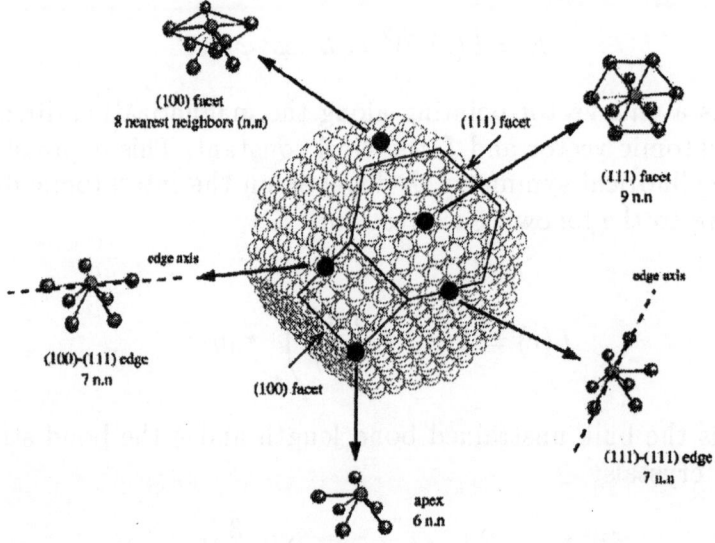

Figure 4. The different atomic positions at the surface of a perfect truncated octahedron containing 1289 atoms. This polyhedron is the equilibrium shape of fcc clusters. It exhibits 8 (111) and 6 (100) facets.

on a given atom by summing pair interactions $L(\vec{m}.\vec{e})^2$ with its nearest neighbors. As an example, we estimate this magnetic anisotropy for different atomic positions in a truncated octahedron (Fig. 4). This polyhedron is the equilibrium shape of fcc cobalt or nickel clusters according to the Wulff theorem [54]. The different positions are depicted in Fig. 4 and the corresponding magnetic anisotropies summarized in Table 1.

First of all, for bulk and apex positions, the resulting anisotropy energy is a constant. Indeed, in both cases, the local symmetry is cubic. Therefore, the Néel model only accounts for the additional magnetic anisotropy arising from symmetry breaking (i.e. on facets and edges). This is the reason why it has been widely used to estimate surface anisotropy (for example perpendicular magnetic anisotropy) in magnetic thin films [55, 56, 57, 58, 59]. In this example, surface anisotropy is locally in-plane if $L < 0$ and out-of-plane if $L > 0$. Moreover magnetic anisotropy is the strongest on (111) facets. In order to derive surface anisotropy for the whole nanoparticle, all the pair interactions given by Eq. (1) must be added up in the particle which gives the following expression [41]:

$$E_{\text{surface}}(\vec{m}) = \frac{L}{2} \sum_{i,j} \frac{(\vec{m}.\vec{r_{ij}})^2}{\| \vec{r_{ij}} \|^2} \tag{11}$$

where atoms i and j are nearest neighbors. We have replaced $\vec{e}$ by $\vec{r_{ij}}/ \| \vec{r_{ij}} \|$ in Eq. (1) and added a factor $1/2$ not to count twice pair interactions. This anisotropy is a quadratic form of the magnetization and thus can be diagonalized along the eigenvectors $\hat{x}$, $\hat{y}$ and $\hat{z}$:

$$E_{\text{surface}}(\vec{m})/v = K_1^{\text{surface}} m_z^2 + K_2^{\text{surface}} m_y^2 \tag{12}$$

where $K_1^{\text{surface}} < 0$ and $K_2^{\text{surface}} > 0$ are the surface anisotropy constants along the easy and hard axis respectively. They only depend on the Néel constant L and the particle faceting. They are also proportional to the surface to volume ratio and thus vanish for large particles.

In conclusion, in this atomic model which is suitable for the study of nanoparticles containing a few thousand atoms, local symmetries are respected. Moreover surface anisotropy is given by a single parameter (the Néel constant) which may include surface strains through its dependence on the interatomic distance. Equations (9) and (10) are only available for free metal surfaces whereas L may be different in magnitude and sign for interfaces. In this case, it should be either estimated from the surface anisotropy constant K_S measured on multi-layers or theoretically calculated using first principles calculations [60]. From any particle geometry and the Néel constant, the calculation of surface anisotropy is straightforward using Eq. (11) and its diagonal-ization leads to surface anisotropy constants as well as easy and hard magnetization axes.

Finally one can mention that magnetic axes for surface and shape anisotropy may not be the same. Indeed summations (3) and (11) are different. Shape anisotropy is related to the particle morphology and surface anisotropy to its faceting. However, except in very particular cases, they are quite similar. Thus if $L < 0$ and surface anisotropy is in-plane, shape and surface anisotropies add up whereas they give opposite contributions if $L > 0$. In addition to shape and magne-tocrystalline anisotropy, clusters may undergo an external stress which relaxes in the particle volume and leads to magnetoelastic anisotropy. This contribution will be neglected in the following [61].

3.1.4. *Method to determine the magnetic anisotropy*

All the anisotropy energies discussed above can be developed in a power series of $m_x^a m_y^b m_z^c$ with $p = a + b + c = 2, 4, 6, \ldots$ giving the order of the anisotropy term. The magnetic anisotropy energy $E_0(\vec{m})$ is:

$$
\begin{aligned}
E_0(\vec{m})/v &= K_1 m_z^2 + K_2 m_y^2 \\
&+ K_4(m_{x'}^2 m_{y'}^2 + m_{x'}^2 m_{z'}^2 + m_{y'}^2 m_{z'}^2) \\
&+ K_6 m_{x'}^2 m_{y'}^2 m_{z'}^2 + \ldots
\end{aligned}
\tag{13}
$$

We divided $E_0(\vec{m})$ by the magnetic volume v in order to express anisotropy constants K_i in J/m^3 as usual. In a cubic particle, x', y' and z' are the [100], [010] and [001] directions. (xyz) is a coordinate system related to the particle shape and may be different from $(x'y'z')$.

Thiaville proposed a geometrical method to calculate the particle's energy and to determine the switching field for all angles of the applied magnetic field yielding the critical surface of switching fields which is analogous to the Stoner–Wohlfarth astroid. In this model, the switching field $\mu_0 \vec{H}_{sw}$ can be written as:

$$
\mu_0 \vec{H}_{sw} = \frac{2}{M_s} \left(\lambda \vec{m} + \frac{1}{2} \partial_\theta (E_0/v) \vec{e}_\theta \right.
\tag{14}
$$

$$
\left. + \frac{1}{2\sin\theta} \partial_\varphi (E_0/v) \vec{e}_\varphi \right)
$$

where (θ, φ) are spherical angles in an arbitrary coordinate system, and

$$
\begin{aligned}
\lambda = &-\frac{1}{4}\left(\partial_{\theta\theta}(E_0/v) + \frac{\cos\theta}{\sin\theta}\partial_\theta(E_0/v) + \frac{1}{\sin^2\theta}\partial_{\varphi\varphi}(E_0/v)\right) \\
&+\frac{1}{4}\left(\left(\partial_{\theta\theta}(E_0/v) - \frac{\cos\theta}{\sin\theta}\partial_\theta(E_0/v) - \frac{1}{\sin^2\theta}\partial_{\varphi\varphi}(E_0/v)\right)^2\right. \\
&\left.+4\left(\partial_\theta\left(\frac{1}{\sin\theta}\partial_\varphi(E_0/v)\right)\right)^2\right)^{1/2}
\end{aligned}
\tag{15}
$$

∂_θ and $\partial_{\theta\theta}$ (resp. ∂_φ and $\partial_{\varphi\varphi}$) are first and second partial derivatives with respect to θ (resp. φ).

The main interest of Thiaville's calculation is that measuring the critical surface of the switching field allows one to find the effective anisotropy of the nanoparticle. The knowledge of the latter is important for temperature-dependent studies and quantum tunneling investigations. Knowing precisely the particle's shape and the crystallographic axis allows one to determine the different contributions to the effective anisotropy.

3.2. EXPERIMENTAL EVIDENCE FOR MAGNETIZATION REVERSAL BY UNIFORM ROTATION

In order to demonstrate experimentally the uniform rotation mode, the angular dependence of the magnetization reversal has often been studied [2]. However, a comparison of theory with experiment is difficult because magnetic particles often have a nonuniform magnetization state that is due to rather complicated shapes and surfaces, crystalline defects, and surface anisotropy. In general, for many particle shapes the demagnetization fields inside the particles are nonuniform leading to nonuniform magnetization states [2]. An example is presented in Fig. 5 which compares typical hysteresis loop measurements of an elliptical Co particle, fabricated by electron beam lithography, with the prediction of the Stoner–Wohlfarth model. Before magnetization reversal, the magnetization decreases more strongly than predicted because the magnetic configuration is not collinear as in the Stoner–Wohlfarth model, but instead presents deviations mainly near the particle surface. The angular dependence of the switching field agrees with the Stoner–Wohlfarth model only for angles $\theta \leq 30°$ where nonlinearities and defects play a less important role [25, 62].

Studies of magnetization reversal processes in ultrathin magnetic dots with in-plane uniaxial anisotropy showed also switching fields

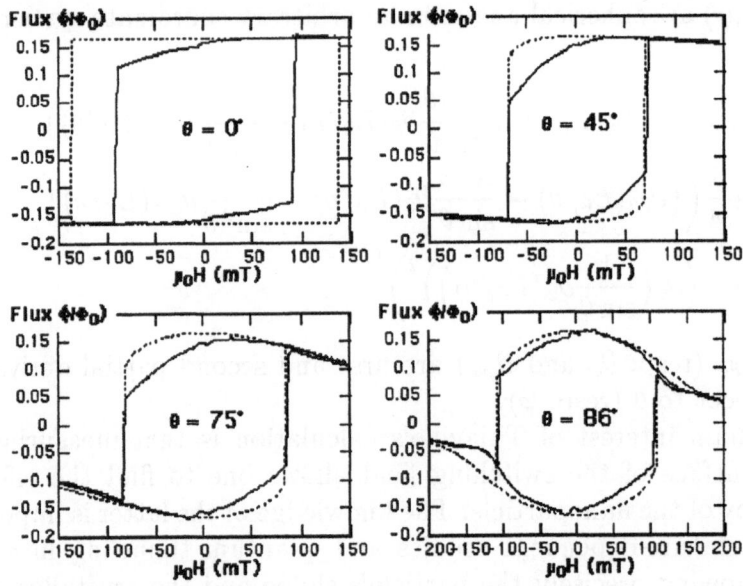

Figure 5. Hysteresis loops of a nanocrystalline elliptic Co particle of $70 \times 50 \times 25$ nm^3. The dashed line is the prediction of the Stoner–Wohlfarth model of uniform rotation of magnetization. The deviations are due to nonuniform magnetization states.

that are very close to the Stoner–Wohlfarth model, although magnetic relaxation experiments clearly showed that nucleation volumes are by far smaller than an individual dot volume [63]. These studies show clearly that switching field measurements as a function of the angles of the applied field cannot be taken unambiguously as a proof of a Stoner–Wohlfarth reversal.

The first clear demonstration of the uniform reversal mode has been found with Co nanoparticles [38], and BaFeO nanoparticles [39], the latter having a dominant uniaxial magnetocrystalline anisotropy. The three-dimensional angular dependence of the switching field measured on BaFeO particles of about 20 nm could be explained with the Stoner–Wohlfarth model taking into account the shape anisotropy and hexagonal crystalline anisotropy of BaFeO [40]. This explication is supported by temperature- and time-dependent measurements yielding activation volumes which are very close to the particle volume (Section 4).

We present here the first measurements on individual cobalt clusters of 3 nm diameter containing about a thousand atoms (Figs. 6 and 7) [41]. In order to achieve the needed sensitivity, Co clusters preformed in the gas phase are directly embedded in a co-deposited thin Nb film

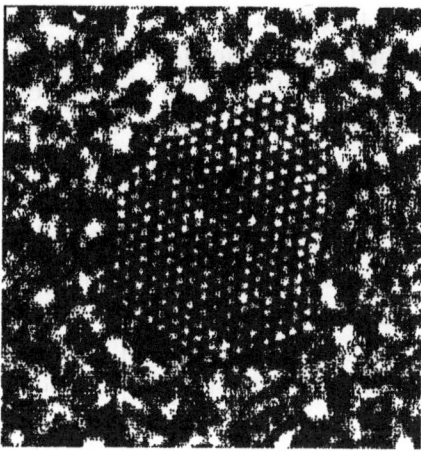

Figure 6. High-resolution transmission electron microscopy observation along a [110] direction of a 3 nm cobalt cluster exhibiting an fcc structure.

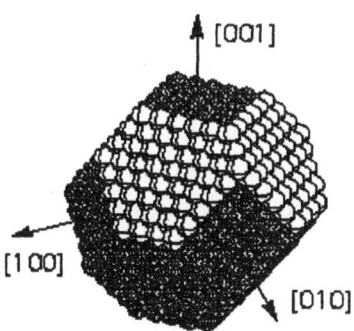

Figure 7. Scheme of a typical cluster shape with light-gray atoms belonging to the 1289-atom truncated octahedron basis and dark-gray atoms belonging to the (111) and (001) added facets.

that is subsequently used to pattern micro-SQUIDs. A laser vaporization and inert gas condensation source is used to produce an intense supersonic beam of nanosized Co clusters which can be deposited in various matrices under ultrahigh-vacuum (UHV) conditions. Due to the low-energy deposition regime, clusters do not fragment upon impact on the substrate [64]. The niobium matrix is simultaneously deposited from a UHV electron gun evaporator leading to continuous films with a low concentration of embedded Co clusters [65]. These films are used to pattern planar microbridge-DC-SQUIDs by electron beam lithography.

The latter ones allow us to detect the magnetization reversal of a single Co cluster for an applied magnetic field in any direction and in the temperature range between 0.03 and 30 K (Section 2.2). However, the desired sensitivity is only achieved for Co clusters embedded into the microbridges where the magnetic flux coupling is high enough. Due to the low concentration of embedded Co clusters, we have a maximum of 5 noninteracting particles in a microbridge which is 300 nm long and 50 nm wide. We can separately detect the magnetization switching for each cluster. Indeed they are clearly different in intensity and orientation because of the random distribution of the easy magnetization directions. The *cold mode* method [42] in combination with the *blind* method [42] allows us to detect separately the magnetic signal for each cluster.

High-resolution transmission electron microscopy observations showed that the Co clusters are well-crystallized in a fcc structure (Fig. 6) with a sharp size distribution [65]. They mainly form truncated octahedrons (Fig. 7) [41].

Figure 8 displays a typical measurement of switching fields in three dimensions of a 3 nm Co cluster at $T = 35$ mK. This surface is a three-dimensional picture directly related to the anisotropy involved in the magnetization reversal of the particle (Section 3.1). It can be reasonably fitted with the generalized Stoner and Wohlfarth model [48] (Section 3.1). We obtain the following anisotropy energy:

$$E_0(\vec{m})/v = -K_1 m_z^2 + K_2 m_x^2 - K_4(m_{x'}^2 m_{y'}^2 + m_{x'}^2 m_{z'}^2 + m_{y'}^2 m_{z'}^2) \quad (16)$$

where K_1 and K_2 are the anisotropy constants along z and x, the easy and hard magnetization axes, respectively. K_4 is the fourth-order anisotropy constant and the $(x'y'z')$ coordinate system is deduced from (xyz) by a 45° rotation around the z axis. We obtained $K_1 = 2.2 \times 10^5$ J/m^3, $K_2 = 0.9 \times 10^5$ J/m^3, and $K_4 = 0.1 \times 10^5$ J/m^3. The corresponding theoretical surface is shown in Fig. 9. Furthermore, we measured the temperature dependence of the switching field distribution (Section 4.3.2). We deduced the blocking temperature of the particle $T_B \approx 14$ K, and the number of magnetic atoms in this particle: $N \approx 1500$ atoms (Section 4.3.2). Detailed measurements on about 20 different particles showed similar three-dimensional switching field distributions with comparable anisotropy ($K_1 = (2.0 \pm 0.3) \times 10^5$ J/m^3, $K_2 = (0.8 \pm 0.3) \times 10^5$ J/m^3, and $K_4 = (0.1 \pm 0.05) \times 10^5$ J/m^3) and size ($N = 1500 \pm 200$ atoms).

In the following, we analyze various contributions to the anisotropy energy of the Co clusters. Fine structural studies using EXAFS mea-

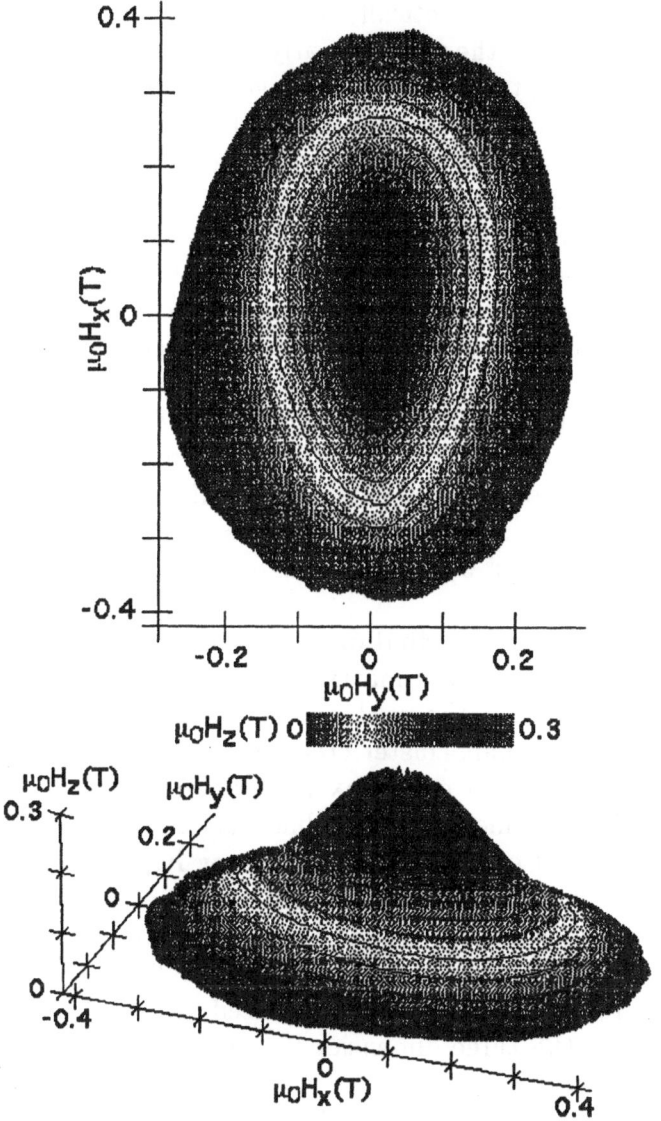

Figure 8. Top view and side view of the experimental three-dimensional angular dependence of the switching field of a 3 nm Co cluster at 35 mK. This surface is symmetrical with respect to the H_x–H_y plane, and only the upper part ($\mu_0 H_z >$ 0 T) is shown. Continuous lines on the surface are contour lines on which $\mu_0 H_z$ is constant.

surements [65] were performed on 500 nm thick niobium films con-taining a very low concentration of cobalt clusters. They showed that niobium atoms penetrate the cluster surface to almost two atomic monolayers because cobalt and niobium are miscible elements. Further magnetic measurements [65] on the same samples showed that these two atomic monolayers are magnetically dead. For this reason, we estimated the shape anisotropy of the typical nearly spherical deposited cluster in Fig. 7 after removing two atomic monolayers from the surface. By calculating all the dipolar interactions inside the particle assuming a bulk magnetic moment of $\mu_{at} = 1.7\mu_B$, we estimated the shape anisotropy constants: $K_1 \approx 0.3 \times 10^5$ J/m^3 along the easy magnetization axis and $K_2 \approx 0.1 \times 10^5$ J/m^3 along the hard magnetization axis. These values are much smaller than the measured ones which means that E_{shape} is not the main cause of the second-order anisotropy in the cluster.

The fourth-order term $K_4 = 0.1 \times 10^5$ J/m^3 should arise from the cubic magnetocrystalline anisotropy in the fcc cobalt clusters. However, this value is smaller than the values reported in previous works [66, 55]. This might be due to the different atomic environment of the surface atoms with respect to that of bulk fcc Co. Taking the value of the bulk [66, 55], $K_{bulk} = 1.2 \times 10^5$ J/m^3 only for the core atoms in the cluster, we find $K_{MC} \approx 0.2 \times 10^5$ J/m^3, which is in reasonable agreement with our measurements.

We expect that the contribution of the magnetoelastic anisotropy energy K_{MC} coming from the matrix-induced stress on the particle is also small. Indeed, using the co-deposition technique, niobium atoms cover uniformly the cobalt cluster creating an isotropic distribution of stresses. In addition, they can relax preferably inside the matrix and not in the particle volume because niobium is less rigid than cobalt. We believe therefore that only interface anisotropy $K_{surface}$ can account for the experimentally observed second-order anisotropy terms. Niobium atoms at the cluster surface might enhance this interface anisotropy through surface strains and magnetoelastic coupling. This emphasizes the dominant role of the surface in nanosized systems.

In conclusion, the three-dimensional switching field measurements of individual clusters give access to their magnetic anisotropy energy. A quantitative understanding of the latter is still difficult, but it seems that the cluster–matrix interface provides the main contribution to the magnetic anisotropy. Such interfacial effects could be promising to control the magnetic anisotropy in small particles in order to increase their blocking temperature up to the required range for applications.

4. Influence of temperature on the magnetization reversal

The thermal fluctuations of the magnetic moment of a single-domain ferromagnetic particle and its decay towards thermal equilibrium were introduced by Néel [12, 13] and further developed by Bean and Livingston [67, 68] and Brown [69, 70, 71]. The simplest case is an assembly

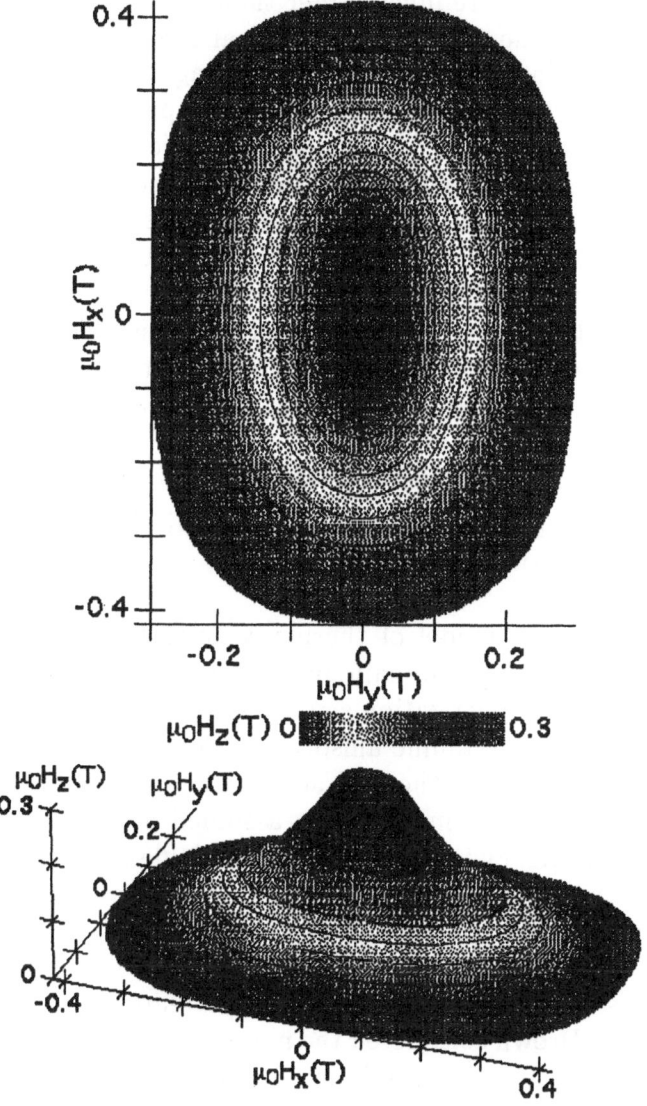

Figure 9. Top view and side view of the theoretical switching field surface considering second- and fourth-order terms in the anisotropy energy.

of independent particles having no magnetic anisotropy. In the absence of an applied magnetic field, the magnetic moments are randomly oriented. The situation is similar to paramagnetic atoms where the temperature dependence of the magnetic susceptibility follows a Curie behavior, and the field dependence of magnetization is described by a Brillouin function. The only difference is that the magnetic moments of the particles are much larger than those of the paramagnetic atoms. Therefore, the quantum mechanical Brillouin function can be replaced by the classical limit for larger magnetic moments, namely the Langevin function. This theory is called superparamagnetism. The situation changes, however, as soon as magnetic anisotropy is present which establishes one or more preferred orientations of the particle's magnetization (Section 3). In the following, we present an overview of the simplest model describing thermally activated magnetization reversal of single isolated nanoparticles which is called the Néel–Brown model. After a brief review of the model (Section 4.1), we present experimental methods to study the thermally activated magnetization reversal (Section 4.2). Finally, we discuss some applications of the Néel–Brown model (Section 4.3).

4.1. NÉEL–BROWN MODEL OF THERMALLY ACTIVATED MAGNETIZATION REVERSAL

In Néel and Brown's model of thermally activated magnetization reversal, a magnetic single-domain particle has two equivalent ground states of opposite magnetization separated by an energy barrier which is due to shape and crystalline anisotropy. The system can escape from one state to the other by thermal activation over the barrier. Just as in the Stoner–Wohlfarth model, they assumed uniform magnetization and uniaxial anisotropy in order to derive a single relaxation time. Néel supposed further that the energy barrier between the two equilibrium states is large in comparison to the thermal energy $k_B T$ which justified a discrete orientation approximation [12, 13]. Brown criticized Néel's model because the system is not explicitly treated as a gyromagnetic one [69, 70, 71]. Brown considered the magnetization vector in a particle to wiggle around an energy minimum, then jump to the vicinity of the other minimum, then wiggle around there before jumping again. He supposed that the orientation of the magnetic moment may be described by a Gilbert equation with a random field term that is assumed to be white noise. On the basis of these assumptions, Brown was able to derive a Fokker–Planck equation for the distribution of magnetization orientations. Brown did not solve his differential equation. Instead he tried some analytic approximations and an asymptotic expansion for the case of the field parallel or perpendicular to the easy axis of magnetization. More recently, Coffey et al. [72, 73] found by numerical methods an exact solution of Brown's differential equation for uniaxial anisotropy and an arbitrary applied field direction. They also derived an asymptotic general solution for the case of large energy barriers in comparison to the thermal energy $k_B T$. This asymptotic solution is of particular interest for single-particle measurements and is reviewed in the following.

For a general asymmetric bistable energy potential $E = E(\vec{m}, \vec{H})$ [Eq. (1)] with the orientation of magnetization $\vec{m} = \vec{M}/M_s$ (M_s is the spontaneous magnetization), $\vec{H}$ is the applied field, and with minima at $\vec{n}_1$ and $\vec{n}_2$ separated by a potential barrier containing a saddle point at $\vec{n}_0$ (with the $\vec{n}_i$ coplanar), and in the case of $\beta(E_0 - E_i) \gg 1$ where $\beta = 1/k_B T$, and $E_i = E(\vec{n}_i, \vec{H})$, Coffey et al. showed that the longest relaxation time [2] is given by the following equation which is valid in the intermediate to high damping limit (IHD) defined by $\alpha\beta(E_0 - E_i) > 1$ [74]:

$$\tau^{-1} = \frac{\Omega_0}{2\pi\omega_0} \left[\omega_1 e^{-\beta(E_0 - E_1)} + \omega_2 e^{-\beta(E_0 - E_2)}\right] \qquad (17)$$

where ω_0 and Ω_0 are the saddle and damped saddle angular frequencies:

$$\omega_0 = \frac{\gamma}{M_s}\sqrt{-c_1^{(0)}c_2^{(0)}} \qquad (18)$$

$$\Omega_0 = \frac{\gamma}{M_s}\frac{\alpha}{1+\alpha^2}\left[-c_1^{(0)} - c_2^{(0)} + \sqrt{(c_2^{(0)} - c_1^{(0)})^2 - 4\alpha^{-2}c_1^{(0)}c_2^{(0)}}\right] \qquad (19)$$

ω_1 and ω_2 are the well angular frequencies:

$$\omega_i = \frac{\gamma}{M_s}\sqrt{c_1^{(i)}c_2^{(i)}} \qquad (20)$$

with $i = 1$ and 2. $c_1^{(j)}$ and $c_2^{(j)}$ ($j = 0, 1, 2$) are the coefficients in the truncated Taylor series of the potential at well and saddle points—that is, the curvatures of the potential at well and saddle points. γ is the gyromagnetic ratio, $\alpha = \nu\gamma M_s$ is the dimensionless damping factor and ν is the friction in Gilbert's equation (ohmic damping).

In the low damping limit (LD), defined by $\alpha\beta(E_0 - E_i) < 1$, the longest relaxation time is given by [75, 76]

$$\tau^{-1} = \frac{\alpha}{2\pi}\left[\omega_1\beta(E_0 - E_1)e^{-\beta(E_0 - E_1)} + \omega_2\beta(E_0 - E_2)e^{-\beta(E_0 - E_2)}\right] \qquad (21)$$

---

[2] The inverse of the longest relaxation time is determined by the smallest non-vanishing eigenvalue of the appropriate Fokker–Planck equation [72, 73]. All other eigenvalues can be neglected in the considered asymptotic limit of $\beta(E_0 - E_i) \gg 1$.

In this case, the energy dissipated in one cycle of motion in the well is very small in comparison to the thermal energy k_BT.

Experimentally, relaxation is observed only if τ is of the order of magnitude of the measuring time of the experiment. This implies for all known single-particle measurement techniques that $\beta(E_0 - E_i) \gg 1$; that is, the asymptotic solutions [Eqs. (17) and (21)] are always a very good approximation to the exact solution of Brown's Fokker–Planck equation [77]. Due to an applied field, $\beta(E_0 - E_1) \gg \beta(E_0 - E_2)$ (taking E_2 as the metastable minimum) might be true. Then the first exponential in Eqs. (17) and (21) can be neglected.

Concerning the possible values of α, we remark that little information is available. Typical values should be between 0.01 and 5 [11], meaning that in practice $\alpha\beta(E_0 - E_i)$ can be $\gg 1$, $\ll 1$, or ≈ 1. Thus the distinction between Eqs. (17) and (21) becomes important.

Finally, we note that $c_1^{(j)}$ and $c_2^{(j)}$ $(j = 0, 1, 2)$ can be found experimentally by measuring the critical surface of the switching field and applying the calculation of Thiaville (Section 3.1) [49].

4.2. EXPERIMENTAL METHODS FOR THE STUDY OF THE NÉEL–BROWN MODEL

As discussed in the previous section, in the Néel–Brown model of thermally activated magnetization reversal a magnetic single-domain particle has two equivalent ground states of opposite magnetization separated by an energy barrier due to, for instance, shape and crystalline anisotropy. The system can escape from one state to the other either by thermal activation over the barrier at high temperatures or by quantum tunneling at low temperatures. At sufficiently low temperatures and at zero field, the energy barrier between the two states of opposite magnetization is much too high to observe an escape process. However, the barrier can be lowered by applying a magnetic field in the opposite direction to that of the particle's magnetization. When the applied field is close enough to the switching field at zero temperature H_{sw}^0, thermal fluctuations are sufficient to allow the system to overcome the barrier, and the magnetization is reversed.

In the following, we discuss three different experimental methods for studying this stochastic escape process which are called waiting time, switching field, and telegraph noise measurements.

4.2.1. *Waiting time measurements*

The waiting time method consists in measuring the probability that

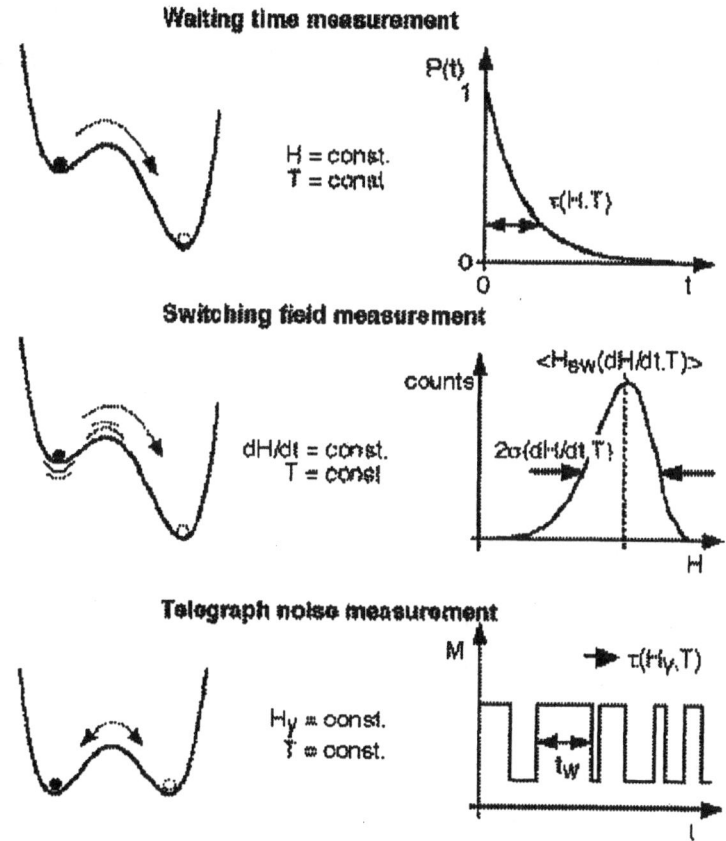

Figure 10. Schema of three methods for studying the escape from a potential well: waiting time and telegraph noise measurements give direct access to the switching time probability $P(t)$, whereas switching field measurements yield histograms of switching fields.

the magnetization has not switched after a certain time. In the case of an assembly of identical and isolated particles, it corresponds to measurements of the relaxation of magnetization. However, in most particle assemblies, broad distributions of switching fields lead to logarithmic decay of magnetization, and the switching probability is hidden behind the unknown distribution functions [11]. For individual particle studies, waiting time measurements give direct access to the switching probability (Fig. 10). At a given temperature, the magnetic field H is increased to a waiting field H_w near the switching field H_{sw}^0. Next, the elapsed time until the magnetization switches is measured. This process is repeated several hundred times, yielding a waiting time histogram. The integral of this histogram and proper normalization yields the probability that the magnetization has not switched after a time t. This probability is measured at different waiting fields H_w and temperatures in order to explore several barrier heights and thermal activation energies.

According to the Néel–Brown model, the probability that the magnetization has not switched after a time t is given by:

$$P(t) = e^{-t/\tau} \tag{22}$$

and τ (inverse of the switching rate) can be expressed by an Arrhenius law of the form:

$$\tau^{-1}(\varepsilon) = B\varepsilon^{a+b-1} e^{-A\varepsilon^a} \tag{23}$$

where $\varepsilon = (1 - H/H_{sw}^0)$ and A, B, a, and b depend on damping, temperature, energy barrier height [(17)–(21)], curvatures at well and saddle points, and reversal mechanism (thermal or quantum) (cf. Table 1 of [78]). For simplicity, experimentalists have often supposed a constant pre-exponential factor τ_0^{-1} instead of $B\varepsilon^{a+b-1}$.

The adjustment of Eq. (22) to the measured switching probabilities yields a set of mean waiting times $\tau^{-1}(H_w, T)$. In order to adjust the Néel–Brown model to this set of data, we propose the following relation that can be found by inserting $\varepsilon = (1 - H_w/H_{sw}^0)$ into Eq. (23):

$$H_w = H_{sw}^0 \left(1 - \left[\frac{1}{A} \ln\left(\tau B\varepsilon^{a+b-1}\right)\right]^{1/a}\right) \tag{24}$$

When plotting the H_w values as a function of $\left[T \ln\left(\tau B\varepsilon^{a+b-1}\right)\right]^{1/a}$, all points should gather on a straight line (master curve) by choosing the proper value for the constants B, a, and b [a and b should be given by (17)–(21)]. A can be obtained from the slope of the master.

The number of exploitable decades for τ values is limited for waiting time measurements: Short-time (milliseconds) experiments are limited by the inductance of the field coils [3] and long-time (minutes) studies by the stability of the experimental setup. Furthermore, the total acquisition time for a set of $\tau^{-1}(H_w, T)$ is rather long (weeks). Thus a more convenient method—namely, the switching field method—is needed for single-particle measurements.

---

[3] A solution to this problem might be a superposition of a constant applied field and a small pulse field.

4.2.2. *Switching field measurements*

For single-particle studies, it is often more convenient to study magnetization reversal by ramping the applied field at a given rate and measuring the field value as soon as the particle magnetization switches. Next, the field ramp is reversed and the process repeated. After several hundred cycles, switching field histograms are established, yielding the mean switching field $\langle H_{sw} \rangle$ and the width σ_{sw} (rms deviation). Both mean values are measured as a function of the field sweeping rate and temperature (Fig. 10).

From the point of view of thermally activated magnetization reversal, switching field measurements are equivalent to waiting time measurements because the time scale for the sweeping rate is typically more than 8 orders of magnitude greater than the time scale of the exponential prefactor, which is in general around 10^{-10} s. We can therefore apply the Néel–Brown model described above. The mathematical transformation from a switching time probability [Eqs. (22)–(23)] to a switching field probability was first given by Kurkijärvi [79] for the critical current in SQUIDs. A more general calculation was evaluated by Garg [78]. In many cases, the mean switching field $\langle H_{sw} \rangle$ can be approximated by the first two terms of the development of Garg [78]:

$$\langle H_{sw}(T,v) \rangle \approx H_{sw}^0 \left(1 - \left[\frac{1}{A} \ln \left(\frac{H_{sw}^0 B}{va A^{1-b/a}} \right) \right]^{1/a} \right) \qquad (25)$$

where the field sweeping rate is given by $v = dH/dt$. The width of the switching field distribution σ_{sw} can be approximated by the first term of Garg's development:

$$\sigma_{sw} \approx H_{sw}^0 \frac{\pi}{\sqrt{6}a} \left(\frac{1}{A} \right)^{1/a} \left[\ln \left(\frac{H_{sw}^0 B}{va A^{1-b/a}} \right) \right]^{(1-a)/a} \qquad (26)$$

In the case of a constant pre-exponential factor τ_0^{-1}, the calculation of $\langle H_{sw} \rangle$ and σ_{sw} is simpler and is given by Eqs. (4) and (5) in Ref. [38], respectively.

Similar to the waiting time measurements, a scaling of the model to a set of $\langle H_{sw}(T,v) \rangle$ values can be done by plotting the $\langle H_{sw}(T,v) \rangle$ values as a function of $\left[T \ln \left(\frac{H_{sw}^0 B}{va A^{1-b/a}} \right) \right]^{1/a}$. All points should gather on a straight line by choosing the proper value for the constants.

The entire switching field distribution $P(H)$ can be calculated iteratively by the following equation [79]:

$$P(H) = \tau^{-1}(H)v^{-1}\left[1 - \int_0^H P(H')dH'\right] \qquad (27)$$

4.2.3. *Telegraph noise measurements*

In order to study the superparamagnetic state [4] of a single-particle, it is simply necessary to measure the particle's magnetization as a function of time. We call this telegraph noise measurement as stochastic fluctuations between two states are expected. According to the Néel–Brown model, the mean time τ spent in one state of magnetization is given by an Arrhenius law of the form of Eq. (23). Because τ increases exponentially with decreasing temperature, it is very unlikely that an escape process will be observed at low temperature. However, applying a constant field in the direction of a hard axis (hard plane) of magnetization reduces the height of the energy barrier (Fig. 10). When the energy barrier is sufficiently small, the particle's magnetization can fluctuate between two orientations which are close to a hard axis (hard plane) of magnetization. The time spent in each state follows an exponential switching probability law as given by Eqs. (22) and (23) with $a = 2$.

4.3. EXPERIMENTAL EVIDENCE FOR THE NÉEL–BROWN MODEL

The Néel–Brown model is widely used in magnetism, particularly in order to describe the time dependence of the magnetization of collections of particles, thin films, and bulk materials. However, until recently, all the reported measurements, performed on individual particles, were not consistent with the Néel–Brown theory. This disagreement was attributed to the fact that real samples contain defects, ends, and surfaces that could play an important, if not dominant, role in the physics of magnetization reversal. It was suggested that the dynamics of reversal occurs via a complex path in configuration space, and that a new theoretical approach is required to provide a correct description of thermally activated magnetization reversal even in single-domain

---

[4] At zero applied field, a magnetic single-domain particle has two equivalent ground states of opposite magnetization separated by an energy barrier. When the thermal energy k_BT is sufficiently high, the total magnetic moment of the particle can fluctuate thermally, like a single spin in a paramagnetic material. Such magnetic behavior of an assembly of independent single-domain particles is called superparamagnetism [11, 67, 68].

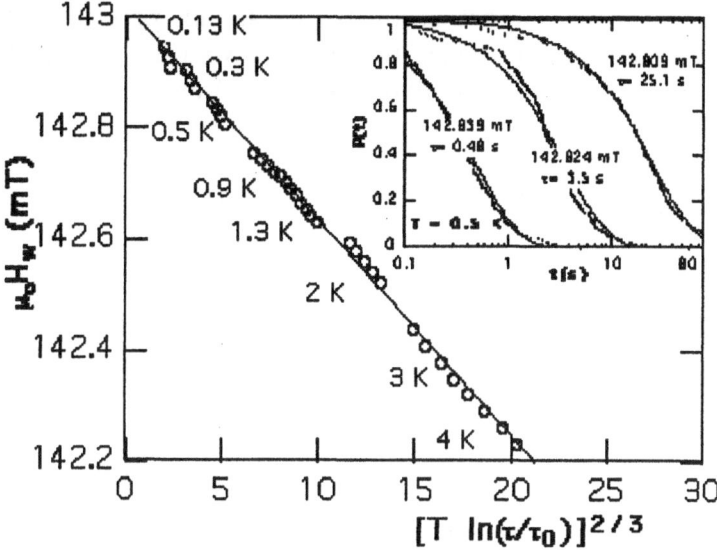

Figure 11. Scaling plot of the mean switching time $\tau(H_w, T)$ for several waiting fields H_w and temperatures (0.1 s $< \tau(H_w, T) < 60$ s) for a Co nanoparticle. The scaling yields $\tau_0 \approx 3 \times 10^{-9}$s. Inset: Examples of the not-switching probability of magnetization as a function of time for different applied fields and at 0.5 K. Full lines are data fits with an exponential function: $P(t) = e^{-t/\tau}$.

ferromagnetic particles [22, 62]. Similar conclusions were drawn from numerical simulations of the magnetization reversal [80, 81, 82, 83, 84].

A few years later, micro-SQUID measurements on individual Co nanoparticles showed for the first time a very good agreement with the Néel–Brown model by using waiting time, switching field, and telegraph noise measurements [38, 39, 41]. It was also found that sample defects, especially sample oxidation, play a crucial role in the physics of magnetization reversal.

In the following subsections, we review some typical results concerning nanoparticles (Section 4.3.1) and clusters (Section 4.3.2). In Section 4.4, we point out the main deviations from the Néel–Brown model which are due to defects.

4.3.1. *Application to nanoparticles*

One of the important predictions of the Néel–Brown model concerns the exponential not-switching probability $P(t)$ [Eq. (22)] which can be measured directly via waiting time measurements (Section 4.2.1): at a given temperature, the magnetic field is increased to a waiting field H_w which is close to the switching field. Then, the elapsed time is measured until the magnetization switches. This process is repeated several hundred times, in order to obtain a waiting time histogram. The integral of this histogram gives the not-switching probability $P(t)$ which is measured at several temperatures T and waiting fields H_w. The inset of Fig. 11 displays typical measurements of $P(t)$ performed on a Co nanoparticle. All measurements show that $P(t)$ is given by an exponential function described by a single relaxation time τ.

The validity of Eq. (23) is tested by plotting the waiting field H_w as a function of $[T \ln(\tau/\tau_0)]^{2/3}$.[5] If the Néel–Brown model applies, all points should collapse onto one straight line (master curve) by choosing the proper values for τ_0. Figure 11 shows that the data set $\tau(H_\mathrm{w}, T)$ falls on a master curve provided that $\tau_0 \approx 3 \times 10^{-9}$ s. The slope and intercept yield the values $E_0 = 214{,}000$ K and $H_\mathrm{sw}^0 = 143.05$ mT. The energy barrier E_0 can be approximately converted to a thermally "activated volume" by using $V = E_0/(\mu_0 M_S H_\mathrm{sw}^0) \approx (25 \text{ nm})^3$ which is very close to the particle volume estimated by SEM. This agreement is another confirmation of a magnetization reversal by uniform rotation. The result of the waiting time measurements are confirmed by switching field and telegraph noise measurements [38, 39]. The field and temperature dependence of the exponential prefactor τ_0 is taken into account in Ref. [73].

4.3.2. *Application to Co clusters*

Figure 12 presents the angular dependence of the switching field of a 3 nm Co cluster measured at different temperatures. At 0.03 K, the measurement is very close to the standard Stoner–Wohlfarth astroid. For higher temperatures the switching field becomes smaller and smaller. It reaches the origin at about 14 K, yielding the blocking temperature $T_\mathrm{B} = 14$ K of the cluster magnetization. T_B is defined as the temperature for which the waiting time Δt becomes equal to the relaxation time τ of the particle's magnetization at $\vec{H} = \vec{0}$. T_B can be used to estimate the total number N_tot of magnetic Co atoms in the cluster. Using an Arrhenius-like law [Eq. (23)] which can be written as $\Delta t = \tau = \tau_0 \exp(K_{at} N_\mathrm{tot}/k_\mathrm{B} T_\mathrm{B})$, where τ_0^{-1} is the attempt frequency typically between 10^{10} and 10^{11} Hz [85], K_{at} is an effective anisotropy energy per atom and k_B is the Boltzmann constant. Using the expression of the switching field at $T = 0$ K and for $\theta = 0$:

---

[5] $a = 3/2$ because the field was applied at about $20°$ from the easy axis of magnetization.

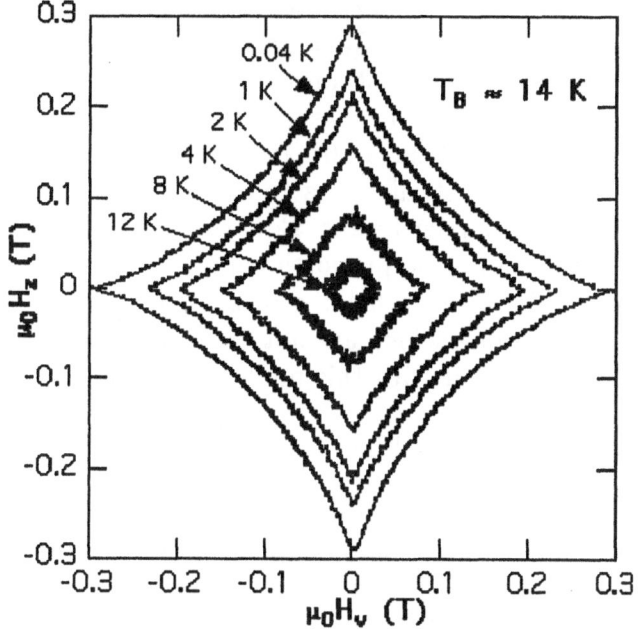

Figure 12. Temperature dependence of the switching field of a 3 nm Co cluster, measured in the plane defined by the easy and medium hard axes ($H_y - H_z$ plane in Fig. 8). The data were recorded using the blind mode method with a waiting time of the applied field of $\Delta t = 0.1$ s. The scattering of the data is due to stochastics and is in good agreement with Eq. (26).

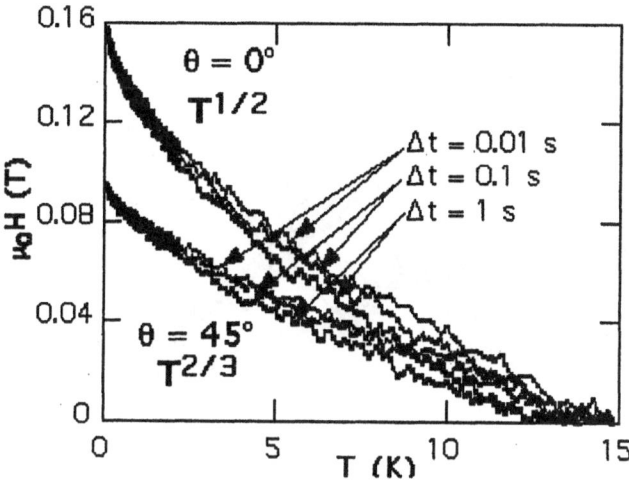

Figure 13. Temperature dependence of the switching field of a 3 nm Co cluster, measured at $0°$ and $45°$. The data were recorded using the blind mode method with different waiting times Δt of the applied field. The scattering of the data is due to stochastics and is in good agreement with Eq. (26).

$\mu_0 H_{sw} = 2K_{at}/\mu_{at} = 0.3T$ (Fig. 12), the atomic moment $\mu_{at} = 1.7\mu_B$, $\Delta t = 0.01$ s, $\tau_0 = 10^{-10}$ s, and $T_B = 14$ K, we deduce $N_{tot} \approx 1500$, which corresponds very well to a 3 nm Co cluster (Fig. 6).

Figure 13 presents a detailed measurement of the temperature dependence of the switching field at 0° and 45° and for three waiting times Δt. This measurement allows us to check the predictions of the field dependence of the barrier height. Equation (25) predicts that the mean switching field should be proportional to $T^{1/a}$ where a depends on the direction of the applied field: $a = 2$ for $\theta = 0°$ and 90°, and $a = 3/2$ for all other angles which are not too close to $\theta = 0°$ and 90°. We found a good agreement with this model (Fig. 14).

4.4. DEVIATIONS FROM THE NÉEL–BROWN MODEL

Anomalous magnetic properties of oxidized or ferrimagnetic nanoparticles have been reported previously by several authors [86, 87]. These properties are, for example, the lack of saturation in high fields and shifted hysteresis loops after cooling in the presence of a magnetic field. These behaviors have been attributed to uncompensated surface spins of the particles and surface spin disorder [88, 89].

Concerning our single-particle studies, we systematically observed aging effects which we attribute to an oxidation of the surface of the sample, forming antiferromagnetic CoO or NiO [62, 90]. We found that the antiferromagnetic coupling between the core of the particle or wire and its oxidized surface changed the dynamic reversal properties. For

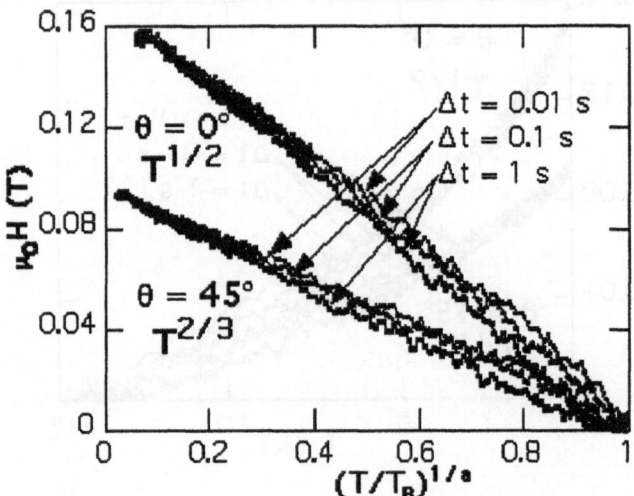

Figure 14. Temperature dependence of the switching field of a 3 nm Co cluster as in Fig. 13 but plotted as a function of $(T/T_B)^{1/a}$ with $T_B \doteq 14$ K and $a = 2$ or 3/2 for $\theta = 0°$ and 45°, respectively, and for three waiting times Δt.

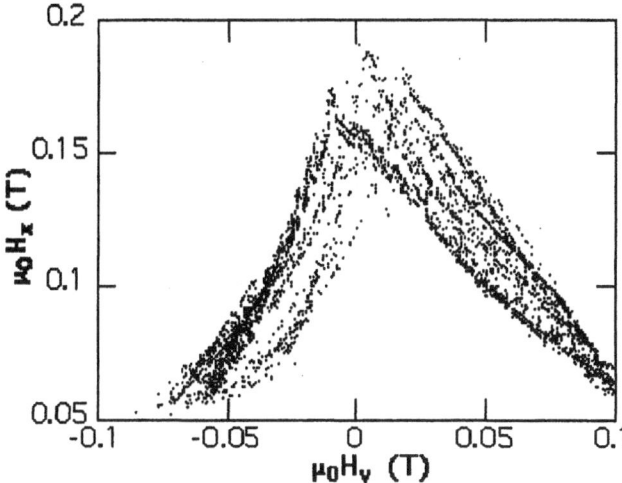

Figure 15. Angular dependence of switching fields of a 3 nm Fe cluster having (probably) a slightly oxidized surface. Each point corresponds to one of the 10,000 switching field measurements. The huge variations of the switching field might be due to exchange bias of frustrated spin configurations.

instance, we repeated the measurements of the magnetization reversal of a Ni wire two days after fabrication, six weeks after, and finally after three months [90]. Between these measurements, the wire stayed in a dry box. The quasi-static micro-SQUID measurements did reveal only small changes. The saturation magnetization measured after six weeks was unchanged and was reduced by one to two percent after three months. The angular dependence of the switching field changed also only slightly. The dynamic measurements showed the aging effects more clearly, as evidenced by a nonexponential probability of not switching, an increase of the width of the switching field distributions, and a decrease of the activation energy. We measured a similar behavior on lithographic fabricated Co particles with an oxidized border [62].

Figure 15 presents the angular dependence of switching fields of a 3 nm Fe cluster having a slightly oxidized surface. Huge variation of the switching fields can be observed which might be due to exchange bias of frustrated spin configurations at the surface of the cluster (Fig. 16).

We propose that the magnetization reversal of a ferromagnetic particle with an antiferromagnetic surface layer is mainly governed by two mechanisms which are both due to spin frustration at the interface between the ferromagnetic core and the antiferromagnetic surface layer(s). The first mechanism may come from the spin frustration differing slightly from one cycle to another, thus producing a varying

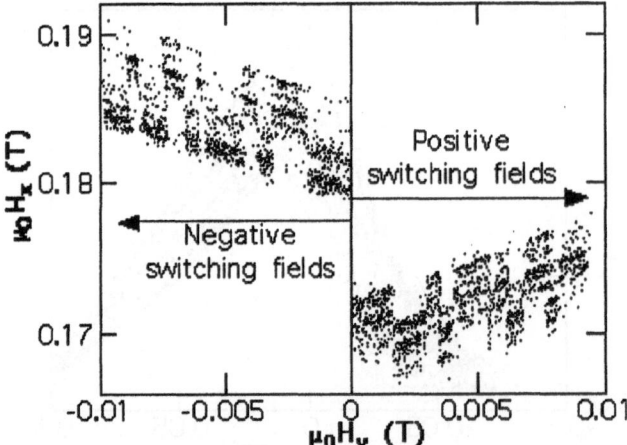

Figure 16. Details of the angular dependence of switching fields of a 3 nm Fe cluster having (probably) a slightly oxidized surface. Each point corresponds to one of the 3000 switching field measurements. Stochastic fluctuations between different switching field distributions are observed. The "mean" hysteresis loop is shifted to negative fields.

energy landscape. These energy variations are less important at high temperatures when the thermal energy ($k_B T$) is much larger. However, at lower temperature the magnetization reversal becomes sensitive to the energy variations. During the hysteresis loop the system chooses randomly a path through the energy landscape which leads to broad switching field distributions. A second mechanism may become dominant at high temperatures: the magnetization reversal may be governed by a relaxation of the spin frustration, hence by a relaxation of the energy barrier. This relaxation is thermally activated—that is, slower at lower temperatures.

5. Conclusion

Nanometer-sized magnetic particles have generated continuous interest as the study of their properties has proved to be scientifically and technologically very challenging. In the last few years, new fabrication techniques have led to the possibility of making small objects with the required structural and chemical qualities. In order to study these objects, new local measuring techniques were developed such as magnetic force microscopy, magnetometry based on micro-Hall probes or micro-

SQUIDs. This led to a new understanding of the magnetic behavior of nanoparticles.

In this chapter we reviewed the most important theories and experimental results concerning the magnetization reversal of single-domain particles and clusters. Special emphasis is laid on single-particle measurements avoiding complications due to distributions of particle size, shape, and so on. Measurements on particle assemblies have been reviewed in Ref. [11].

The understanding of the magnetization reversal in nanostructures requires the knowledge of many physical phenomena, and nanostructures are therefore particularly interesting for the development of new fundamental theories of magnetism and in modeling new magnetic materials for permanent magnets or high density recording.

Acknowledgements

The author is indebted to A. Benoit, E. Bonet Orozco, V. Bouchiat, I. Chiorescu, M. Faucher, K. Hasselbach, M. Jamet, D. Mailly, B. Pannetier, and C. Thirion for their experimental contributions and the developement of the micro-SQUID technology. The author acknowledges the collaborations with J.-Ph. Ansermet, B. Barbara, B. Doudin, V. Dupuis, O. Fruchart, H. Pascard, and A. Perez.

References

1. A. Hubert and R. Schäfer, *Magnetic Domains: The Analysis of Magnetic Microstructures* (Springer-Verlag, Berlin Heidelberg New York, 1998).
2. Aharoni, *An Introduction to the Theory of Ferromagnetism* (Oxford University Press, London, 1996).
3. I. M. L. Billas, A. Châtelain, and W. A. de Heer, J. Magn. Magn. Mat. **168**, 64 (1997).
4. S. E. Apsel, J. W. Emmert, J. Deng, and L. A. Bloomfield, Phys. Rev. Lett. **76**, 1441 (1996).
5. A.-L. Barra, P. Debrunner, D. Gatteschi, Ch. E. Schulz, and R. Sessoli, EuroPhys. Lett. **35**, 133 (1996).
6. M.A. Novak and R. Sessoli, in *Quantum Tunneling of Magnetization-QTM'94*, Vol. 301 of *NATO ASI Series E: Applied Sciences*, edited by L. Gunther and B. Barbara (Kluwer Academic Publishers, London, 1995), pp. 171–188.
7. J. R. Friedman, M. P. Sarachik, J. Tejada, and R. Ziolo, Phys. Rev. Lett. **76**, 3830 (1996).
8. L. Thomas, F. Lionti, R. Ballou, D. Gatteschi, R. Sessoli, and B. Barbara, Nature (London) **383**, 145 (1996).

9.   C. Sangregorio, T. Ohm, C. Paulsen, R. Sessoli, and D. Gatteschi, Phys. Rev. Lett. **78**, 4645 (1997).

10. W. Wernsdorfer and R. Sessoli, Science **284**, 133 (1999).

11. J. L. Dormann, D. Fiorani, and E. Tronc, Adv. Chem. Phys. **98**, 283 (1997).

12. L. Néel, Ann. Geophys. **5**, 99 (1949).

13. L. Néel, C. R. Acad. Science **228**, 664 (1949).

14. A. H. Morrish and S. P. Yu, Phys. Rev. **102**, 670 (1956).

15. J. E. Knowles, IEEE Trans. Mag. **MAG-14**, 858 (1978).

16. A. Tonomura, T. Matsuda, J. Endo, T. Arii, and K. Mihama, Phys. Rev. B **34**, 3397 (1986).

17. H. J. Richter, J. Appl. Phys. **65**, 9 (1989).

18. S. J. Hefferman, J. N. Chapman, and S. McVitie, J. Magn. Magn. Mat. **95**, 76 (1991).

19. C. Salling, S. Schultz, I. McFadyen, and M. Ozaki, IEEE Trans. Mag. **27**, 5185 (1991).

20. N. Bardou, B. Bartenlian, C. Chappert, R. Megy, P. Veillet, J. P. Renard, F. Rousseaux, M. F. Ravet, J. P. Jamet, and P. Meyer, J. Appl. Phys. **79**, 5848 (1996).

21. T. Chang and J. G. Chu, J. Appl. Phys. **75**, 5553 (1994).

22. M. Ledermann, S. Schultz, and M. Ozaki, Phys. Rev. Lett. **73**, 1986 (1994).

23. J. Bansmann, V. Senz, L. Lu, A. Bettac, and K.H. Meiweis-Broer, J. Electron Spectrosc. Relat. Phenom., **106**, 221 (2000).

24. K.H. Meiweis-Broer, Phys. Bl. **55**, 21 (1999).

25. W. Wernsdorfer, K. Hasselbach, D. Mailly, B. Barbara, A. Benoit, L. Thomas, and G. Suran, J. Magn. Magn. Mat. **145**, 33 (1995).

26. K. Hong and N. Giordano, J. Magn. Magn. Mat. **151**, 396 (1995).

27. J. E. Wegrowe, S. E. Gilbert, D. Kelly, B. Doudin, and J.-Ph. Ansermet, IEEE Trans. Mag. **34**, 903 (1998).

28. A.D. Kent, S. von Molnar, S. Gider, and D.D. Awschalom, J. Appl. Phys. **76**, 6656 (1994).

29. J.G.S. Lok, A.K. Geim, J.C. Maan, S.V. Dubonos, L. Theil Kuhn, and P.E. Lindelof, Phys. Rev. B **58**, 12201 (1998).

30. T. Schweinböck, D. Weiss, M. Lipinski, and K. Eberl, J. Appl. Phys. **87**, 6496 (2000).

31. V. Gros, Shan-Fab Lee, G. Faini, A. Cornette, A. Hamzic, and A. Fert, J. Magn. Magn. Mat. **165**, 512 (1997).

32. W. J. Gallagher, S. S. P. Parkin, Yu Lu, X. P. Bian, A. Marley, K. P. Roche, R. A. Altman, S. A. Rishton, C. Jahnes, T. M. Shaw, and Gang Xiao, J. Appl. Phys. **81**, 3741 (1997).

33. J.-E. Wegrowe, D. Kelly, A. Franck, S. E. Gilbert, and J.-Ph. Ansermet, Phys. Rev. Lett. **82**, 3681 (1999).

34. L.F. Schelp, A. Fert, F. Fettar, P. Holody, S. F. Lee, J. L. Maurice, F. Petroff, and A. Vaures, Phys. Rev. B **56**, R5747 (1997).

35. S. Guéron, M. M. Deshmukh, E. B. Myers, and D. C. Ralph, Phys. Rev. Lett. **83**, 4148 (1999).

36. W. Wernsdorfer, K. Hasselbach, A. Benoit, B. Barbara, D. Mailly, J. Tuaillon, J. P. Perez, V. Dupuis, J. P. Dupin, G. Guiraud, and A. Perez, J. Appl. Phys. **78**, 7192 (1995).

37. W. Wernsdorfer, B. Doudin, D. Mailly, K. Hasselbach, A. Benoit, J. Meier, J.-Ph. Ansermet, and B. Barbara, Phys. Rev. Lett. **77**, 1873 (1996).

38. W. Wernsdorfer, E. Bonet Orozco, K. Hasselbach, A. Benoit B. Barbara, N. Demoncy, A. Loiseau, D. Boivin, H. Pascard, and D. Mailly, Phys. Rev. Lett. **78**, 1791 (1997).

39. W. Wernsdorfer, E. Bonet Orozco, K. Hasselbach, A. Benoit, D. Mailly, O. Kubo, H. Nakano, and B. Barbara, Phys. Rev. Lett. **79**, 4014 (1997).

40. E. Bonet, W. Wernsdorfer, B. Barbara, A. Benoit, D. Mailly, and A. Thiaville, Phys. Rev. Lett. **83**, 4188 (1999).

41. M. Jamet, W. Wernsdorfer, C. Thirion, D. Mailly, V. Dupuis, P. Mélinon, and A.Pérez, Phys. Rev. Lett. **86**, 4676 (2001).

42. W. Wernsdorfer, Adv. Chem. Phys. **118**, 99 (2001).

43. J. Clarke, A. N. Cleland, M. H. Devoret, D. Esteve, and J. M. Martinis, Science **239**, 992 (1988).

44. M. Ketchen, D. J. Pearson, K. Stawiasz, C-H. Hu, A. W. Kleinsasser, T. Brunner, C. Cabral, V. Chandrashekhar, M. Jaso, M. Manny, K. Stein, and M. Bhushan, IEEE Applied Superconductivity **3**, 1795 (1993).

45. W. Wernsdorfer, Ph.D. thesis, Joseph Fourier University, Grenoble, 1996.

46. E. C. Stoner and E. P. Wohlfarth, Philos. Trans. London Ser. A **240**, 599 (1948), reprinted in IEEE Trans. Magn. **27**, 3475 (1991).

47. L. Néel, C. R. Acad. Science **224**, 1550 (1947).

48. A. Thiaville, J. Magn. Magn. Mat. **182**, 5 (1998).

49. A. Thiaville, Phys. Rev. B **61**, 12221 (2000).

50. M. Jamet, W. Wernsdorfer, C. Thirion, V. Dupuis, P. Mélinon, A.Pérez, and D. Mailly, Phys. Rev. B **69**, 024401 (2004).

51. L. J. Lewis, P. Jensen, and J. L. Barrat, Phys. Rev. B **56**, 2248 (1997).

52. N. Combe, P. Jensen, and A. Pimpinelli, Phys. Rev. Lett. **85**, 110 (2000).

53. L. Néel, J. Phys. Radium **15**, 376 (1954).

54. G. Wulff, Z. Krist. **34**, 449 (1901).

55. D. S. Chuang, C. A. Ballentine, and R. C. O'Handley, Phys. Rev. B **49**, 15084 (1994).

56. C. Chappert and P. Bruno, J. Appl. Phys. **64**, 5736 (1988).

57. R. Jungblut, M. T. Johnson, J. aan de Stegge, A. Reinders, and F. A. J. den Broeder, J. Appl. Phys. **75**, 6424 (1994).

58. G. Bochi, O. Song, and R. C. O'Handley, Phys. Rev. B **50**, 2043 (1994).

59. P. J. H. Bloemen F. A. J. den Broeder, W. Hoving, J. Magn. Magn. Mat. **93**, 562 (1991).

60. J. M. MacLaren and R. H. Victora, J. Appl. Phys. **76**, 6069 (1994).

61. V. Dupuis, M. Jamet, L. Favre, J. Tuaillon-Combes, P. Mélinon, and A. Pérez, J. Vac. Sci. and Technol. A **21**, 1519 (2003).

62. W. Wernsdorfer, K. Hasselbach, A. Benoit, G. Cernicchiaro, D. Mailly, B. Barbara, and L. Thomas, J. Magn. Magn. Mat. **151**, 38 (1995).

63. O. Fruchart, J.-P. Nozieres, W. Wernsdorfer, and D. Givord, Phys. Rev. Lett. **82**, 1305 (1999).

64. A. Perez, P. Melinon, V. Depuis, P. Jensen, B. Prevel, J. Tuaillon, L. Bardotti, C. Martet, M. Treilleux, M. Pellarin, J.L. Vaille, B. Palpant, and J. Lerme, J. Phys. D **30**, 709 (1997).

65. M. Jamet, V. Dupuis, P. Mélinon, G. Guiraud, A. Pérez, W. Wernsdorfer, A. Traverse, and B. Baguenard, Phys. Rev. B **62**, 493 (2000).

66. C. H. Lee, Hui He, F. J. Lamelas, W. Vavra, C. Uher, and Roy Clarke, Phys. Rev. B **42**, 1066 (1990).
67. C. P. Bean, J. Appl. Phys. **26**, 1381 (1955).
68. C. P. Bean and J. D. Livingstone, J. Appl. Phys. **30**, 120S (1959).
69. W. F. Brown, J. Appl. Phys. **30**, 130S (1959).
70. W. F. Brown, J. Appl. Phys. **34**, 1319 (1963).
71. W. F. Brown, Phys. Rev. **130**, 1677 (1963).
72. W. T. Coffey, D. S. F. Crothers, J. L. Dormann, Yu. P. Kalmykov, , and J. T. Waldron, Phys. Rev. B **52**, 15951 (1995).
73. W. T. Coffey, D. S. F. Crothers, J. L. Dormann, Yu. P. Kalmykov, E. C. Kennedy, and W. Wernsdorfer, Phys. Rev. Lett. **80**, 5655 (1998).
74. W. T. Coffey, D. S. F. Crothers, J. L. Dormann, Yu. P. Kalmykov, E. C. Kennedy, and W. Wernsdorfer, J. Phys. Cond. Mat. **10**, 9093 (1998).
75. I. Klik and L. Gunther, J. Stat. Phys. **60**, 473 (1990).
76. I. Klik and L. Gunther, J. Appl. Phys. **67**, 4505 (1990).
77. W. T. Coffey, Adv. Chem. Phys. **103**, 259 (1998).
78. A. Garg, Phys. Rev. B **51**, 15592 (1995).
79. J. Kurkijärvi, Phys. Rev. B **6**, 832 (1972).
80. H. L. Richards, S. W. Sides, M. A. Novotny, and P. A. Rikvold, J. Appl. Phys. **79**, 5749 (1996).
81. J. M. Gonzalez, R. Ramirez, R. Smirnov-Rueda, and J. Gonzalez, J. Appl. Phys. **79**, 6479 (1996).
82. D. Garcia-Pablos, P. Garcia-Mochales, and N. Garcia, J. Appl. Phys. **79**, 6021 (1996).
83. D. Hinzke and U. Nowak, Phys. Rev. B **58**, 265 (1998).
84. E. D. Boerner and H. Neal Bertram, IEEE Trans. Mag. **33**, 3052 (1997).
85. M. Respaud, J. M. Broto, H. Rakoto, A. R. Fert, L. Thomas, and B. Barbara, Phys. Rev. B **57**, 2925 (1998).
86. A. E. Berkowitz, J. A. Lahut, I. S. Jacobs, L. M. Levinson, and D.W. Forester, Phys. Rev. Lett. **34**, 594 (1975).
87. J. T. Richardson, D. I. Yiagas, B. Turk, J. Forster, and M. V. Twigg, J. Appl. Phys. **70**, 6977 (1991).
88. R. H. Kodama, A. E. Berkowitz, E. J. McNiff, Jr., and S. Foner, Phys. Rev. Lett. **77**, 394 (1996).
89. R.H. Kodama, J. Magn. Magn. Mat. **200**, 359 (1999).
90. W. Wernsdorfer, K. Hasselbach, A. Benoit, B. Barbara, B. Doudin, J. Meier, J.-Ph. Ansermet, and D. Mailly, Phys. Rev. B **55**, 1155 (1997).

Index

Ageing, 113, 225
Amorphous particles, 239
Antiferromagnetic particles, 57, 206

Barium ferrites particles, 108

Cobalt clusters, 270
Cobalt oxide particles, 211
Coercivity, 49, 54, 64, 230
Core-shell model, 241

Density functional theory, 2
Domain walls, 31

Exchange bias, 27, 113, 233

Ferrimagnetic particles, 105, 198, 201
Ferromagnetic particles, 48, 57, 59,
    65, 100, 219
Ferromagnetic resonance, 141, 172, 177
Frustration, 105

Glassy behaviour, 107

.Interface effects, 27, 53, 113, 233
Iron clusters, 293
Iron films, 24, 31
Iron/iron oxide, 228
Iron oxide particles, 101, 102,199, 205
Iron particles, 219

Magnetic relaxation, 114, 117, 118,
    224, 233, 248
Magnetocrystalline anisotropy, 7, 12,
    14, 270
Magnetostriction, 17
Monte Carlo simulations, 45, 89,
    121, 197, 243, 249

Nanowires, 12
Néel-Brown model, 282, 288
Nickel oxide particles, 206

Oxide particles, 101, 102, 123, 189,
    198, 199, 201, 205, 206
Oxidized particles, 52

Shape anisotropy, 269
Single particle measurements, 263
Small angle neutron scattering, 50, 69
Specific heat, 124
Spin-orbit coupling, 6
Spin-wave resonance, 153
Spin-wave theory, 85
Stoner-Wolfarth model, 268
Surface anisotropy, 91, 141, 202, 271
Switching field, 93, 287